Lecture Notes in Computer Science　　8260

Commenced Publication in 1973
Founding and Former Series Editors:
Gerhard Goos, Juris Hartmanis, and Jan van Leeuwen

T0240555

Lecture Notes in Computer Science   8200

Commenced Publication in 1973
Founding and Former Series Editors:
Gerhard Goos, Juris Hartmanis, and Jan van Leeuwen

Editorial Board

David Hutchison
  Lancaster University, UK
Takeo Kanade
  Carnegie Mellon University, Pittsburgh, PA, USA
Josef Kittler
  University of Surrey, Guildford, UK
Jon M. Kleinberg
  Cornell University, Ithaca, NY, USA
Alfred Kobsa
  University of California, Irvine, CA, USA
Friedemann Mattern
  ETH Zurich, Switzerland
John C. Mitchell
  Stanford University, CA, USA
Moni Naor
  Weizmann Institute of Science, Rehovot, Israel
Oscar Nierstrasz
  University of Bern, Switzerland
C. Pandu Rangan
  Indian Institute of Technology, Madras, India
Bernhard Steffen
  TU Dortmund University, Germany
Madhu Sudan
  Microsoft Research, Cambridge, MA, USA
Demetri Terzopoulos
  University of California, Los Angeles, CA, USA
Doug Tygar
  University of California, Berkeley, CA, USA
Gerhard Weikum
  Max Planck Institute for Informatics, Saarbruecken, Germany

Marc Fischlin   Stefan Katzenbeisser (Eds.)

Number Theory and Cryptography

Papers in Honor of Johannes Buchmann
on the Occasion of His 60th Birthday

Springer

Volume Editors

Marc Fischlin
Technische Universität Darmstadt & CASED
Mornewegstraße 30
64293 Darmstadt, Germany
E-mail: marc.fischlin@gmail.com

Stefan Katzenbeisser
Technische Universität Darmstadt & CASED
Mornewegstraße 32
64293 Darmstadt, Germany
E-mail: katzenbeisser@seceng.informatik.tu-darmstadt.de

Cover illustration: Johannes Buchmann is a passionate photographer.
His pictures of mannequin faces were displayed at Schloss Dagstuhl in 2009.
© Johannes Buchmann

ISSN 0302-9743   e-ISSN 1611-3349
ISBN 978-3-642-42000-9   e-ISBN 978-3-642-42001-6
DOI 10.1007/978-3-642-42001-6
Springer Heidelberg New York Dordrecht London

CR Subject Classification (1998): F.2, F.1, C.2, E.3, G.1, K.6.5, D.4.6

LNCS Sublibrary: SL 1 – Theoretical Computer Science and General Issues

© Springer-Verlag Berlin Heidelberg 2013

Typesetting: Camera-ready by author, data conversion by Scientific Publishing Services, Chennai, India

Printed on acid-free paper

Springer is part of Springer Science+Business Media (www.springer.com)

Johannes Buchmann

Preface

Johannes Buchmann was the first professor in Darmstadt to work on cryptography when he joined the university in 1996. Now, slightly more than 15 years later, Darmstadt hosts one of the largest centers for IT security in the world. In between, many success stories happened, various research projects were initiated, and several colleagues working in the areas of cryptography and computer security were hired (including the editors of this volume). All these achievements can be traced back to Johannes' efforts.

This year we are celebrating Johannes' 60th birthday. It is a perfect occasion to honor Johannes' academic achievements. Johannes is the (co-)author of more than 100 scientific publications. He has supervised more than 60 Ph.D. theses, and various of his students now hold positions in academic institutions around the world. Johannes' book *Introduction to Cryptography*, published by Springer Verlag, has been very influential in teaching cryptography around the world, as witnessed by several translations. Johannes' scientific work has been acknowledged over and over again. In 1993 he was awarded the Leibniz Prize of the German Research Foundation (DFG), the most important research award in Germany. He has received prestigious awards such as the Beckurts-Prize (2006), the German IT Security Award (2008) and the Tsungming Tu-Alexander von Humboldt Research Award (2012), the highest scientific award in Taiwan. Johannes is a member of the Academy of Science of Berlin-Brandenburg, the National Academy of Science and Engineering (acatech) and the German National Academy of Sciences Leopoldina.

A Festschrift can hardly be compared to awards like the ones mentioned above. And yet, a Festschrift has a very distinguished property: it acknowledges the academic achievements of the honoree by means of scientific contributions. That is, fellow researchers endow their scientific works and their ideas as a tribute to their colleague. At the same time, because of its special nature, a Festschrift also serves as a venue to tell personal anecdotes, characterizing important aspects of the accomplishments of the honored person.

This Festschrift reflects the many research areas in which Johannes has been active over the past years, and yet it cannot comprise all aspects of his research to full extent. Being a mathematician, at the early stage of his career in the 80's Johannes made important contributions to the field of algebraic number theory. The first part of this book is dedicated to this research area. With his background it comes as no surprise that Johannes soon turned his attention to cryptography, which emerged as a new application area of these mathematical foundations in the 80's and 90's. This dedication to cryptography, a branch of research that is nowadays rooted in computer science, continues to last. Johannes made important contributions to estimating the hardness of security assumptions which form the basis of cryptographic mechanisms and to the appropriate choices of key sizes. The second part of this book is devoted to this field.

With cryptography being increasingly used in modern communication, Johannes' interest in the more practical aspects of cryptography has grown as well. This is reflected in the third part of this book that discusses the efficiency and security of cryptographic schemes when implemented in hardware. Recently Johannes became interested in the application of cryptography and security techniques to real-world scenarios, including the ability to make cryptography usable for the broad public. The final contributions of this volume therefore cover these aspects. All in all, we believe that this volume constitutes an interesting overview about Johannes' scientific work, and a rightful tribute to his achievements.

When the idea of the Festschrift came to our minds, we easily compiled a list of potential contributors. We merely had to think of close colleagues, former students, and other well-respected researchers in the areas Johannes has worked on, and the list was instantaneously filled with names. The positive feedback of all the researchers we approached was overwhelming, such that assembling this volume became a joyful procedure. This, of course, was only possible in close collaboration with all authors who contributed to this Festschrift. We are very thankful to them. We would also like to thank Springer Verlag for giving us the possibility to publish this volume in their Lecture Notes in Computer Science Series.

Happy birthday, Johannes!

November 2013   Marc Fischlin
                                                 Stefan Katzenbeisser

Greeting by the Department

The Computer Science department would like to preface this honorary publication with very special and heartfelt congratulations to Johannes Buchmann. With Johannes Buchmann, the department is honoring a member from its midst who was one of the department's most defining personalities in its more than forty years of history.

Like almost no other, Johannes Buchmann is both a researcher and a teacher with heart and soul, and with utmost commitment. Even in this later stage of his career he keeps breaking new ground with great enthusiasm — particularly of course in cryptographic research. To mention just one example: Elliptic curve cryptography, to which he contributed significantly, was still in the midst of its development, when he already started to target quantum cryptography. Besides enjoying the highest reputation internationally, his achievements procured him numerous awards in Germany, such as memberships in prestigious scientific academies. This publication will honor his outstanding research achievements in more detail in the following chapters. Further, it is a great pleasure for the department to honor Johannes Buchmann as one of its most committed and gifted teachers. He never made a secret out of being proud to stem from a family of teachers and also to have pursued this profession for a while himself. Whenever we experience a taste of his pedagogical skills, we envy his PhD candidates and students for being able to benefit from his talent and enthusiasm.

When we, his colleagues, think of Johannes Buchmann, we do not first and foremost think of him as the excellent researcher and teacher, but as the leader and creator, the persuader and strategist. Without him, Darmstadt would not even have come close to being the internationally recognized IT security research landmark it is now. At Technische Universität Darmstadt alone, approximately one third of the computer science department's professors work on core topics of security, and even more carry out parts of their research in this area. This status, unique in Germany and possibly even in Europe, could not be attained overnight. It was Johannes Buchmann who had the courage to imagine such a vision and the tenacious dedication to realize it. This involved overcoming resistance on all levels, convincing decision makers, and attracting the most distinguished colleagues to come and work in Darmstadt. Such accomplishments cannot be achieved without plenty of determination and great skill. Understandably enough, none of the department's members are comfortable with an undertaking if it comes to their attention that Johannes Buchmann is not backing it. Thereby he is not seen as a ruthless power seeker, but as a brilliant rhetorician and crystal clear thinker. He knows how to refine facts in a precise and concise manner, and to present decisive arguments — who can blame him for giving priority to the arguments he himself considers to be the most important.

Despite all of these preeminent talents, Johannes Buchmann has not succumbed to the temptation of abusing them for his own well-being. On the contrary, he is fighting for his field of research, which he is deeply convinced to be of utmost importance for our society. He is also fighting for our department as a whole, for two different reasons: On one hand, he is convinced of information technology's key role for our future; he wants to foster cutting edge information research in Germany and to ensure that our junior scientists are optimally educated and trained. On the other hand, Johannes Buchmann feels committed towards us as his colleagues and the scientific community, a fact which frequently touched us directly and amazed us when witnessing it. His commitment is fueled by deep ethical convictions — Johannes Buchmann is a mentor and advisor devoting a substantial part of his life to developing and promoting others. Most of us were able to benefit from his mentorship and advice more or less intensively in the past — and with this the circle closes and the first sentence of this greeting bears its substance: We bring truly sincere good wishes, borne by many grateful colleagues and a department paying the deepest respects for Johannes Buchmann's life achievements.

November 2013   Max Mühlhäuser
                                    Prodekan, Computer Science Department
                                        Technische Universität Darmstadt

Table of Contents

Hardware Security

Privacy and Security

Application Security

Laudatio in Honour of Professor Dr. Johannes Buchmann on the Occasion of His 60th Birthday

Hugh C. Williams

Dept. of Mathematics and Statistics University of Calgary

I am very pleased and honoured to be asked to provide this laudatio for Johannnes Buchmann, a good friend, colleague and collaborator for over 28 years.

Johannes was born on November 20, 1953 in Cologne and attended university there at the University of Cologne, where he obtained his doctorate in 1982 under the supervision of the late Hans-Joachim Stender. He went on in 1985 to become a research assistant at the University of Düsseldorf and completed his Habilitation there, with Michael Pohst as advisor, in 1988. From 1988 until 1996 he served as Professor of Computer Science at the University of the Saarland, and since 1996 he has been Professor of Computer Science and Mathematics at the Technical University in Darmstadt.

There can be no doubt that Johannes is internationally recognized as one of the leading figures in areas of computational number theory, cryptography and information security. He has published in refereed journals or conference proceedings over 160 scientific papers, spanning a very wide spectrum of interests including computational algebraic number theory, encryption schemes, signature schemes, cryptanalysis, internet privacy, security architecture, and post quantum cryptography. His work in these various areas is outstanding. In particular, I regard his work in post quantum cryptography, a subject in which he is a pioneer, to be of particular quality and importance. I must confess, however, that the work he has done on applying algebraic number theory to cryptography remains my personal favourite. Perhaps this is because this was the area on which we collaborated so many years ago.

Over the past 30 years, cryptography and cryptographic protocols have become a key element of information systems, protecting data and communications to ensure confidentiality, integrity and authenticity of data. The systems that are used for these purposes rely for their security on the (presumed) difficulty of specific mathematical problems such as integer factorization and the modular discrete logarithm problem. It is these schemes on which the security of all internet traffic relies.

The possibility of quantum computing becoming practical would change this picture dramatically. Some (very small) quantum computers have already been constructed, and the principles of fault-tolerant quantum error correction indicate that the difficult challenge of translating this experience to efficiently scalable quantum computers is a technological one, a problem which some specialists claim might be solved within the next 30 years. If realized, the difficulty of the problems on which the security of our current cryptosystems relies would diminish considerably, rendering these systems useless. Much of Johannes's work is focussed on the development of cryptographic schemes that would be resistant to attacks by quantum computers. This is

M. Fischlin and S. Katzenbeisser (Eds.): Buchmann Festschrift, LNCS 8260, pp. 1–2, 2013.

what post quantum cryptography is all about, and given our present level of reliance on the internet, it is vitally important that this research continues. Research in this area often involves elegant applications of diverse mathematics, including algebraic geometry, combinatorics, coding theory, information theory, lattice theory and probability.

Johannes has been awarded a number of prizes for his work, including the Leibnitz Prize of the Deutsche Forschungsgemeinschaft, the Karl-Heinz-Beckurts Prize, the IT-Security Prize of the Horst Görtz-Foundation and the Tsungming Tu Prize of the National Science Council of Taiwan. Also, the University of Debrecen has conferred on him an honorary doctorate. Furthermore, he is a member of several German academies of science. All of these provide ample testimony to the impact of his work and his international reputation for scholarly excellence. He is frequently asked to present his work as a plenary speaker at international conferences and workshops and he has served as an editor of several international, high quality journals. He currently sits on the editorial board of three of these, and from 1990 until 2009 sat on the editorial board of the *Journal of Cryptology*, arguably the premier journal devoted to academic cryptography.

I should also mention that Johannes has written several books; one of these, *Einführung in die Kryptographie*, an introductory textbook in cryptography, was so successful that it has gone through five editions and has been translated into seven languages. He has also written or coauthored books on such diverse subjects as binary quadratic forms, internet privacy, coding theory, and post quantum cryptography.

He has supervised over 220 Bachelor, Master's and Diploma theses and over 60 PhD theses, a most remarkable record of teaching and advising. Many of his PhD students have contributed to top quality research and have gone on to positions that require highly desired skills, obtained through the training that they received. Several have been awarded academic appointments in prestigious universities.

All of this is evidence of his outstanding research and teaching track record; however, I would be most remiss if I did not comment on another of his attributes, one that is very hard to find among academics: leadership. As indicated above, Johannes has enjoyed considerable success in building up and directing his research group CDC at Darmstadt, but he also served as the Dean of the Department of Computer Science at TU Darmstadt and then went on to become Vice President of the university for six years (2001-2007).

In view of Johannes's long, consistent and distinguished record of scholarly achievement and academic leadership, it is clear that we are not just celebrating his birthday, but also his exceptional academic career. Happy Birthday, Johannes!

Have a Break – Have a Security Centre:
From DZI to CASED

Harald Baier

da/sec – Biometrics and Internet Security Research Group
Hochschule Darmstadt, Darmstadt, Germany
and
Center for Advanced Security Research Darmstadt (CASED)
harald.baier@cased.de

Abstract. If you want to be successful you must follow a vision and be able to inspire your environment. In my long shared history with Johannes Buchmann since July 1998 I learned that he is very effective in seeding a suitable idea and managing an eligible ambience to let the thought become successful reality. One prominent example is the establishment of the predecessor of CASED, the Darmstädter Zentrum für IT-Sicherheit (DZI). In this article I summarise the different steps from the idea to found a security centre in Darmstadt to the current form of CASED. Furthermore I show the significance of the pizzeria Da Nino at the beginning of this process.

1   Introduction

Once you have submitted your PhD thesis and prepare for your defence you have to decide about your near future (and probably your distant one, too). The main tracks you may follow are either a scientific career or an employment in industry. After about 3 years at Johannes Buchmann's cdc chair at Technische Universität Darmstadt (*cdc* represents his three research areas *c*ryptography, *d*istributed systems and *c*omputer algebra), my decision period started in March 2002.

Johannes was aware of my uncertainty. Already in previous situations, where I had to come to an important decision, Johannes invited me for lunch to *Da Nino*, an Italian restaurant vis-à-vis of our former office site in Alexanderstraße 10. We had a mutual break, it was some day in April 2002, but I do not remember the actual date. He reported me the idea to found a security centre at Technische Universität Darmstadt (TUD) and offered me to serve as managing director of the centre.

Johannes emphasised two points: first he referred to Bochum as an example that an institutionalisation can quicken the IT security activities in a university. In the previous years Horst Görtz, the founder of the IT security corporation Utimaco Safeware AG, supported the establishment of an IT security centre at the Ruhr-Universität Bochum, which is still very successful as *Horst Görtz Institut für IT-Sicherheit* (HGI). Bochum managed to attract outstanding people

M. Fischlin and S. Katzenbeisser (Eds.): Buchmann Festschrift, LNCS 8260, pp. 3–18, 2013.

like Hans Dobbertin (who unfortunately died in 2006), Christof Paar and Jörg Schwenk.

Second he pointed to the excellent boundary conditions for such a centre at TUD, because Claudia Eckert previously accepted an offer for a professorship of IT security at TUD, which also involves the management of the Fraunhofer Institute for Secure Information Technology (SIT). And like in Bochum Horst Görtz sponsored her chair.

Claudia's two jobs relied on a formal cooperation between TUD and SIT, and the Fraunhofer-Gesellschaft provided a funding to TUD until 2007, which may be used to establish a security centre. Additionally the success of further IT security institutions (or departments) in Darmstadt like the famous A8 of the Fraunhofer Institute for Computer Graphics Research (IGD) or the Competence Centre for Applied Security Technology (CAST) would stimulate and benefit from an IT security centre at TUD. His vision was to bring these institutions together to become the leading IT security centre in Germany (and maybe in Europe or even in the world).

I do not remember my immediate answer, but I guess that I asked Johannes for some respite. Effectively I accepted his offer and got involved in establishing the IT security centre at TUD, the Darmstädter Zentrum für IT-Sicherheit (DZI). As of today the DZI is replaced by its successor CASED, one of the leading IT security centres in the world. Hence Johannes' initial vision became reality.

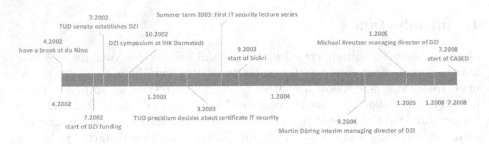

Fig. 1. The timeline of the evolution of DZI to CASED

In this article I exhibit key milestones in the history of the DZI as depicted in Figure 1. First I describe in Section 2 the *birth and initial steps* of DZI as a scientific centre at TUD. Then Section 3 shows the *adolescence* of the DZI, which mainly comprises the year 2003. Section 4 gives some insights into different aspects of our *public relation concept*, e.g., the juicy history of the DZI logo. Finally, I sketch in Section 5 the *handover* to Martin Döring and Michael Kreutzer and the final step to CASED.

2   Birth and Childhood of DZI

About one month after our joint lunch break at Da Nino, I defended my PhD on May 07th 2002. Beforehand Johannes had informed Claudia about his plans to offer the management director position of the planned IT security centre at TUD to me and invited her to be part of the examination board of my defence to get knowledge of my scientific work. She accepted and the following day we had an appointment at SIT to share our plans about the IT security centre and our roles. Since that point in time it was confirmed that Claudia, Johannes and myself would work on establishing the DZI.

The first two important issues were the formal establishment of DZI as scientific centre of TUD as described in Section 2.1 and the public celebration at IHK Darmstadt in cooperation with the Hessian symposium science-economy (see Section 2.2), the sponsor of the inception event.

2.1   Formal Foundation

Our aim was to establish DZI as fast as possible, i.e., during summer term 2002. Formally we had the choice between two organisational forms: an institute and a centre. An institute, however, is part of only one department, which contradicted our understanding of IT security as an interdisciplinary topic. Thus we decided to found DZI as a scientific centre including various departments of TUD. I remember well, when Johannes started gathering people by contacting them by phone and said *four participants is quite good for one afternoon.*

After all, we managed to convince people from 5 departments to participate in DZI. Each of these departments sent us their letters of participation. On July 31st 2002 the senate of TUD accepted DZI as scientific centre of TUD.

The press release to announce the DZI inception enumerated four working fields: Education (e.g., establishing an IT security track, support of further education in cooperation with the CAST forum), research (e.g., interdisciplinary IT security research, support of third-party funding proposals), public relations of the IT security activities at TUD, and finally technology transfer from TUD to the outside world. If you are familiar with CASED and its goals, you will immediately recognise these areas.

The primary 20 professors from 5 departments were:

- FB01 (Law and Economics): J. Marly
- FB04 (Mathematics): B. Kümmerer, J. Lehn, N. Schappacher, T. Streicher, W. Schindler
- FB05 (Physics): G. Alber, W. Elsäßer, Th. Walther
- FB18 (Electrical Engineering and Information Technology): H. Eveking, A. Schürr, R. Steinmetz
- FB20 (Computer Science): A. Buchmann, J. Buchmann, C. Eckert, J. Encarnação, S. Huss, M. Mühlhäuser, T. Takagi, Chr. Walther

2.2   Public Celebration of Inception

After the formal inception of DZI in July 2002 we prepared the public celebration of its establishment. In order to stress the technology transfer goal of DZI we decided in favour of a symposium in cooperation with the Hessian forum science-economy and the local Chamber of Industry and Commerce (IHK). The Hessian forum science-economy provided the funding of the symposium. It is mainly supported by the Hessian State Ministry of Higher Education, Research and the Arts (HMWK) and the Hessian Ministry of Economics, Transport, Urban and Regional Development (HMWVL). Hence we strenghened the relation to our federal state government and forced the perception of Darmstadt as city of IT security in Wiesbaden.

Referent	Titel	Zeit
Dr. H. Hirschler (Staatssekretär im HMWVL)	Grußwort	13:30 – 13:40
Joachim F. Krahl (Vizepräsident IHK Darmstadt)	Grußwort	13:40 – 13:50
Prof. J. Buchmann (Vizepräsident TUD)	Ziele und Aufgaben des DZI	13:50 – 14:00
S. Engel–Flechsig (CEO Radicchio)	Visionen und Anwendungen von E–Commerce	14:00 – 14:40
Dr. S. Paulus (Director Product Management Security, SAP)	Wie Sicherheit zum Einsatz von E–Business–Prozessen verhilft	14:40 – 15:20
	Pause	15:20 – 15:50
Prof. C. Eckert (Leiterin des Fraunhoferinstituts Sichere Telekooperation)	Sicherer E–Commerce	15:50 – 16:30
O. Jüptner (Senior Consultant, IBH)	IT–Sicherheit im Mittelstand	16:30 – 17:10
R. Lauenroth (IBM Solution Manager)	E–Payment–Systeme	17:10 – 17:50
	Pause	17:50 – 18:15
	Podiumsdiskussion	18:15 – 19:00
	Gemeinsames Abendessen	19:00 – 20:30

Fig. 2. The agenda of the inception symposium on October 10th 2002

The symposium took place in the afternoon of October 10th 2002. Its title was *E-commerce – but secure*. A screenshot of the original agenda is listed in

Sichere Anwendungen
· sichere Mobiltechnologie
· Sicherheit in Mediendaten
· Cyberlaw

Zukunftstechnologie
· Quantenkryptographie
· Quantenkommunikation

Forschungs-
schwerpunkt
IT-Sicherheit
Interdisziplinär,
Gesamt-Systeme

Sichere Hardware
· effiziente Krypto-Module
· Chipkarten-Entwicklung

Sicherheitstechnologie
· Kryptographie
· Public-Key Infrastruktur
· Protokolle und Verifikation

Fig. 3. Advertising slide of DZI competencies

Figure 2. After some short welcoming speeches from the HMWVL and the IHK, Johannes gave a short introduction into the goals of DZI. Then 5 speakers gave their presentations to different aspects of e-commerce, including a talk given by Claudia Eckert on secure e-commerce. The symposium ended with a panel discussion and a joint dinner. About 100 participants including both experts and interested laymen were present.

3   Adolescence of DZI

Once the DZI was born and we had celebrated its nativity, we had to work on achieving the goals of the DZI as listed in Section 2.1. This adolescence period mainly took place in 2003. Coordination of the IT security research at TUD was one key item of the DZI. I give details on the TUD internal instrument called *research foci* in Section 3.1. Furthermore I list some DZI acquired projects. The DZI supported education in IT security in cooperation with the CAST forum through its certificate IT security as described in Section 3.2. Section 3.3 recapitulises information on people working at DZI and its premises, mainly focused on the period until 2004.

3.1   TUD Research Focus IT Security and First Projects

In this section I describe the successful application of DZI to become a TUD research focus. Then I mention the first DZI acquired projects: a guideline of digital signatures, a usability and security study on public key infrastructure applications, and the first large-scale project SicAri.

Johannes served during 2002 to 2007 as vice president of research at TUD. During this period he established an internal funding instrument called *research focus* at TUD (in German *Forschungsschwerpunkt*).

A research focus is represented by TUD researchers, who work on a topic of excellence at TUD. Through the research focus they apply for further funding to support their growth (e.g., through administrative infrastructure like secretary, technical infrastructure, premises). The research focus should attract further external funding. The typical initial funding period was 2 years.

In a first step, the TUD presidium evaluated an application as research focus. In case of a positive assessment the senate of TUD took the final decision. The first research focus at TUD had been biotechnology, established in April 2003. Exactly one year after my PhD defence, the TUD senate decided about three further research foci on May 07th 2003: IT security, computational engineering, and technical flow and burn-up.

Claudia Eckert defended the proposal and used the advertising slide as shown in Figure 3. The application was mainly based on the acquisition of the security platform project SicArI as described below (at that point in time, the final letter was a capital one), further education in cooperation with CAST, and the establishment of a certificate IT security for students (see Section 3.2). Interestingly, SicArI was mentioned as *fundamental research* funded as *Collaborative Research Centre* by the German Research Foundation (DFG). Below we will see that the actual project called SicAri was funded by the German Ministry of Education and Research (BMBF).

5.4   Einrichtung des Forschungsschwerpunkts „IT-Sicherheit"

Da der Vizepräsident den Forschungsschwerpunkt mit prägt, übernimmt der Präsident die Sitzungsleitung.

Frau Professorin Eckert stellt den Forschungsschwerpunkt anhand von Folien vor, die die Anlage FN 12/03 ergänzen und der Ergebnisniederschrift als Anlage 5 beigefügt sind. Die beantragten Personal- und Sachmittel liegen bei 50 K€ p.a.; außerdem werden vier Räume benötigt. Der Präsident berichtet, dass das Präsidium den Antrag nach den Kriterien überprüft und positiv bewertet hat; es empfiehlt eine zweijährige Förderung mit jeweils 50 K€ und die benötigten Räume nach Möglichkeit bei der Belegung des Zintl-Instituts vorzusehen.

Der Senat befürwortet mit einer Enthaltung die Einrichtung des Forschungsschwerpunkts „IT-Sicherheit" und die vorgeschlagene Unterstützung mit jeweils 50 K€ und vier Räumen für zwei Jahre.

Fig. 4. The senate protocol of the research focus IT security

Finally Claudia was successful and the DZI acquired 100,000 EUR over two years to setup a secretary and to extend our premises to 4 rooms once the department of computer science moves to the Zintl building (today called Piloty building). Figure 4 shows the senate protocol part of the establishment of the IT security research focus.

The first DZI acquired project was a rather small one. In September 2002 the Investment Bank of Hesse (Investitionsbank Hessen, IBH) organised workshops to advertise IT security to Hessian small businesses. I gave talks on secure data transmission. The IBH invited me to write a guideline on digital signatures (Digitale Signatur – Leitfaden zum Einsatz digitaler Signaturen). I convinced my former colleague Tobias Straub to support me in writing the technical part and his wife Judith Klink to address the legal aspects of that topic. We wrote the guideline in the time period December 2002 until February 2003. Amazingly the guideline is still available in September 2013 via the web presence of Hessen IT[1].

The second DZI project was a study on the usability of PKI products. It was sponsored by Microsoft Germany. Johannes was contacted by Microsoft, after some short discussion with Microsoft about the form of the study and the funding we agreed upon this project. The study was mainly organised and written by Tobias Straub and myself. We had been supported by students in form of a seminar or as student workers. We tested different PKI product classes, e.g., mail clients, browsers, web servers, and login applications. The study evolved from March 2003 until June 2003. The final document comprised 197 pages, a quite large volume for a lobbying document in Berlin.

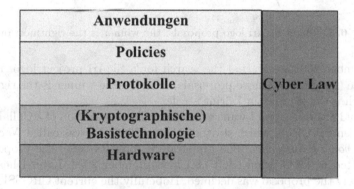

Fig. 5. The well-knwon SicAri layered architecture

The third and most prominent DZI project in 2003 was SicAri (a security architecture and its tools for ubiquitous Internet utilisation). The project was headed by Claudia and Johannes. We initially planned to fund SicAri through the BMBF call *Internetökonomie*, where it unfortunately did not meet the topic. Then we planned to submit the SicAri proposal to the DFG (see Claudia's senate presentation on the research focus above), however, in the meantime the BMBF pointed us to a fitting funding scheme *FUTUR*, and we decided to submit it again to the BMBF.

---

[1] http://www.hessen-it.de/mm/DigitaleSignatur.pdf

After an initial positive evaluation of our proposal, we had to defend it on July 10th 2003 at the BMBF premises in Bonn. As Claudia was part of an expert group, which held a meeting at that day in BMBF, Johannes had to present SicAri. During our journey to Bonn, we jointly finalised the presentation on the back seats of the TUD staff car. The layered architecture image as depicted in Figure 5 was shown quite often in the context of SicAri.

In all the TUD acquired about 2.8 million EUR for a 4 year period starting in October 2003. The SicAri consortium consisted besides the DZI of several chairs and insitutions from TUD (e.g., Johannes Buchmann, Claudia Eckert, Sorin Huss, Max Mühlhäuser, Viola Schmid, Ralf Steinmetz, ITO). In addition the Fraunhofer institutes IGD and SIT were involved as well as cv cryptovision GmbH (Gelsenkirchen), FlexSecure GmbH (Darmstadt), MediaSec Technologies GmbH (Essen), NEC Europe Ltd. (Heidelberg), Philips Semiconductors GmbH (Hamburg), T-Nova GmbH (Darmstadt) und usd.de ag (Langen/Hessen). However, the involvement of NEC was a non-trivial task.

Fig. 6. Different SicAri logo proposals, the winner is the rightmost one

In December 2003 we started the search for a SicAri project logo. Figure 6 shows three (out of much more proposals). Finally, the winner is the right-hand draft, which is due to Martin Döring's wife.

At the end of this section I want to mention the preparation of a Collaborative Research Centre (CRC), which started in June 2003 and was called *Nachhaltige Sicherheit von IT-Systemen*. The preparation took its time, the proposal was not submitted to DFG when I left DZI in September 2004. Later I heard that unfortunately the proposal was declined. Hopefully the current CROSSING submission will fill the gap of a security focused CRC in Darmstadt.

3.2   Certificate IT Security

In this section I sketch the educational contribution of DZI for IT security interested students at TUD. One goal of DZI was to extend the IT security education. We (that were mainly Johannes, Claudia, Christoph Busch and myself) discussed different options, e.g., the establishment of a dedicated IT security master programme. However, due to the administrative overhead of such a solution, we decided to simply adapt an existing further educational model of TUD to a student certificate.

The International Institute in Lifelong Learning I^3L^3 is responsible to organise the further educational offerings at TUD. Beate Kriegler is managing this institution, and in 2000 she started a certificate IT security in further education, which still exists and which is supported by CAST. The model is quite easy: in its original form, external people came to TUD on Tuesday afternoon and attended the regular lectures at TUD. For one specific lecture a dedicated exercise course was provided exclusively for the certificate participants including some refreshments. For instance Bodo Möller was responsible for Johannes' exercise course in winter 2001, which we internally called *Bodos Brötchenübungen*. Additionally the participants were allowed to take part in the CAST workshops during the respective term. Once they successfully passed 8 modules (2 mandatory ones, the rest are electives), they obtain their certificate.

To provide an offering for students, we simply adapted the I^3L^3 certificate to the special needs of students, that is they had to pass a seminar and practical, respectively. As a certificate there was no need to pass a faculty board or the senate, we simply had to apply for the certificate at the presidium of TUD. In March 2003 the presidium approved the student certificate IT security and we could advertise it in summer term 2003.

And it worked fine. Already in its starting phase I got a number of requests. And the certificate was a good pool to acquire highly motivated and skilled students. Alexander von Bardeleben and Manuel Hartl, two student workers of DZI, passed successfully the certificate and were valuable in installing the technical infrastructure of DZI after its movement to the Piloty buildung in March 2004.

You may consider this certificate as the predecessor of today's master IT security at TUD.

3.3 Staff and Premises

This section enumerates core people working at DZI itself, that is I do not aim to mention the exciting large number of external persons who supported the DZI in its life period. If I forgot somebody, I would be glad if he/she pardons me.

Fig. 7. DZI directors and managing directors

I start with the DZI directors: Claudia Eckert and Johannes Buchmann as shown in Figure 7. Their pictures were gathered in 2002 to be shown on the DZI website. The evolvement of DZI within the faculty and the university was mainly driven by them. Their creative spirit and the constructive atmosphere was very impressing to me. And it is consequent that both were the driving force during the CASED application.

In all the DZI had three managing directors: it started with me on July 01st 2002. I left DZI on August 31st 2004 due to my appointment at Fachhochschule Bingen. Johannes and Claudia started their search for my successor in May 2004. After an announcement and the subsequent selection procedure, it was clear that Michael Kreutzer was the successful candidate. However, he was still in the final period of his PhD work. The trade-off was that he would join DZI on January 01st 2005. As with Johannes and Claudia, Michael is one of the designers of CASED.

Hence we had to fill a gap of 4 months. Fortunately Martin Döring, the SicAri staff member at cdc, was willing to serve as interim managing director. I am still very grateful for his support.

Due to the research focus funding we were able to install a secreteriat. However, before our relocation to the Piloty building the DZI premises had been a one-office 'centre' in Alexanderstraße 10. Hence we simply did not have an office for a secretary. After the move Dörte Lührs became the first DZI secretary on July 09th 2004. She supported the DZI until June 2006. Since then she is involved in the public relations department of TUD. Cornelia Reitz followed Dörte, and Cornelia is still serving in the CASED office.

I had valuable support of student workers. I would first like to mention Alexander von Bardeleben, Manuel Hartl and Jochen Becker, who organised and installed the technical environment at Piloty (before the move I could make use of the cdc environment). Manuel later worked for some years at FlexSecure, Johannes' spin off in Eberstadt. Hence he was devoted to his IT security activities. Additionally Golriz Chehrazi and Michele Boivin supported me in different tasks, e.g., organising events or public relations.

As final person I would like to point to Vangelis Karatsiolis, who was the first scientific staff member of the DZI itself in a project since June 2004. He was involved in the TUD card project to deploy a smart card to members of TUD.

With respect to the DZI premises, the start was comparable to the famous garage stories of HP, Apple or Microsoft. In the beginning the DZI worked in exactly one room in Alexanderstr. 10 (S1|15) on the third floor. Then on March 24th 2004 the DZI moved jointly with the faculty of computer science to the Piloty building, which previously was called Zintl. There we had 4 large offices in track E. Amazingly March 24th 2004 was the day of my interview in Bingen.

4   Public Relations

The external presentation of the IT security activities at TUD was one key goal of the DZI. It is not the aim of this section to present any detail of our concept, instead I highlight some activities. I start in Section 4.1 with the evolution of the DZI logo. Then Section 4.2 gives us a feeling of the web layout back in 2002. Finally I present in Section 4.3 some information on our first lecture series, which addressed the broad public.

4.1 Logo

A logo is an important identifier of an institution and aims at being simple, but somehow related to the working fields of the people. Although computer scientists are far away from being experts in designing an appropriate logo, we started in October 2002 to create drafts ourselves.

Fig. 8. Unsuccessful logo proposals of the DZI

The first logo proposal depicted on the left side in Figure 8 was due to Markus Ruppert and myself. It is kept simple. The five dots represent the five departments supporting the DZI. Nevertheless Johannes declined to accept that proposal. His objection was that the formation of the dots resembles too much the symbol of blind people.

The second proposal in Figure 8 is based on the classical key-lock-principle. It was due to Emre Karaca, a student worker of the DZI. Our key trouble with the logo was its obvious self-made property. Besides please note that DZI is entitled as Darmstädter *Institut* für IT-Sicherheit.

Finally, the third unsuccessful proposal is printed on the right in Figure 8. I do not remember the name of its creator, but it is obviously too complicated to be accepted as a logo. However, key aspects of the DZI like the departments of law, mathematics and physics are visualised, encircled by different interaction channels of different IT devices.

And the winner is ... At some point in time, Christoph Busch noticed our internal logo competition and our inappropriate drafts. He acquired the support of his wife Nana, who submitted the actual logo as shown in Figure 9 on January 07th 2003, that is about 3 months after the logo search started. In his mail Christoph pointed to four aspects of the DZI visualised in the logo: *Ich bin gespannt, wer alle vier Themen "Was wollte die Kuenstlerin damit sagen....?" herausfindet.* I leave this task to the reader.

Finally I point to a quite interesting fact: like in the case of the SicAri logo (see Page 6), the final DZI logo is due to a wife of a DZI related scientist.

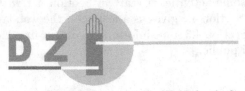

Darmstädter Zentrum für IT-Sicherheit

Fig. 9. The actual DZI logo, drafted by Nana Busch

4.2 Web Site and Newsletter

Important public relations instruments in the pre-Facebook and -Twitter age
had been a web page and newsletter, respectively. Therefore the DZI made use
of these information channels, too.

During the second half of 2002 we prepared the web presence of the DZI.
Honestly speaking I do not remember, who was involved, but I guess it was
mainly self-made. On St Nicholas' Day in 2002 (i.e., December 06th 2002) I
announced per mail the launch of the web site to the members of the DZI. Its
URL was www.dzi.tu-darmstadt.de.

Fig. 10. The initial web site of DZI

The web site aimed at providing essential information on the activities of DZI, hence its structure was straightforward. As one item the web site gave details on the involved persons. Figure 10 shows parts of Johannes' initial self-description.

From a today's point of view its layout and choice of colours appear 'interesting'. Nevertheless Johannes was enthusiastic and answered my announcement late in the evening the same day as given in Figure 11.

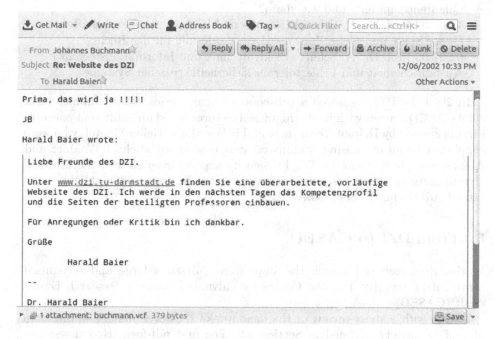

Fig. 11. Johannes' mail answer to the launch of the DZI web site

Besides a web site we started to inform about the DZI activities per newsletter on April 15th, 2003. The first one was sent in text format per mail. Later, since December 2004 (when I had already left the DZI), the newsletter was provided in pdf. Figure 12 shows a part of this first pdf-newsletter, which outlines the staff situation at DZI.

4.3   First Lecture Series 'Sind Sie Sicher?'

About 1 year after its establishment the DZI organised in the summer term 2003 for the first time a lecture series in the scope of IT security. The initial concept was to present the different fields of activity of its members to the general public. Therefore the presentations were given in German by 6 DZI representatives of different departments.

The lecture series was held with respect to the slogan *'Sind Sie sicher?'*. The lectures were announced in the local press and highly accepted by the citizens of Darmstadt. The agenda was as follows:

- 23.04.03 Prof. Dr. C. Eckert (Informatik, Fraunhofer-Institut SIT):
  Sind Sie sicher? Einführung in die IT-Sicherheit
- 07.05.03 Dr. Chr. Busch (Fraunhofer-IGD):
  Biometrische Verfahren: Sicherheit durch Körpermerkmale
- 21.05.03 Prof. Dr. J. Buchmann (Mathematik, Informatik):
  Digitale Signaturen: Grundlagen und Anwendungen
- 04.06.03 Prof. Dr. Th. Walther (Physik):
  Quantencomputer - und was dann?
- 18.06.03 Prof. Dr. V. Schmid (Rechts- und Wirtschaftswissenschaften):
  Cyper Space und Cyber Law als Herausforderung für das Recht
- 02.07.03 Prof. Dr. A. Schürr (Elektrotechnik und Informationstechnik):
  Ausfallsicherheit und Fehlertoleranz sicherheitskritischer Systeme

In 2004 the DZI organised a subsequent lecture series. Since summer term 2009 CASED started with its distinguished lectures based on a different concept. First organised by Helmut Veith (now at TU Wien) and Heiko Mantel, who were supported by an organising committee, it is now up to Michael Waidner and Ahmad Sadeghi to invite leading IT security experts from all over the world to give insights into their respective scientific research. A current highlight is the talk of Adi Shamir in September 2013.

5   From DZI to CASED

In this final section I sketch the steps from DZI to a large and recognised centre of IT security, i.e., the Center for Advanced Security Research Darmstadt (CASED).

I start with a short review of the handover of the three managing directors of DZI as already outlined in Section 3.3. The first pdf-formatted newsletter of the DZI from December 2004 gave the key information on the handover of the managing directors. Figure 12 shows the relevant part of this newsletter. It announced that I had left the DZI on August 31st 2004 to work as a professor of mathematics at Fachhochschule Bingen. Additionnally the start of Michael Kreutzer on January 01st 2005 was declared as well as the intermediate activity of Martin Döring. Once again I am very grateful for Martin's willingness to serve 4 months as managing director of DZI.

An aphorism due to Carl Hilty says: *Zum wirklichen Erfolge im Leben, d.h. zur Erreichung der höchstmöglichen menschlichen Vollkommenheit und wahren, nutzbringenden Tätigkeit gehört notwendig ein öfterer äußerer Mißerfolg.* Before the big success CASED was established Hilty's statement became painful reality in three prominent cases.

First as mentioned in Section 3.1 the long prepared Collaborative Research Centre was declined. Although widely supported by TUD key persons like Claudia Eckert, Johannes Buchmann, and Ralf Steinmetz it was rejected.

TECHNISCHE
UNIVERSITÄT
DARMSTADT

Darmstädter Zentrum für IT-Sicherheit

D Z I – Newsletter
Nr. 1, Dezember 2004

Darmstädter Zentrum für IT-Sicherheit
Hochschulstraße 10, D-64289 Darmstadt,
Tel. 0 61 51 / 16 – 61 65 (Geschäftsführer)
Tel. 0 61 51 / 16 – 48 95 (Sekretariat), Fax 0 61 51 / 16 – 48 25
http://www.dzi.tu-darmstadt.de

Sehr geehrte Leserinnen und Leser,

wir freuen uns, Ihnen den ersten DZI-Newsletter vorstellen zu können. Der Newsletter wird zunächst in loser Folge erscheinen, soll in absehbarer Zeit jedoch quartalsmäßig und später in zweimonatigen Abständen erscheinen. Anregungen und Kritik von Ihrer Seite sind stets willkommen.

▶ Personalia

- Herr Dr. Harald **Baier**, seit der Gründung am 31. Juli 2002 Geschäftsführer des DZI, hat die TUD zum 31. August 2004 verlassen, um ab September 2004 eine Professur für Mathematik an der Fachhochschule Bingen anzutreten. Wir gratulieren ihm zu diesem Erfolg und bedanken uns für seine wertvolle Mitarbeit.

  Ab Januar 2005 wird Herr Michael **Kreutzer** das Amt des Geschäftsführers des DZI übernehmen. Herr Kreutzer promoviert zur Zeit über das Thema „Dienstfindung in mobilen multihop Adhoc-Netzen" an der Universität Freiburg bei Herrn Prof. Dr. Günter Müller. Herr Kreutzer hat bereits verschiedene Projekte betreut und ist weiterhin als Gutachter, Dozent und Co-Autor tätig.

  Von September bis Dezember 2004 übernimmt Herr Martin **Döring** die kommissarische Geschäftsführung des DZI.

Fig. 12. The first pdf newsletter of DZI

Second LOEWE is said to be an answer of our local state Hesse to the failure of Hessian universities in the federal competition of the excellence of universities. It is interesting to know if a comparable initiative like LOEWE would have been established if Hessian universities were successful in the federal contest. But without LOEWE the centre CASED probably would not have been founded.

Finally a very painful personal experience for Johannes, which from a today's perspective is very important for CASED: In July 2007 the election committee of TUD voted for Hans Jürgen Prömel as president of TUD, not for Johannes. If Johannes were elected as president, he would not have been an operational driving force in the application of CASED and its director for more than three years. Hence CASED would be quite different from its actual organisation, if it existed at all.

Since 2007 the CASED proposal was written mainly by Claudia Eckert, Johannes Buchmann, Michael Kreutzer, Christoph Busch, and Max Mühlhäuser. Fortunately, it was selected as a LOEWE research centre and started formally on July 01st 2008.

Fig. 13. The final newsletter of DZI

CASED announced different professorships at TUD and HDA (see Figure 13), and for me the wheel has turned full circle. I applied for the Internet security position at HDA and actually got the offer from HDA. As during my first period in Darmstadt I collaborate with my academic father.

Since 2008/2009 CASED is growing significantly, and I am looking tensely to the near future, what will happen in a possible third funding period of CASED and later. To sum up the initial goals of DZI as formulated 11 years ago have fully become reality through CASED. I am very glad to be part of it and for my joint time with Johannes.

Operating Degrees for XL vs. F_4/F_5 for Generic $\mathcal{MQ}$ with Number of Equations Linear in That of Variables

Jenny Yuan-Chun Yeh, Chen-Mou Cheng, and Bo-Yin Yang

Academia Sinica, Taipei, Taiwan
{jenny,doug,by}@crypto.tw

Abstract. We discuss the complexity of $\mathcal{MQ}$, or solving multivariate systems of m equations in n variables over the finite field $\mathbb{F}_q$ of q elements. $\mathcal{MQ}$ is an important hard problem in cryptography. In particular, the complexity to solve overdetermined $\mathcal{MQ}$ systems with randomly chosen coefficients when $m = cn$ is related to the provable security of a number of cryptosystems.

In this context there are two basic approaches. One is to use XL ("eXtended Linearization") with the solving step tailored to sparse linear algebra; the other is of the many variations of Jean-Charles Faugère's F_4/F_5 algorithms.

Although F_4/F_5 has been the de facto standard in the cryptographic community, it was proposed (Yang-Chen, 2004) that XL with Sparse Solver may be superior in some cases, particularly the generic overdetermined case with $m/n = c + o(1)$.

At the Steering Committee Meeting of the Post-Quantum Cryptography workshop in 2008, Johannes Buchmann listed several key research questions to all post-quantum cryptographers present. One problem in $\mathcal{MQ}$-based cryptography, he noted, is "if the difference between the operating degrees of XL(-with-Sparse-Solver) and F_4/F_5 approaches can be accurately bounded for random systems."

We answer in the affirmative when $m/n = c + o(1)$, using Saddle Point analysis:

1. For instances with randomly drawn coefficients, the degrees of operation of XL and F_4/F_5 has the most pronounced differential in the large-field, "barely overdetermined" ($m - n = c$) cases, where the discrepancy is $\propto \sqrt{n}$.
2. In most other types of random systems with $m/n = c + o(1)$, the expected difference in the operating degrees of XL and F_4/F_5 is constant which can be evaluated mathematically via asymptotic analysis.

Our conclusions are partially backed up using tests with Maple, MAGMA, and an XL implementation featuring Block Wiedemann as the sparse-matrix solver.

Keywords: sparse solver, Gröbner basis, XL, MQ, asymptotic analysis, F_4, F_5.

1 Introduction

$\mathcal{MQ}$ (Multivariate Quadratic), or finding variables $\mathbf{x} = (x_1, x_2, \ldots, x_n) \in (\mathbb{F}_q)^n$ from quadratic equations $p_1(\mathbf{x}) = p_2(\mathbf{x}) = \cdots = p_m(\mathbf{x}) = 0$, is an important hard problem. Instances of $\mathcal{MQ}$ appear in cryptographic situations such as a key step of many attacks known collectively as algebraic cryptanalysis.

J.-C. Faugère's F_4/F_5 algorithms are excellent system-solving algorithms both for $\mathcal{MQ}$ and for even more generalized problems with higher-degree polynomials with

M. Fischlin and S. Katzenbeisser (Eds.): Buchmann Festschrift, LNCS 8260, pp. 19–33, 2013.
© Springer-Verlag Berlin Heidelberg 2013

general applicability — that is, they work well for a large variety of systems including random ones — and are recognized as the de facto standard in the crypto community. Good commercially available implementations of generic F_5 being still sadly lacking, the F_4 implementation in MAGMA [21] is the usual yardstick against which equation-solving is measured [18].

If we limit ourselves to a somewhat theoretic context, cryptographers would like to find the best estimate of complexity of solving a random $\mathcal{MQ}$ when $m/n = c + o(1)$. It is generally believed [4] that the probability of any sub-exponential algorithm solving such random systems becomes negligible as the pameters increase. Such an algorithm would be the most important generic attack for what is known as multivariate quadratic PKCs (cf. [5]), and (iii) it determines the security of several provably secure constructions such as QUAD [4].

1.1 Questions

Despite a near monopoly of Faugère's F_4/F_5 algorithms in the crypto community, other algorithms has been proposed over F_4/F_5 in various contexts:

1. it has been noted that SAT solvers excel in specific cases;
2. other variants of Gröbner Basis methods, such as MutantXL [22, 23] or GGV [17], have been claimed to have better general performance; and
3. it has was suggested that the complexity of $\mathcal{MQ}$ might be better estimated via XL with sparse-matrix solvers when $m/n = c + o(1)$.

While the superiority of "Sparse XL" variants has been suggested since 2004 [28, 30], the issue of their merit has never been comprehensively settled.

In determining what circumstances favor Sparse XL over F_4/F_5, and vice versa, it is clear that systems which have a smaller difference between operating degrees of XL and F_4/F_5— smaller fields, generic ("semi-regular") systems, and more rather than less overdetermined — are better for XL. By late 2008, system-solving experts understood that if this difference is small, then XL with Sparse matrices will dominate F_4/F_5 to hold as the problem sizes get larger for generic $\mathcal{MQ}$ instances with n $m/n = c + o(1)$, provided that that certain heuristic conditions continue to hold.

With the Second Post-Quantum Cryptography Workshop at the University of Cincinnati (Oct. 17-19, 2008, Cincinnati Ohio, USA), a meeting of the Steering Committee of the workshop series was held during which Johannes Buchmann named some key research questions in PQCrypto, one being *whether the difference between the operating degrees of XL and* F_4/F_5 *approaches can be accurately bounded for random systems.*

1.2 Results

We are able to show, using saddle point asymptotic analysis that

1. in many cases of cryptographic interest, the difference in the degrees of operation for XL and F_4/F_5 is bounded tightly by a constant;
2. as n increases, the expected value of the difference approach a constant which can be rigorously determined;

3. the difference is at most 1 for many types of generic systems with $m/n = c + o(1)$, which means that the degrees of operation are usually the same for XL and F_4/F_5;
4. a specific case where the difference is unbounded is the large-field case with $f = m - n =$constant, which is mathematically expected and explainable.

Example: For most generic large-field systems with $m/n = 2$, the degree of operation for XL and F_4/F_5 are equal — actually about 80% of the time — and in the remaining cases the difference is 1.

Example: For direct attacks on QUAD-like ciphers (where provable security reduces to an $m/n = 2$ generic $\mathcal{MQ}$), the degree of operation for XL and F_4/F_5 also differs by 1 about half of the time, and are equal the rest of the time.

1.3 Prior and Related Work

Matrix Techniques in Gröbner-Basis Computations: Most modern system-solvers compute Gröbner Bases. In the 1965 original Buchberger algorithm [8], we take equations two at a time and eliminate around their lead terms by some ordering strategy. Lazard [19] noted that since each successive step involves linear combinations of the original equations, we save work by computing and storing a batch of monomial-equation products. Further, by making each equation a row in a matrix, we enable the use of efficient and well-studied elimination algorithms in linear algebra. This is the initial appearance of the algorithm now known as XL, and leaves open the use of sparse matrix algorithms.

Initial Appearances of XL with Sparse Solver: In 2004, Yang et al mentioned the possibility that despite a higher operating degree, XL with a sparse matrix solver would be better than F_4/F_5 with a conventional solver (such as Strassen with $\omega \approx 2.8$) and can when $q = 2$ it can potentially outdo a brute force search when $m = n$ [28]. Later that year, it was noted that in by adding the "F" ("fix", or guessing) approach [29], FXL with a sparse solver would be the method with the best time complexity and discusses how to compute the optimal number of guesses. In 2006 we see an initial implementation of such an algorithm in [31], now using standard Wiedemann as the solver.

Actual Use of Sparse XL for Cryptanalysis: 2006 marks the initial cryptanalytic paper [30] where the sparsity of the matrices plays a role. In this work it was noted that using Wiedemann allows an attack to be carried out with a practical computer in Sparse XL but not in MAGMA [21] of the time, due to the smaller memory footprint.

Parallelization of Sparse XL: In [15], the authors use a tailored XL algorithm with a parallelized standard (not block) Wiedemann using a large computer and OpenMP in defeating a Rainbow/TTS scheme with a suboptimal structure.

In [24], the authors implement an XL algorithm using block Wiedemann for 32 equations and 32 variables using just 8GB of main memory (this runs out of memory in MAGMA [21] at the time, and as late as 2012).

In [9] we find a block Wiedemann optimized for XL for a variety of different fields, including $\mathbb{F}_{16}$, $\mathbb{F}_2$, and $\mathbb{F}_{31}$, using both contiguous and MPI pragmas.

Legitimatization of Sparse XL: [3] introduces an algorithm termed BooleanSolve but is effectively the same as XL using a sparse solver and guessing (fixing). The formula for evaluating monomials is also one that would be used for FXL, not one corresponding to "the Hybrid Approach" [6] which advocates guessing with F_4/F_5 instead of XL.

[3] after nearly a decade of denial of neglect against XL still does a poor job of describing prior art, but it effectively vindicates Sparse XL and affirms the asymptotic superiority of Sparse XL than F_4/F_5 for generic systems. While we cannot pretend to read minds one practical reason for this late concession is precisely the fact pointed out in this work, i.e., that the degree of operation for XL and F_4/F_5 are often the same and has a bounded difference in most random cases.

1.4   Future Work

There are several issues identified by our study.

- Complexities previously evaluated with F_4/F_5 may need to be recomputed with Sparse XL variants. [3] does this to some extent but is incomplete in this aspect.
- An obvious improvement to Sparse XL for many situations would be Sparse F_4 (XL2). However, an straightforward implementation of that approach would be wasteful in that it throws away previously performed work with each raised degree. A better combination between a Wiedemann-like solver and F_4 would instantly lead to great advances.
- The number of columns that actually appear in the final F_4/F_5 matrix, a parameter that would determine a cut-off size when XL with sparse matrices catches up to F_4/F_5, is still yet to be determined conclusively.

2   History and the Status Quo of XL vs. F_4/F_5

Notations: We will denote by $\mathbf{x}^{\mathbf{b}}$ the monomial $x_1^{b_1} x_2^{b_2} \cdots x_n^{b_n}$, and its degree by $|\mathbf{b}| = \sum_{i=1}^{n} b_i$. We will choose a degree of operation D, and let $\mathcal{T} = \mathcal{T}^{(D)} = \{\mathbf{x}^{\mathbf{b}} : |\mathbf{b}| \le D\}$ be the set of degree-D-or-lower monomials. Multiply each equation p_i by all monomials $\mathbf{x}^{\mathbf{b}} \in \mathcal{T}^{(D-\deg p_i)}$ to form the set of relations $\mathcal{R} = \mathcal{R}^{(D)} = \{\mathbf{x}^{\mathbf{b}} p_j(\mathbf{x}) = 0 : 1 \le j \le m, |\mathbf{b}| \le D - \deg p_j\}$ at degree $\le D$. $T := |\mathcal{T}^{(D)}|$ is the number of terms, and we will use the combinatorial notation $[t^k]s(t)$ for the coefficient of t^k in the Maclaurin series expansion of the function $s(t)$ in t, so $T = [t^D]\left((1 - t^q)^n/(1 - t)^{n+1}\right)$. We denote also by ω the exponent in the complexity of matrix multiplication/inversion. The infimum of this complexity exponent has recently been shown to be as low as 2.3727 [26], but for practical purposes is likely to be $\log_2 7 \approx 2.8$.

Basic XL: Solve $\mathcal{R}^{(D)}$ as a linear system in monomials $\mathbf{x}^{\mathbf{b}} \in \mathcal{T}^{(D)}$, with complexity $\propto \left(T^{(D)}\right)^{\omega}$.

The original XL article [12] mentioned the possibility of the Macaulay matrix reducing to a univariate equation. This happens when $I = \mathrm{span}\mathcal{R} \ge T - D$, then brute force or Berlekamp's algorithm will find the solution. However, in the overdetermined case we usually either see that $T - I = 0$ with a self-contradictory system or 1 with exactly 1 solution. Indeed, XL', or reducing to r equations in r variables when $1 < T - I < \binom{r+D}{D}$, is not known to makes a difference in any practical case known [27, 29].

XL with Sparse Matrices: The Macaulay matrix $\mathcal{M}^{(D)}$ has total weight $\sim Rn^2/2$, where $R = |\mathcal{R}^{(D)}| = m|\mathcal{T}^{(D-2)}|$ is the number of equations. Hence, the linear system may be solved via (Block) Wiedemann or some similar sparse matrix solver [11, 25] in $\sim \frac{3}{2}RTn^2$ multiplications. A heuristic variant [30] discards rows randomly to come down to only T rows and then solve using (Block) Wiedemann, using only $\approx \frac{3}{2}T^2n^2$ multiplications. [30] notes that this produces a single solution for most "random" overdetermined systems. If the nullity $\ell > 1$, then perhaps we dropped an essential equation, or if the system started with more than one solution. Here we must check below at every vectors of a subspace with an entry of 1 in the slot correspond to the monomial 1 ("normalized"), about $q^{\ell-1}$ points.

Why would randomly tossing rows work? Heuristically, N random vectors in $(\mathbb{F}_q)^N$ span the entire space with non-zero probability $\approx 1 - \frac{1}{q-1}$ even as $N \nearrow \infty$. Empirically, in many runs failures are even fewer and farther in between, and most singular matrices are further only of small nullity (1 or 2).

XL2 (a.k.a. MutantXL [14]): Consider starting XL at degree D and performing some kind of elimination on the equations $\mathcal{R}^{(D)}$ to attempt eliminating all the highest-degree monomials. If we fail, then raise the degree by multiplying each remaining row by every variable and repeat; if we succeed, then we have found lower-degree equations ("mutants") which can be multiplied by monomials to form new equations without raising the operating degree. In this case, with more elimination and degree-raising stages, we will usually continue to termination [27] without the degree increasing again.

F_4/F_5: It suffices to know that these are "better" versions of XL2 where the matrix-building and row-operation sequences are run according to certain rules to avoid redundancy, but F_4/F_5/XL2 all run at the same degree [27], which is D_{reg} for semi-regular systems. The time complexity will be bound by $\left(T^{(=D)}\right)^\omega$, where $T^{(=D)}$ is the number of degree-D terms, so $\binom{n}{D}$ for $q = 2$, $\binom{n+D}{D}$ for large q.

How Does Sparsity Matter? If the dimension of the matrix does not differ by a large factor, then eventually the log-complexity of XL (with Sparse Matrix Solvers) would be $2/\omega$ that of F_4/F_5, *but only if the latter cannot work with very sparse matrices.* The main determining factor for the dimension of the matrix would then be the degree of operation, which is the subject of this study.

2.1 Degrees of Operation

Small Fields Even though the claimed "proof" of the formula is mistaken, most experts expect XL to operate at the degree indicated by the heuristic formula [28, Theorem 2]:

$$D_0 := \min\{d : [t^d]\left((1-t)^{-n-1}(1-t^q)^n(1-t^2)^m(1-t^{2q})^{-m}\right) \leq 0\}.$$

Analogously, we expect F_4/F_5 to operate at what is known as "degree of regularity" (cf. [1] for $q = 2$):

$$D_{\text{reg}} := \min\{d : [t^d] \left((1 - t)^{-n} (1 - t^q)^n (1 - t^2)^m (1 - t^{2q})^{-m}\right) < 0\}.$$

Large Fields $F_4/F_5/XL2$ and XL operate [13, 28] at (respectively)

$$D_{\text{reg}} := \min\{d : [t^d] \left((1 - t)^{m-n}(1 + t)^m\right) < 0\},$$

and

$$D_0 := \min\{d : [t^d] \left((1 - t)^{m-n-1}(1 + t)^m\right) < 0\}.$$

The former is assumed to be true from the definition of semiregularity; the latter is proved assuming the Maximal Rank Conjecture.

As functions of n and m, it is obvious that almost always $D_0(n, m) = D_{\text{reg}}(n + 1, m)$. Exceptions occur when $[t^{D_0}] \left(((1 - t)^{m-n-1}(1 + t)^m\right) = 0$, the most common case being $m - n = 2$ and n odd.

Operating Degree as a Root of a Function via Integrals: The degree d coefficient of the Maclaurin series of $f(t)$ is given by a contour integral $S_f(d) := (2\pi i)^{-1} \oint \left(f(z) z^{-(d+1)}\right) dz$. The power of the first nonpositive (or resp. negative) coefficient of $f(t)$ is then the smallest integer no less (resp. greater) than the smallest positive real root of S_f. Here we will denote by $\widehat{D_0}$ and $\widehat{D_{\text{reg}}}$ the smallest positive roots of the corresponding integral functions, hence $D_0 = \lceil \widehat{D_0} \rceil$ and $D_{\text{reg}} = \lfloor \widehat{D_{\text{reg}}} \rfloor + 1$.

Known Asymptotic Results [1, 2, 29]: For $m = (c + o(1))n$ where c is a constant and for any $\mathbb{F}_q$, we have D_0 and D_{reg} also equal to $(w + o(1))n$, where w depends only on c and q. $\lg T$ will also asymptotically proportional to n. For any q and ω (i.e., algorithm of elimination), one guesses up to a $c \sim m/n$ to optimize the number of field multiplications one makes.

2.2   A Note on Why Is Not XL Better?

[27, 29] advocated XL with Sparse solvers as asymptotically better than F_4/F_5 in the generic case, especially as the memory size gets larger. Indeed, one might imagine that F_4/F_5 is no match for XL with sparse matrices if the former works with degree 3 or $\log_2 7$ complexity in matrix size, and the latter degree $2 + o(1)$. Yet, the question is, if XL is fundamentally better, why is there not such a report?

Linear Algebra Implementation: It is understandable that the per-multiplication cost in a sparse matrix solver is larger than that in solving a dense linear system, because linear algebra with dense systems is a well-known subject. Even linear algebra in finite fields are optimized very well using tricks like [20] (often known erroneously as the Method of 4 Russians, due to Lupanov-Kronrod).

The Choice of Parameters. However, the difference in speed caused by implementation issues is usually a constant or at most polynomial factor. We will show later that the decisive factor was a different one: Typically such investigations examined a large-field, $m - n$ =constant case, which as we will see below is exactly the worse case for XL among generic systems with $m/n = c + o(1)$.

2.3   Matrix Operations in XL vs. F_4/F_5 if Both $\frac{D_0}{n}$ and $\frac{D_{\text{reg}}}{n} = w + o(1)$

Difference in Operating Degree Means Difference in Size: If $D_0 > D_{\text{reg}}$, then the XL matrix dimension increases with respect to $F_4/F_5/\text{XL2}$ by an extra factor of

$$\frac{(n + D_{\text{reg}})(n + D_{\text{reg}} + 1) \cdots (n + D_0 - 1)}{(D_{\text{reg}} + 1)(D_{\text{reg}} + 2) \cdots D_0} \approx \left(1 + \frac{1}{w}\right)^{D_0 - D_{\text{reg}}}.$$

We could say each increment of $D_0 - D_{\text{reg}}$ costs XL (versus F_4) a factor of $\geq 3\times$.

Other Factors: The Sparsity and the structure of the extended Macaulay matrix works in favor of XL with sparse solvers in terms of better memory footprint and lower complexity. Everything else will be operating against XL. Here we list some differences of XL vs. F_4/F_5.

Memory Use and Storage for Matrix: In F_4/F_5, recent lectures (ECC 2011 [16] and earlier, Polynomial Equations Solving workshop at KTH, Stockholm, Sweden) gave density of the non-zero entries involved in the matrix steps as about $d^{-1}\sqrt{(6/\pi n)}$ for random systems, where d is the degree of the polynomials, so the matrix is fairly dense, with total weight $\sim T^2/\text{poly}(n)$.

As for the XL Sparse version, each row in the Extended Macaulay Matrix has essentially the same number of entries, which is $\sim n^2/2$. There are various ways to compress a sparse system. To take it to the extremes, one could simply store the original equations and generate all of the Macaulay matrix on the fly. In practice, one need to store column indices in each block of rows to avoid recomputation. The total Macaulay matrix storage is thus $Tn \text{ polylog}(n)$.

However, there is one operational detail which sometimes offsets some of the advantages of XL which is that when parallelizing, memory needs to be handled in cache lines and each core needs full vectors data each load in any parallelized solver like Block Wiedemann, and storage is needed for source and destination. So the memory footprint for Block Wiedemann vectors is $2T\nu s_v$ where ν is the number of cores and s_v is the vector length to fit cache lines. For small-to-medium cases this is often larger than the matrix size.

Extra Columns: The number of terms in XL is larger by a factor of $T^{(D)}/T^{(=D)} = \binom{n+D}{D}/\binom{n+D-1}{D} = \frac{n+D}{n} \sim (1 + w + o(1))$ even if the two degrees are equal – but the constant is not far removed from 1.

A more important issue: some terms may be completely eliminated (and with it the associated columns and pivots) during F_4/F_5 and never appear again. To give an example, there are 2^{21} monomials of degree 9 where F_4/F_5 solves $\mathcal{MQ}$ with $(n, m) =$

$(17, 19)$, but the MAGMA output indicates that the matrix is only about 2^{34} bytes, which indicates that the matrix is much emptier or smaller than the 2^{42} bytes that a raw extrapolation would indicate even taking into account the sparsity estimate given above. This phenomenon was also described in [22, 23] ("the partial enlargement strategy"), and noted in passing by F_4/F_5 investigators.

Extraneous Rows: In F_4, there are some extraneous rows; in F_5, there are no extraneous rows (that will reduce to zero) generated at the cost of some restrictions on linear algebra; in XL, there are *many* extra rows in the Macaulay matrix, but any random row-tossing scheme would cut it down to a square matrix. We thus expect the matrix dimensions to be relatively close at the same degree.

How and When XL Might be Better Than F_4/F_5. We expect that the ratio of the per-multiplication cost in the linear algebra is going to be more or less constant with good programming; it is also something that is harder for us to control. However, it is easier to find cases that favor XL over F_4/F_5 if we choose cases where there *is not* a large difference in the operating degrees.

The attacks in [30] dealt with random $\mathcal{M}Q$ where $m = 2n$, which is related to the provable security of QUAD, is one such example. The XL-with-Block-Wiedemann implementation of [9] on a 32-core Xeon E5620 2.4GHz mini-cluster, takes 577 seconds with $(n, m) = (24, 48)$ over $\mathbb{F}_{16}$. MAGMA-2.17 on a Xeon X7550 2.0GHz takes 68628 seconds when $(n, m) = (24, 48)$. This is a good case for XL — the operating degrees are the same; if we change to the parameters $(n, m) = (23, 46)$, XL operates at a higher degree (6 vs 5). This is consistent with the above impression that XL does better where the difference $D_0 - D_{\text{reg}}$ is small, in particular zero. In the remainder of this article, we try to find out which parameters tend to satisfy this property.

3   Degrees of Operation for F_4/F_5 vs. XL and Asymptotics

In this section, we examine the difference in the operating degree of F_4/F_5 vs XL. Clearly $\lceil \widehat{D_0 - D_{\text{reg}}} \rceil \geq D_0 - D_{\text{reg}} \geq \lfloor \widehat{D_0 - D_{\text{reg}}} \rfloor \geq 0$. We show that $\widehat{D_0 - D_{\text{reg}}}$ is asymptotically large in large-field, almost-square systems only. We then discuss how this reflects on the practical complexities of XL vs F_4/F_5. All the details would be included in a future full version.

3.1   Large Fields ($q > D$), Barely Overdetermined ($m/n = 1 + o(1)$) Cases

We observe empirically that $D_0 - D_{\text{reg}}$ is seldom zero. If $m, n \to \infty$ while $m - n = f > 1$ fixed, then the degree of regularity D_{reg} for a system of m quadratic equations in n variables is asymptotically given by $\widehat{D_{\text{reg}}} = \frac{m}{2} - h_{f,1} \cdot \sqrt{\frac{m}{2}} \cdot (1 + o(1))$ [2], where $h_{f,1} = \sqrt{2f + 1} + O(f^{-1/6})$ is the largest zero of the Hermite polynomial of order f. Hence $\widehat{D_0 - D_{\text{reg}}} = (h_{f,1} - h_{f-1,1}) \sqrt{\frac{m}{2}} (1 + o(1))$.

Practical Implication for XL vs. F_4/F_5: It is difficult for XL with Sparse solvers to catch up to F_4/F_5, because $\widehat{D_0} - \widehat{D_{\text{reg}}} \geq 1$, and at some point (which for $m - n = 2$ is (20, 22)) becomes always 2 or more, which means that the number of monomials in XL is $> 10\times$ or more that of F_4/F_5, without taking into account the pivots that disappear from the later stages of F_4/F_5. In fact, the only practical way that XL would be better might be because the matrix would be too large to handle in F_4/F_5.

3.2   Large Fields ($q > D$), QUAD-Like ($m/n \sim \alpha > 1$) Case

The asymptotic expansion for $\widehat{D_{\text{reg}}}$ for large q is given by [2]:

$$\widehat{D_{\text{reg}}} = (\alpha - \frac{1}{2} - \sqrt{\alpha(\alpha - 1)})n + \frac{-a_1}{2(\alpha(\alpha - 1))^{\frac{1}{6}}}n^{\frac{1}{3}} - \left(2 - \frac{2\alpha - 1}{4(\alpha(\alpha - 1))^{\frac{1}{2}}}\right) + O(\frac{1}{n^{1/3}}).$$

So in a typical case for QUAD, $m = 2n$ ($\alpha = 2$), and

$$\widehat{D_{\text{reg}}}(n, 2n) \approx 0.0858n + 1.0415n^{1/3} - 1.4697 + O(n^{-1/3}).$$

It is worth noting that in asymptotic analysis of the root of a function using coalescent saddle points [10], due to the characteristics of the Airy integral expansions, typically the expansion is a series in $n^{-1/3}$ missing the second term, or $f(n) := a_0 n^\beta + a_2 n^{\beta - 2/3} + a_3 n^{\beta - 1} + a_4 n^{\beta - 4/3} + \cdots$, so, on first thought we would expect a difference in the next-to-leading term, or $\widehat{D_0}(n, 2n) - \widehat{D_{\text{reg}}}(n, 2n) = a_2 n^{1/3} + O(1) \nearrow \infty$ as $n \to \infty$, *except that it doesn't but rather approaches a constant near* 0.207.

Indeed, we can carry through the same Coalescent Saddles computations to find that

$$\widehat{D_0}(n, 2n) \approx 0.0858n + 1.0415n^{1/3} - 1.2626 + o(1), \tag{1}$$

$$\text{or} \approx \quad \widehat{D_{\text{reg}}}(n, 2n) + 0.2071 + o(1). \tag{2}$$

We verified the 0.207 asymptotic for semiregular systems over the range $n = 10 \cdots 120000$. Practically, this number starts at around 1/4 in the practical range and decreases toward 0.207. The upshot is that **when $m/n \approx 2$, more than three quarters of the time XL and XL2/F_4/F_5 runs at the same degree.**

Practical Implications for XL vs. F_4/F_5: $D_0(n, 2n) - D_{\text{reg}}(n, 2n) \leq 1$ for almost all n. Furthermore, if we regard the linear and $n^{1/3}$ terms as supplying a random fractional part between 0 and 1, *we can expect that exactly* 20.7% *among all n have operating degrees $D_0(n, 2n)$ and $D_{\text{reg}}(n, 2n)$ differ by 1*. This puts XL in a (relatively speaking) good position compared to F_4/F_5. Degree increases (or drops) in XL/F_4/F_5 matter a lot because the number of monomials increases by a factor that is often between $4\times$ to $6\times$ but asymptotically a factor of $\approx (1 + 0.0858)/(0.0858) \approx 12.66$.

If we fix q and increase n, the system ceases to be "large-field" in that eventually $D_{\text{reg}} > q$. However, for practical attacks we expect the degree drop $D_0 - D_{\text{reg}}$ to be limited to 1 (see following sections). Empirically size of the matrix in the MAGMA F_4 is also somehow larger for the same T for $m/n = 2$ cases than for $m/n \sim 1$ cases. This is again understandable heuristically in the sense that n is larger but D is smaller if we compare an $m/n = 2$ instance with an equal-T instance where $m/n \approx 1$, which means far fewer eliminated columns and pivots.

We conclude that in QUAD ($m/n = 2$) type instances, estimation of cryptographic complexities must take into account attacks using XL as opposed to F_4/F_5, if not using the former outright.

An Explanation of 0.207: Why does it happen here that the $n^{1/3}$ term coefficient does not change? There is a good reason for that. A heuristic "proof" is that we can write the uniform asymptotic expansion as follows

$$\widehat{D_{\text{reg}}} = \left(1 - \frac{\alpha^{-1}}{2} - \sqrt{(1 - \alpha^{-1})}\right) m + \frac{-a_1 \cdot (\alpha^{-1})^{2/3}}{2(1 - \alpha^{-1}))^{\frac{1}{6}}} m^{\frac{1}{3}} - \left(2 - \frac{2 - \alpha^{-1}}{4(1 - \alpha^{-1})^{\frac{1}{2}}}\right) + O(\frac{1}{m^{\frac{1}{3}}}).$$

Hence, if we write $\widehat{D_{\text{reg}}}(n, \alpha n) = f(\alpha^{-1}, m)$, then

$$\widehat{D_0}(n, 2n) - \widehat{D_{\text{reg}}}(n, 2n) = f\left(\left(\frac{1}{2} + \frac{1}{2n}\right), 2n\right) - f\left(\frac{1}{2}, 2n\right)$$

$$\approx \frac{1}{2n} \cdot \frac{\partial f}{\partial(\alpha^{-1})}\bigg|_{\alpha^{-1} = \frac{1}{2}, m = 2n} = \left(\frac{\sqrt{2} - 1}{2}\right) + o(1),$$

which explains why $D_0(n, 2n+k) - D_{\text{reg}}(n, 2n+k)$ is also on average 0.207 as soon as n gets somewhat large. We verified this for integers from $k = -10, \ldots, 10$. Similarly, we can verify that for $m/n \approx 1.5$ and 2.5, the degree drop converges to 0.367 and 0.145, respectively. This analysis can be made rigorous (and similarly for the heuristic "proof" below) with some complex analysis.

The Distinctiveness of the $m/n = 1 + o(1)$ *case:* The reason that the Large Fields ($q > D$), Barely Overdetermined case is so different is that $\alpha = 1$ is a singularity and hence there is no way to take a differentiative at that point with respect to α.

3.3 Small-Field Cases

For $\mathbb{F}_2$, $\mathbb{F}_3$, and $\mathbb{F}_4$ and all $\alpha = m/n > 1$, $D_0 - D_{\text{reg}} \leq 1$ for all practical cases.

$\mathbb{F}_2$, $m = n$ **case:** this resembles the large-field, $m/n = 2$ case in behavior. In part this is because we can think of the "field equations" $x_i^2 = x_i$ as n more equations. Note that as the smallest field, the behavior of $\mathbb{F}_2$ is not truly representative of all small fields, but $\mathbb{F}_2$ is so important in cryptography we simply have to use it as the example. If we carry out the requisite coalescent saddle points computations, we find that

$$\widehat{D_0} - \widehat{D_{\text{reg}}} = 0.2339 + o(1).$$

Just like for large q cases, this implies that (for all practical purposes) $D_0 - D_{\text{reg}} \leq 1$ and is only non-vanishing on less than a quarter of possible n's on average. We verified that this average is roughly correct by using Maple to compute the series up to $n = 10000$.

Practical Implications: In the practical range, we expect that Sparse XL will do better than F_4/F_5 as is first claimed in [28], verified by [24] and conceded in [3]. *However, this is not a good case for Sparse XL because XL will be comfortably outrun by brute force searches.* In the parallelized Block Wiedemann implementations of XL of [9], it was seen that for 35 variables in 35 equations takes 45571s on a test machine with 64 2.3GHz AMD Bulldozer cores and 256GB of contiguous main memory. The same machine would use < 1s on a brute-force search [7].

$\mathbb{F}_2$, $m = 2n$ **case:** Similar to the previous case, we did Coalescent Saddle Point analysis to find that

$$\widehat{D_0} - \widehat{D_{reg}} = 0.1169 + o(1).$$

This implies that $D_0 - D_{reg} \leq 1$ in practice for all D_0, and the proportion of n where $D_0 - D_{reg} = 1$ is a scant 15% within or less for $n \leq 10000$.

This is a more interesting case to be talking about XL because of two reasons. One is that this is the security assumption for QUAD stream ciphers and variants. The other is that it is easier for XL to beat brute force. Unfortunately, it is not *that* easy. For an example, at $n = 96$, $m = 192$, Sparse XL is projected to take $2^{94.6}$ field multiplications. Each field multiplication takes about 1/20 of a cycle. But if we consider the memory footprint, XL still loses badly to a brute-force attack. If we take memory size into account, to use XL in $\mathbb{F}_2$, we should continue to guess until m/n is close to 3.

Other Small Fields. We can verify that for $\mathbb{F}_3$ and $\mathbb{F}_4$ n-variables-n-equation systems have $D_0 - D_{reg} \leq 1$, just like $\mathbb{F}_2$. In fact,

$$\widehat{D_0}(n, n; 3) - \widehat{D_{reg}}(n, n; 3) = 0.3660 + o(1)$$
$$\widehat{D_0}(n, 2n; 3) - \widehat{D_{reg}}(n, 2n; 3) = 0.1650 + o(1)$$
$$\widehat{D_0}(n, n; 4) - \widehat{D_{reg}}(n, n; 4) = 0.4940 + o(1)$$
$$\widehat{D_0}(n, 2n; 4) - \widehat{D_{reg}}(n, 2n; 4) = 0.1912 + o(1)$$

Using maple, one can check for $m = n$ that for roughly 38% and 53% of all $n < 5000$ that $D_0 - D_{reg} = 1$ for $\mathbb{F}_3$ and $\mathbb{F}_4$ respectively, and the other times the two degrees are equal. What this means is that generic equations in smaller fields generally favor XL over F_4/F_5. However (although a little surprising to begin with), when facing a system with $m = n$ in these small fields we need to check whether brute force is best also.

Heuristic Evaluation of $\widehat{D_0} - \widehat{D_{reg}}$ for Small Fields: Without entering into the complex analysis required to prove all of the above rigorously, we will compute an example the asymptotic behavior of $\widehat{D_0}(n, 2n; 2) - \widehat{D_{reg}}(n, 2n; 2)$. Here $\widehat{D_{reg}}(n, 2n; 2)$ and $\widehat{D_0}(n, 2n; 2)$ are respectively the smallest positive root of

$$S_1(d) := \frac{1}{2\pi i} \oint \frac{(1+z)^n \, dz}{z^{d+1} (1+z^2)^{2n}}, \quad \text{and} \quad S_2(d) := \frac{1}{2\pi} \oint \frac{(1+z)^n \, dz}{(1-z) z^{d+1} (1+z^2)^{2n}}.$$

Let $w = d/n$ and consider S_1 and S_2 as special cases of the following contour integral

$$S(n; w; \alpha, \beta, \gamma) := \oint \frac{dz}{2\pi i z} \left(\frac{(1+z)^\alpha}{z^w (1+z^2)^\beta (1-z)^\gamma} \right)^n.$$

To evaluate this we need the following equation in z to have double roots:

$$\frac{-w}{z} + \frac{\alpha}{1+z} - \frac{2\beta z}{1+z^2} + \frac{\gamma}{1-z} = 0.$$

If we let $w := F(\alpha, \beta, \gamma)$ represent the smallest positive real $w = d/n$ that allows double roots for z, then we see that $F(\alpha, \beta, 0)$ is the coefficient of the $\Theta(n)$ term in the asymptotic expansion of $\widehat{D_{\mathrm{reg}}}(\alpha n, \beta n; 2)$ and $\widehat{D_0}(\alpha n, \beta n; 2)$, and if we skip all the analysis, we eventually get to

$$\widehat{D_0}(n, 2n; 2) - \widehat{D_{\mathrm{reg}}}(n, 2n; 2) = \left(F(1, 2, \frac{1}{n}) - F(1, 2, 0) \right) n = \left. \frac{\partial F}{\partial \gamma} \right|_{\alpha=1, \gamma=0} + o(1),$$
(3)

which we may evaluate with implicit differentiation to obtain the 0.1169 above.

3.4   Direct Attacks on QUAD

In a direct attack against QUAD we face this problem: take random polynomials $\mathbf{P} = (P_1, \ldots, P_n)$ and $\mathbf{Q} = (Q_1, \ldots, Q_n)$ in the variables $\mathbf{x} = (x_1, \ldots, x_n)$. Solve ℓn equations for $\mathbf{x}$ using vectors $\mathbf{y}_1, \ldots, \mathbf{y}_\ell$:

$$\mathbf{y_1} = \mathbf{P(x)}, \ \mathbf{y_2} = \mathbf{P(Q(x))}, \ \mathbf{y_3} = \mathbf{P(Q(Q(x)))}, \ldots.$$

This arises from studying the security of the stream cipher QUAD [4]. [30] suggested this direct attack in the known-plaintext setting and verified empirically that this system behaves like random systems (i.e., a system with n random quadratic equations, n random quartic equations, and so on) if one tries to solve it with Gröbner basis methods, including XL.

Our investigations and tests show that the systems created in direct algebraic attacks on QUAD-like systems also have $D_0 - D_{\mathrm{reg}} \leq 1$ if we assume semi-regularity, and hence we expect XL to overtake F_4/F_5 as the best estimate of complexities for moderately large n.

$\mathbb{F}_2$ *cases:* We compare the operating degrees of XL and F_4/F_5 as given by [30, Sec. 4.5]:

$$D_0(\mathtt{QUAD}(2, n, n)) = \min \left\{ D : [t^D] \frac{(1+t)^n}{(1-t)} \left((1+t^2)(1+t^4) \cdots (1+t^{2^\ell}) \right)^{-n} \leq 0 \right\},$$

$$D_{\mathrm{reg}}(\mathtt{QUAD}(2, n, n)) = \min \left\{ D : [t^D] \frac{(1+t)^n}{\left((1+t^2)(1+t^4) \cdots (1+t^{2^\ell}) \right)^n} < 0 \right\}.$$

As $\ell \nearrow \infty$, we have $\left((1+t^2)(1+t^4)\cdots(1+t^{2^\ell}) \right) = (1-t^2)^{-1}\left(1-t^{2^\ell}\right) \longrightarrow$
$(1-t^2)^{-1}$, hence

$$D_0(\text{QUAD}(2,n,n)) = D_0(n,2n;\text{large } q), \quad D_{\text{reg}}(\text{QUAD}(2,n,n)) = D_{\text{reg}}(n,2n;\text{large } q).$$

So the operating-degree difference of XL and F_4/F_5 bounded by 1 and average 0.2071.

Large-Field Cases: As found by [30, Sec. 4.4], using more than the quartics does not lead to substantial gains. We will hence restrict ourselves to the direct attack using only quadratics and quartics:

$$D_0(\text{QUAD}(\text{large } q, n, n)) = \min\left\{ D : [t^D]\left((1-t)^{-(n+1)}(1-t^2)(1-t^4) \right)^n < 0 \right\},$$
$$D_{\text{reg}}(\text{QUAD}(\text{large } q, n, n)) = \min\left\{ D : [t^D]\left((1-t)^{-n}(1-t^2)(1-t^4) \right)^n < 0 \right\}.$$

We discover that $D_0(\text{QUAD}(\text{large } q, n, n)) - D_{\text{reg}}(\text{QUAD}(\text{large } q, n, n))$ is zero for roughly half of all n and 1 for the other half. This might seem surprising but again this can be explained as follows: Let $R(\alpha)$ be the smallest positive w that gives a double root to $\frac{d}{dz}(1-z)^\alpha(1+z)^2(1+z^2)z^{-w} = 0$. Then as before $\widehat{D_{\text{reg}}}(\text{QUAD}(\text{large } q, n, n)) = n(R(1) + o(1))$ while $\widehat{D_0}(\text{QUAD}(\text{large } q, n, n)) = n(R(1-\frac{1}{n}) + o(1))$. The implicit function theorem lets us find $R'(1)$ and derive (heuristically, but can be made rigorous):

$$\widehat{D_0}(\text{QUAD}(\text{large } q, n, n)) - \widehat{D_{\text{reg}}}(\text{QUAD}(\text{large } q, n, n)) = 0.4843 + o(1).$$

We note here that the asymptotic result is similar if we include higher-order equations.

4   Discussion and Concluding Remarks

In this paper, we discuss the difference in the degrees of operation of the XL and the F_4/F_5 alforitms (and all similar algorithms such as MutantXL/XL2 [23], or the GGV algorithm of [17]) for multivariate systems with randomly chosen cocfficients, where the ratio of the number of equations to the number of variables is nearly constant. We show that usually the difference is small. In fact, for most cryptographically relevant cases, it is at most one, and the expectation value over many possible set of parameters can be evaluated precisely using asymptotic analysis.

The inevitable conclusion is that for generic/random and mildly overdetermined systems with $m/n = c + o(1)$, XL with sparse matrices may be a better way to find roots than any of the more advanced methods. This vindicates the conjectures of [27, 29] regarding XL with Sparse solvers, and is consistent with the recent article [3] which implicitly assumes a sparse matrix method and XL rather than F_4/F_5.

Future work remains to determine the best way to implement similar methods using Wiedemann type solvers. Here practical study is made difficult as so much details about F_5 and MAGMA-F_4 are unknown, but we believe that we have shed some light on the comparison of XL vs. F_4/F_5 in theory. We are preparing a full version for journal publication.

Acknowledgements. Thanks are due for their sponsorship to Academia Sinica (for an Career Advancement Award to BY) and to the National Science Council of Taiwan for grant NSC100-2628-E-001-004-MY3.

Many more thanks and appreciation are due from BY and CMC to Johannes Buchmann, who over many years had been a good friend who offers sage advice, an esteemed and supportive elders of our trade, as well as a valued colleague and co-author.

References

1. Bardet, M., Faugère, J.-C., Salvy, B.: On the complexity of Gröbner basis computation of semi-regular overdetermined algebraic equations. In: Proceedings of the International Conference on Polynomial System Solving, pp. 71–74 (2004); Previously INRIA report RR-5049
2. Bardet, M., Faugère, J.-C., Salvy, B., Yang, B.-Y.: Asymptotic expansion of the degree of regularity for semi-regular systems of equations. In: Gianni, P. (ed.) MEGA 2005, Sardinia, Italy (2005)
3. Bardet, M., Faugère, J.-C., Salvy, B., Spaenlehauer, P.-J.: On the complexity of solving quadratic boolean systems. Journal of Complexity 29(1), 53–75 (2013) ISSN 0885-064X
4. Berbain, C., Gilbert, H., Patarin, J.: QUAD: A practical stream cipher with provable security. In: Vaudenay, S. (ed.) EUROCRYPT 2006. LNCS, vol. 4004, pp. 109–128. Springer, Heidelberg (2006)
5. Bernstein, D.J., Buchmann, J., Dahmen, E. (eds.): Post Quantum Cryptography, 1st edn. Springer (2008) ISBN 3-540-88701-6
6. Bettale, L., Faugère, J.-C., Perret, L.: Hybrid approach for solving multivariate systems over finite fields. Journal of Mathematical Cryptology 3(3), 177–197 (2010)
7. Bouillaguet, C., Chen, H.-C., Cheng, C.-M., Chou, T., Niederhagen, R., Shamir, A., Yang, B.-Y.: Fast exhaustive search for polynomial systems in F_2. In: Mangard, S., Standaert, F.-X. (eds.) CHES 2010. LNCS, vol. 6225, pp. 203–218. Springer, Heidelberg (2010)
8. Buchberger, B.: Ein Algorithmus zum Auffinden der Basiselemente des Restklassenringes nach einem nulldimensionalen Polynomideal. PhD thesis, Innsbruck (1965)
9. Cheng, C.-M., Chou, T., Niederhagen, R., Yang, B.-Y.: Solving quadratic equations with xl on parallel architectures. In: Prouff, E., Schaumont, P. (eds.) CHES 2012. LNCS, vol. 7428, pp. 356–373. Springer, Heidelberg (2012)
10. Chester, C., Friedman, B., Ursell, F.: An extension of the method of steepest descents. Proceedings of Cambridge Philosophical Society 53, 599–611 (1957)
11. Coppersmith, D.: Solving homogeneous linear equations over GF(2) via block wiedemann algorithm. Mathematics of Computation 62(205), 333–350 (1994)
12. Courtois, N.T., Klimov, A.B., Patarin, J., Shamir, A.: Efficient algorithms for solving overdefined systems of multivariate polynomial equations. In: Preneel, B. (ed.) EUROCRYPT 2000. LNCS, vol. 1807, pp. 392–407. Springer, Heidelberg (2000), http://www.minrank.org/xlfull.pdf
13. Diem, C.: The XL-algorithm and a conjecture from commutative algebra. In: Lee, P.J. (ed.) ASIACRYPT 2004. LNCS, vol. 3329, pp. 323–337. Springer, Heidelberg (2004)
14. Ding, J., Buchmann, J., Mohamed, M.S.E., Mohamed, W.S.A.E., Weinmann, R.-P.: Mutant XL. In: talk at the First International Conference on Symbolic Computation and Cryptography (SCC 2008), Beijing (2008)
15. Ding, J., Yang, B.-Y., Chen, C.-H.O., Chen, M.-S., Cheng, C.-M.: New differential-algebraic attacks and reparametrization of rainbow. In: Bellovin, S.M., Gennaro, R., Keromytis, A.D., Yung, M. (eds.) ACNS 2008. LNCS, vol. 5037, pp. 242–257. Springer, Heidelberg (2008), http://eprint.iacr.org/2008/108

16. Faugère, J.-C.: Solving efficiently structured polynomial systems and applications in cryptology (September 2011), http://ecc2011.loria.fr/slides/faugere.pdf; Talk at ECC 2011, 9:30 AM (September 20, 2011)
17. Gao, S., Guan, Y., Volny, F.: A new incremental algorithm for computing groebner bases. In: Koepf, W. (ed.) ISSAC, pp. 13–19. ACM (2010)
18. Joux, A., Vitse, V.: A variant of the F4 algorithm. In: Kiayias, A. (ed.) CT-RSA 2011. LNCS, vol. 6558, pp. 356–375. Springer, Heidelberg (2011)
19. Lazard, D.: Gröbner-bases, Gaussian elimination and resolution of systems of algebraic equations. In: EUROCAL 1983. LNCS, vol. 162, pp. 146–156. Springer, Heidelberg (March 1983)
20. Lupanov, O.B.: On rectifier and contact-rectifier circuits. Akademii Nauk SSSR 111, 1171–1174 (1956) ISSN 0002àV3264
21. MAGMA project, Computational Algebra Group, University of Sydney. The MAGMA computational algebra system for algebra, number theory and geometry, http://magma.maths.usyd.edu.au/magma/
22. Mohamed, M.S.E., Cabarcas, D., Ding, J., Buchmann, J., Bulygin, S.: MXL_3: An efficient algorithm for computing Gröbner bases of zero-dimensional ideals. In: Lee, D., Hong, S. (eds.) ICISC 2009. LNCS, vol. 5984, pp. 87–100. Springer, Heidelberg (2010)
23. Mohamed, M.S.E., Mohamed, W.S.A.E., Ding, J., Buchmann, J.: *MXL2*: Solving Polynomial Equations over GF(2) using an improved mutant strategy. In: Buchmann, J., Ding, J. (eds.) PQCrypto 2008. LNCS, vol. 5299, pp. 203–215. Springer, Heidelberg (2008)
24. Mohamed, W.S.A., Ding, J., Kleinjung, T., Bulygin, S., Buchmann, J.: PWXL: A parallel Wiedemann-XL algorithm for solving polynomial equations over GF(2). In: Cid, C., Faugère, J.-C. (eds.) Proceedings of the 2nd International Conference on Symbolic Computation and Cryptography, pp. 89–100 (June 2010)
25. Wiedemann, D.: Solving sparse linear equations over finite fields. IEEE Transactions on Information Theory, IT-32(1), 54–62 (1976)
26. Williams, V.V.: Breaking the Coppersmith-Winograd barrier (2011), www.cs.berkeley.edu/~virgi/matrixmult.pdf
27. Yang, B.-Y., Chen, J.-M.: All in the XL family: Theory and practice. In: Park, C.-s., Chee, S. (eds.) ICISC 2004. LNCS, vol. 3506, pp. 67–86. Springer, Heidelberg (2005)
28. Yang, B.-Y., Chen, J.-M.: Theoretical analysis of XL over small fields. In: Wang, H., Pieprzyk, J., Varadharajan, V. (eds.) ACISP 2004. LNCS, vol. 3108, pp. 277–288. Springer, Heidelberg (2004)
29. Yang, B.-Y., Chen, J.-M., Courtois, N.T.: On asymptotic security estimates in XL and Gröbner bases-related algebraic cryptanalysis. In: López, J., Qing, S., Okamoto, E. (eds.) ICICS 2004. LNCS, vol. 3269, pp. 401–413. Springer, Heidelberg (2004)
30. Yang, B.-Y., Chen, O.C.-H., Bernstein, D.J., Chen, J.-M.: Analysis of QUAD. In: Biryukov, A. (ed.) FSE 2007. LNCS, vol. 4593, pp. 290–308. Springer, Heidelberg (2007)
31. Yang, B.-Y., Chen, O.C.-H., Chen, J.-M.: The limit of XL implemented with sparse matrices. In: Workshop Record, PQCrypto Workshop, Leuven (2006), http://postquantum.cr.yp.to/pqcrypto2006record.pdf

Solving Degree and Degree of Regularity
for Polynomial Systems over a Finite Fields

Jintai Ding[1,*] and Dieter Schmidt[2]

[1] University of Cincinnati and Chongqing University
jintai.ding@gmail.com
[2] University of Cincinnati
schmiddr@ucmail.uc.edu

Abstract. In this paper, we try to clarify some of the questions related to a key concept in multivariate polynomial solving algorithm over a finite field: the degree of regularity. By the degree of regularity, here we refer to a concept first presented by Dubois and Gama, namely the lowest degree at which certain nontrivial degree drop of a polynomial system occurs. Currently, it is somehow commonly accepted that we can use this degree to estimate the complexity of solving a polynomial system, even though we do not have systematic empirical data or a theory to support such a claim. In this paper, we would like to clarify the situation with the help of experiments. We first define a concept of solving degree for a polynomial system. The key question we then need to clarify is the connection of solving degree and the degree of regularity with focus on quadratic systems. To exclude the cases that do not represent the general situation, we need to define when a system is degenerate and when it is irreducible. With extensive computer experiments, we show that the two concepts, the degree of regularity and the solving degree, are related for irreducible systems in the sense that the difference between the two degrees is indeed small, less than 3. But due to the limitation of our experiments, we speculate that this may not be the case for high degree cases.

Keywords: Solving degree, degree of regularity, HFE, HFEv, random polynomial system, non-degenerate system.

1 Introduction

1.1 Motivations

One way of attacking a symmetric or asymmetric cryptosystem is by solving a set of multivariate polynomial equations over a finite field. This is a rapidly developing area in cryptography. The security analysis of many cryptosystems is very much affected by the complexity to solve the related polynomial systems. Such an attack is studied most intensively in the context of multivariate public key cryptography, since here the public key is a set of multivariate polynomials.

* Partially sponsored by National Science Foundation of China, Grant #60973131 and U1135004, and the Charles Phelps Taft Research Center.

M. Fischlin and S. Katzenbeisser (Eds.): Buchmann Festschrift, LNCS 8260, pp. 34–49, 2013.
© Springer-Verlag Berlin Heidelberg 2013

The security of such a system depends directly on the complexity of solving the given polynomial equations. Therefore, understanding the complexity to solve multivariate polynomial systems is a critical problem in cryptography. In addition, since polynomial solving is used in many other places, the answer to this question has broad impact on other areas in theory and applications.

Since higher order polynomial equations can be transformed into a set of quadratic polynomial equations, the later can be considered to have the most general form in some sense. We thus focus on the quadratic polynomials

$$p_1(x_1, \ldots, x_n) = \cdots = p_m(x_1, \ldots, x_n) = 0,$$

over a finite field $\mathbb{F}_q$ of order q. We will concentrate on the cases with $m = n$, since these are the hard cases, which occur most frequently in applications.

To solve a multivariate polynomial system, the key method is the Gröbner basis algorithm. In general, solving a set of generic multivariate polynomial equations is a very hard problem. For instance, we know that to solve a set of randomly selected quadratic equations over a finite field with n equations and n variables is NP-complete. It is also known that the complexity of the Gröbner basis algorithms for a generic system is doubly exponential in the number of variables for a field of characteristic zero, as was shown in [15].

In the last two decades, polynomial solving algorithms have undergone a fast development. People realized that computationally (not in terms of storage or memory), the original Gröbner basis algorithm – Buchberger's algorithm – is very inefficient due to the need to perform multivariate polynomial reduction independently on each new S-pairs. The new trend is to improve the algorithm with a far better computational complexity but usually with a much larger storage or memory usage, namely a trade-off between computation cost and storage cost. This is achieved by transforming the polynomial solving process into several steps of solving linear systems. Here the linear systems are derived from the polynomial themselves directly by rewriting each polynomial as a row of a matrix. The reduction of the polynomials is then achieved by Gaussian elimination. The origin of the idea can be traced back to the original XL algorithm (later rediscovered as the XL Method) and was proposed by Lazard [2, 14]. The XL algorithm can be described in simplified terms as follows

1. multiply the equations with monomials to form a collection of relations up to some degree d;
2. linearize (i.e., treat each individual monomial as a variable), and use matrix algorithms (for example Gaussian elimination) over $\mathbb{F}_q$ on the resulting matrix (the *extended Macaulay matrix*).

In this paper, we assume that the Macauley matrix is in the form that rows represent monomials and columns represent polynomials, while the usual Macauley matrix is the other way around.

The newly developed algorithms include F_4, XL algorithm and Mutant XL algorithms [2, 5, 12, 16, 17]. For these algorithms it is clear that the computational complexity is dominated by the step, where performing the linear algebra computation (Gaussian elimination) takes the longest time. On the other hand, if one is more concerned with memory complexity then the step dealing with the largest matrix will determine the space complexity. We call such a step the *solving step*.

The complexity of the solving step is determined by the form of the multivariate polynomials. The number of monomials determine the size of the rows of the corresponding Macauley matrix and the number of polynomials determines the number of columns of the matrix. The number of monomials is determined by the degree of the polynomial. The corresponding matrices we need to deal with for the solving step are in general almost square. Therefore, the complexity of the solving step is determined by the highest degree of the polynomials involved. We call this degree the *solving degree*.

The complexity analysis problem now becomes a problem of finding such a degree. It is clear that the concept of solving degree is very vague, and what we try to do is to find something more mathematically tangible. This leads us to the degree of regularity introduced by Dubois and Gama [11].

We first define the graded ring $B := \mathbb{F}_q[x_1, \ldots, x_n] / \langle x_1^q, \ldots, x_n^q \rangle$ and B_d its degree-d subspace. By B_d^m, we mean the vector space of direct product of m copies of B_d.

Definition 1. *For homogeneous quadratic polynomials* $(\lambda_1, \ldots, \lambda_m) \in B_2^m$, *let* $\psi_d : B_d^m \to B_{d+2}$ *be the map defined as*

$$\psi(b_1, \ldots, b_m) = \sum_{i=1}^m b_i \lambda_i.$$

Then

$$R_d(\lambda_1, \ldots, \lambda_m) := \ker \psi_d$$

defines the subspace of relations

$$\sum_{i=1}^m b_i \lambda_i = 0.$$

Further let $T_d(\lambda_1, \ldots, \lambda_n)$ *be the subspace of trivial relations generated by the elements*

$$\{b(\lambda_i e_j - \lambda_j e_i) \mid 1 \le i < j \le m, b \in B_{d-2}\},$$

and

$$\{b(\lambda_i^{q-1})e_i \mid 1 \le i \le m, b \in B_{d-2(q-1)}\}.$$

Here e_i *means the i-th unit vector consisting of all zeros except 1 at the i-th position.*

$$e_i = (0, \ldots, 0, 1, 0, \ldots, 0).$$

The degree of regularity *of a homogeneous quadratic set is then*

$$D_{reg}(\lambda_1, \ldots, \lambda_m) := \min\{d \mid R_{d-2}(\lambda_1, \ldots, \lambda_m)/T_{d-2}(\lambda_1, \ldots, \lambda_m) \ne \{0\}\}.$$

For a generic polynomial (non-homogeneous) system $p_1 = \cdots = p_m = 0$,

$$D_{reg}(p_1, \ldots, p_m) := D_{reg}((p_1)^h, \ldots, (p_m)^h),$$

where $(p_i)^h$ is the highest degree homogeneous component of p_i.

Clearly the degree of regularity is the lowest degree at which we have a linear combination of multiples of p_i that has a nontrivial cancellation of all of the highest degree components, and therefore a nontrivial degree-drop.

In this definition, we can see that the subspace T_d of trivial syzygies represents a "known-to-be-useless" degree drop in the following sense:

Let

$$p_i = c^{(i)} + \sum_k b_k^{(i)} x_k + \sum_{k \le \ell} a_{k\ell}^{(i)} x_k x_\ell.$$

Let $(p)^h$ represent the homogeneous highest degree part of the polynomial p. Clearly $(p_j)^h (p_i)^h - (p_i)^h (p_j)^h = 0$ is a trivial syzygy, which is equivalent to the combination of degree-4 rows $\left(\sum_{k\ell} a_{k\ell}^{(i)} (x_k x_\ell p_j) \right) - \left(\sum_{k\ell} a_{k\ell}^{(j)} (x_k x_\ell p_i) \right)$ being of degree-3 (or fewer). Equally clearly this "degree-drop" will not give us anything useful since

$$\left(c^{(i)}(p_j) + \sum_k b_k^{(i)}(x_k p_j) + \sum_{k\ell} a_{k\ell}^{(i)}(x_k x_\ell p_j) \right)$$
$$= \left(c^{(j)}(p_i) + \sum_k b_k^{(j)}(x_k p_i) + \sum_{k\ell} a_{k\ell}^{(j)}(x_k x_\ell p_i) \right),$$

given that both give $(p_i p_j)$. Thus we just "found" a linear combination of polynomials we already have at degree 3. So a trivial or *principal* syzygy between the top-degree parts $(p_i)^h$ leads to a *trivial* degree drop useless for generating new equations. We must verify that a degree-drop is nontrivial before we can claim that we have reached the degree of regularity.

This concept is very mathematical (not computational) in the sense that the degree of regularity is invariant under invertible linear transformations in terms of either the variables and the polynomials.

This critical concept, *degree of regularity*, is actually the lowest degree where we find a *nontrivial* degree drop in terms of linear combinations of multiples of the original polynomials that define the system. By now, it is commonly accepted that this degree somehow in general matches the highest degree of polynomials we need to deal with in a polynomial solving algorithm, or in an abused term, the degree at which F_4, Mutant XL, and similar algorithms usually terminate. But we shall see later, this concept of termination degree is not really a good concept that sometimes it can be misleading. Therefore, we will use the new term the solving degree.

The mutant XL algorithm is an improved XL algorithm as follows: for a fixed degree d, multiply each p_i with all monomials of degree $d - \deg p_i$ to create a large collection of relations of degree d, order the monomials and linearize these equations to obtain the Macaulay matrix $\mathrm{Mac}^{(d)}(p_1, \ldots, p_m)$, then try to eliminate the highest degree monomials from $\mathrm{Mac}^{(d)}(p_1, \ldots, p_m)$ to create new relations of degree $d-1$ or lower, and once we find such polynomials with degree drop, we try to use them fully before we move on to the next degree. These new polynomial created from nontrivial degree drop are called mutants. This idea allows us to greatly improve the XL algorithm to become one of the most efficient polynomial solving algorithms.

1.2   Questions about the Terminology

There is some confusion about the term "the degree of regularity". The rank of Macaulay matrices at any given degree, which describes the dimension related to the space of the XL algorithm can be computed with certain generating functions and the strong assumption that there are no nontrivial syzygies. A system where this assumption is valid for any degrees is called *regular*. However this can not happen in the case of a finite field. To deal with such a problem, there is a definition of "semi-regular" system [3]. The degree of regularity in such a setting, is the degree at which the system ceases to behave as if it is regular. The degree of regularity as described in Definition 1 is the degree at which the first appearance of "nontrivial degree drop" is observed, that is the system ceases to behave semi-regular.

A heuristic formula for the degree of regularity of most random systems (including asymptotics) is given by Bardet et al [1, 18]. However, this formula is not at all applicable to most systems with a structure that we are interested in.

Certain simple upper bounds to D_{reg} for the multivariate cryptosystems for HFE, HFE-, HFEv and HFEv- [4, 7, 8, 10] were found, and are shown to be good bounds to find the computational complexity.

1.3   The Contribution of This Paper

The question, we would like to clarify: Is it indeed true that the degree of regularity is a good concept to help us to determine the complexity to solve a given polynomial system?

We would like to first point out that this is not true for just any system. We will use an example of a triangular system to demonstrate this first. This leads to new definitions of degenerate systems and partially degenerate systems, and irreducible systems, and we would like to find out that if it is indeed true that the degree of regularity is a good concept to help us to determine the complexity to solve a given irreducible polynomial system.

We would like to show via experiments that the degree of regularity and the solving degree are closely related. Since we were not able to perform experiments on systems with a large number of variables, we are not sure what the relationship will be. We will also discuss briefly the applicability of the two degrees for higher degree systems.

2   A Degenerate System and an Irreducible System

We will first study the concept of degenerate systems.

2.1   An Example of a Degenerate System

We would like to first show an example, where the degree of regularity and solving degree have a big difference.

The constructions of this example is a type of triangular system. The example is a polynomial system, which looks like the following:

$$p_1(x_1,\ldots,x_n) = f_1(x_1,x_2)$$
$$p_2(x_1,\ldots,x_n) = f_2(x_1,x_2)$$
$$p_3(x_1,\ldots,x_n) = f_3(x_1,\ldots,x_n)$$
$$\ldots$$
$$p_n(x_1,\ldots,x_n) = f_n(x_1,\ldots,x_n)$$

The first two equations involves only the first 2 variables, and the rest are much more complicated polynomials. In this case, what a Gröbner basis solver will actually do is to try to solve the subsystem formed by f_1 and f_2 first, where a nontrivial degree drop will occur, and then try to solve the rest. Therefore in this case, the degree of regularity comes from the f_1 and f_2 system, but the solving degree actually comes from $f_3,\ldots,f_n$. We can expect a big difference, or as big a difference as we wish by manipulating the system.

Below we will give a concrete example for such a system. Before we present the example, we would like to say a few words about how we present the experimental results. We used both mutant XL algorithms and Magma implementation of the Gröbner basis algorithms for our experiments, but those mutant XL implementations are not yet publicly available. Therefore, to enable other researchers to check on our results, we will use only the data from the Magma implementation of the Gröbner basis algorithms and will not publish the data from the mutant XL algorithms, which matches well with the data from the Magma implementation of the Gröbner basis algorithms. In the Magma implementation, the algorithm goes through many steps of computations, and the key computation in each step is the Gaussian elimination which performs a reduction of the polynomials at a fixed degree. We call this degree the *step degree*. The degree of regularity is the first step degree at which the step degree starts to go either flat or down. We will present a graph, where the horizontal direction is the step number and the vertical direction is the step degree. Then we will add two vertical line segments to represent the relative metrics of the matrix size (left line) and the time (right line) to each step. In this way, we can read out easily the solving degree and the step which dominates the whole computation process.

We will show via computer experiments that the degree of regularity and solving degree can be far apart. We use an HFE system (whose definition is given in the section below) and then replace the first two equations by quadratic

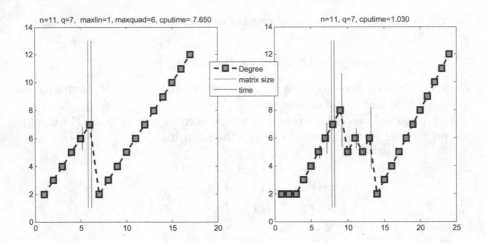

Fig. 1. The degrees at each step. The (relative) size of the matrix is given by the length of the left vertical line and the size of the line on the right gives the (relative) time for each step.
a) A HFE system over $GF(7)$ b) A system of same size but first two equations are replaced by quadratic equations in x_1 and x_2.

equations involving only the first two variables x_1, x_2. From Figure 1, we see that the degree of regularity for the HFE system is 7 and occurred in step 6. Most of the time was spent in this step and the largest matrix was encountered there, so that the solving degree is 7. When the first two equations were replaced by quadratic equations in x_1 and x_2 the degrees encountered by the Gröbner basis algorithm looks different. The degree of regularity is now 2 but the solving degree remains 7 and was encountered in step 8. The reason in the second case that the degree of regularity is low is exactly because the solver is actually first working on solving the system made of the first two equations.

We can actually create systems where the difference between the degree of regularity and the solving degree can be as large as we desire.

Also we know that for linear algebra based Gröbner basis algorithms the change of basis or the mixing of polynomials through invertible linear transformations does not change the degree of regularity and the solving degree. This leads to our definition of a degenerate system.

2.2   Definition of a Degenerate System

To simplify the exposition, we would like to deal with quadratic system where $m = n$ and where the solution space is of zero dimension.

Let us assume that we are dealing with a set of polynomials:

$$p_1(x_1,\ldots,x_n) = \cdots = p_n(x_1,\ldots,x_n) = 0.$$

We would like to first define a set of degenerate system.

Definition 2. *A quadratic system $p_1(x_1,\ldots,x_n) = \cdots = p_n(x_1,\ldots,x_n) = 0$ is degenerate, if we can find m' linearly independent polynomials $h_i(x_1,\ldots,x_{n'})$, $i = 1,\ldots,m'$, $n' < n$ and $m' \geq n'$ such that*

$$h_i(x_1,\ldots,x_{n'}) = \left(\sum a_{ij}(p_j)^h\right) \circ L(x_1,\ldots,x_n).$$

It is clear that the example above is an example of a degenerate system, where $m' = n' = 2$, while $n = 7$.

Now the question is: Is it true that for all non-degenerate systems the degree of regularity and the solving degree are close?

For this, we are actually not sure. The reason for this is that we are still not at all sure what happens for a system that is partially degenerate, namely what happens if we can find such a m' polynomials such that $m' < n'$. Since such a case is very complicated, what we would like to do is to concentrate on what we call an irreducible system.

Definition 3. *A quadratic system $p_1(x_1,\ldots,x_n) = \cdots = p_n(x_1,\ldots,x_n) = 0$ is an irreducible system, if we can not find a non-zero polynomial $\sum a_{ij}(p_j)^h$, such that the corresponding quadratic form is not of full rank (for the case, q=2, it should be full rank-2, when n is even).*

Here we would like to remark that for the case when n is odd and $q = 2$, the full rank can only be $n - 1$ not n. Another remark is that we do not know what happens in the case of the partially degenerate system and we speculate that the differences between the solving degree and the degree of regularity can be anything.

3   Solving Degree and Degree of Regularity

What we will do is to systematically perform testing on irreducible systems to check on the connection of degree of regularity and the solving degree. An easy way to construct an irreducible system is to use a random systems, whose coefficient are generated independently and uniformly. But how can we generate other type of irreducible systems? The trick here is that we will use HFE or a related cryptosystem from multivariate public key cryptography to construct irreducible systems that behave differently from random systems.

3.1   The HFE, HFEv, and IPHFE Cryptosystems

In the standard formulation of a multivariate public-key cryptosystem over a finite field $\mathbb{F}$, the public-key $P : \mathbb{F}^n \mapsto \mathbb{F}^m = T \circ Q \circ S$ is a composition of two invertible affine maps $S : \mathbb{F}^n \mapsto \mathbb{F}^n$ and $T : \mathbb{F}^m \mapsto \mathbb{F}^m$, and a quadratic map (possibly with some parameters) $Q : \mathbb{F}^n \mapsto \mathbb{F}^m$ which is easily invertible when all parameters are given. The maps S and T are part of the secret key, and properties of the central map Q determines most of the properties of the cryptosystem.

Let $\mathbb{F} \cong \mathbb{F}_q$ be a finite field of order q and $\mathbb{K}$ a degree-n extension of $\mathbb{F}$, with a "canonical" isomorphism ϕ identifying $\mathbb{K}$ with the vector space $\mathbb{F}^n$. That is, $\mathbb{F}^n \xrightarrow{\phi} \mathbb{K}$, $\mathbb{K} \xrightarrow{\phi^{-1}} \mathbb{F}^n$. Any function or map F from $\mathbb{K}$ to $\mathbb{K}$ can be expressed *uniquely* as a polynomial function with coefficients in $\mathbb{K}$ and degree less than q^n, namely

$$F(X) = \sum_{i=0}^{q^n-1} a_i X^i, \quad a_i \in \mathbb{K}.$$

Denote by $\deg_{\mathbb{K}}(F)$ the degree of $F(X)$ for any map F. Using ϕ, we can build a new map $F' : \mathbb{F}^n \to \mathbb{F}^n$

$$P(x_1, \ldots, x_n) = (p_1(x_1, \ldots, x_n), \ldots, p_n(x_1, \ldots, x_n)) = \phi^{-1} \circ F \circ \phi(x_1, \ldots, x_n),$$

which is essentially F but viewed from the perspective of $\mathbb{F}^n$. We will denote F' also by F unless there is a chance of confusion.

An $\mathbb{F}$-degree-2 or $\mathbb{F}$-quadratic function from $\mathbb{K}$ to $\mathbb{K}$ can in this framework be seen to be a polynomial all of whose monomials have exponent $q^i + q^j$ or q^i or 0 for some i and j. The general form of this $\mathbb{F}$-quadratic function is $Q(X) = \sum_{i,j=0}^{n-1} a_{ij} X^{q^i+q^j} + \sum_{i=0}^{n-1} b_i X^{q^i} + c.$, the *extended Dembowski-Ostrom polynomial map.* Such a $Q(X)$ with a fixed low $\mathbb{K}$-degree is used to build the HFE multivariate public key cryptosystems, as in the following

$$Q(X) = \sum_{\substack{i,j=0, j \leq i}}^{q^i+q^j \leq D} a_{ij} X^{q^i+q^j} + \sum_{i=0}^{q^i \leq D} b_i X^{q^i} + c;$$

Note that the coefficients are values in $\mathbb{K}$, and all coefficients $a_{ii} = 0$ if $q = 2$, since those are covered by the b-part of the coefficients.

For an overview of multivariate cryptosystems, including all the common modifiers such as "minus", "internal perturbation", and "vinegar" see [6,9]. It gives this formulation of HFEv, which uses the vinegar modification [13], built from the polynomial:

$$Q(X, \bar{X}) = \sum_{i,j} a_{ij} X^{q^i+q^j} + \sum_{i,j} b_{ij} X^{q^i} \bar{X}^{q^j} + \sum_{i,j} \alpha_{ij} \bar{X}^{q^i+q^j} + \sum_i b_i X^{q^i} + \sum_i \beta_i \bar{X}^{q^i} + c$$

$$(1)$$

where the auxiliary variable $\bar{X}$ occupies only a subspace of small rank v in $\mathbb{K} \cong \mathbb{F}^n$. The function Q is quadratic in the components of X and $\bar{X}$, and so is

$P = T \circ Q \circ S$ for affine bijections T and S in $\mathbb{F}^n$ and $\mathbb{F}^{n+v}$. We hope that P is hard to invert to the adversary, while the legitimate user, with the knowledge of (S, T) can compute X by substituting a random $\bar{X}$, then solving for X via root-finding algorithms such as Berlekamp (or Cantor-Zassenhaus, if $q \neq 2$). To limit the effort of Berlekamp, we restrict the maximum degree D of the polynomial.

Another closely related scheme to HFEv is IPHFE (internally perturbed HFE). Suppose in Eq. 1, $\bar{X}$ is not a free variable, but is instead the image of ℓ, a map from $\mathbb{F}^n$ onto $\mathbb{F}^v$. So the central map is really $Q'(X) := Q(X, \ell(X))$.

For our experiments, we will use a more generalized version of HFE, namely we allow the maximum degree of the quadratic part (the terms in the form of $X^{q^i + q^j}$) to be different from the linear part (the terms in the form of X^{q^i}). We call the highest degree of the quadratic part the quadratic degree and the highest degree of the linear part the linear degree of the HFE polynomial.

Here we would like to make one remark that we only look at systems with relative large n, since when n is small, special combinatorial identity could occur.

3.2   The Experimental Results

We will first present a few graphs of systems from many experiments we have done to give the reader a basic idea of what happens in the solving process. All computations were performed on a PC with the 64 bit version of Magma for Unix. The first example (Fig. 2) is an example, where the differences between the corresponding two degrees, the degree of regularity and the solving degree, are the same and occur at the same step.

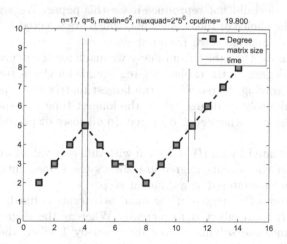

Fig. 2. This is an example of HFE with q=5, n=17, the quadratic degree is 2 and the linear degree 25. This is a case where both degrees are 5 and both occur at the step 4.

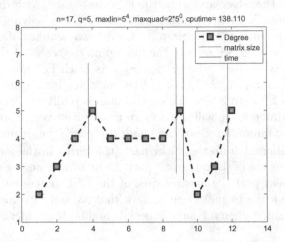

Fig. 3. This is an example of HFE with q=5, n=17, the quadratic degree is 2 and the linear degree to be 625. This is a case where the two degrees have the same value 5 but they occur at different steps.

But the degree of regularity and the solving degree can occur at very different steps despite the fact that they have the same values. Fig. 3 is such an example.

The degree of regularity and the solving degree can occur at very different steps and the difference now is 1. Fig. 4 is an example.

We have seen cases where the degree of regularity and the solving degree differ by 2, but we could not reproduce it for this paper. We are not sure if it was caused by the choice of the coefficients, which are selected at random, or by a programming or another error from our side.

Below are some of the tables from many we made for the degree of regularity and the solving degree. In the tables 'deg-reg' stands for the degree of regularity, 'deg-size' is the solving degree where the largest matrix size was encountered, and 'deg-time' the solving degree where the longest time was spent. The entry 'at step' gives the step where each occurred. In all cases displayed the difference is at most 1.

Table 1 was created by an HFE system with different values of n. The degree of regularity and the solving degree are always the same, but sometimes the largest matrix is encountered at a different step.

In Tables 2 and 3 the degree of the quadratic terms is fixed, but the degree of the linear terms are allowed to increase. Whereas the degree of regularity remains the same, the solving degree increases by 1 when the linear degree reaches a certain threshold. Table 4 shows that internal perturbation of an HFE system has no effect on the difference of the degrees.

For a quadratic system with the coefficients selected at random there will be some difference between the degrees as seen in Table 5 and it also occurs for other finite fields.

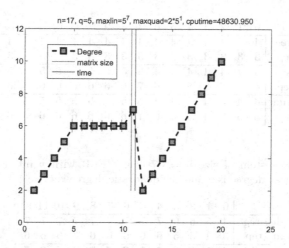

Fig. 4. This is an example of HFE with q=5, n=17 the quadratic degree is degree 10 and the linear degree is 78125. This is a case where the two degrees differ by 1 and are 6 and 7 respectively.

Table 1. HFE systems over $GF(3)$ with quadratic degree 6 and linear degree 9 for different values of n

n	10	11	12	13	14	15	16	17	18	19	20	21	22	23	24	25	26	27	28	29	30	31	32
deg-reg	4	4	4	4	4	4	4	4	4	4	4	4	4	4	4	4	4	4	4	4	4	4	4
at step	3	3	3	3	3	3	3	3	3	3	3	3	3	3	3	3	3	3	3	3	3	3	3
deg-size	4	4	4	4	4	4	4	4	4	4	4	4	4	4	4	4	4	4	4	4	4	4	4
at step	3	3	3	3	3	3	3	7	3	8	8	3	8	8	7	3	7	3	8	3	8	7	8
deg-time	4	4	4	4	4	4	4	4	4	4	4	4	4	4	4	4	4	4	4	4	4	4	4
at step	3	3	3	3	3	3	3	3	3	3	3	3	3	3	3	3	3	3	3	3	3	3	3

Table 2. HFE systems of size $n = 17$ over $GF(5)$ with quadratic degree 2 and the linear degree 5^{r_1}

r_1	0	1	2	3	4	5	6	7	8	9	10	11	12
deg-reg	5	5	5	5	5	5	5	5	5	5	5	5	5
at step	4	4	4	4	4	4	4	4	4	4	4	4	4
deg-size	5	5	5	5	5	5	5	5	6	6	6	6	6
at step	4	11	4	4	12	4	4	4	9	8	8	8	8
deg-time	5	5	5	5	5	5	5	5	6	6	6	6	6
at step	4	4	4	4	9	8	8	9	9	8	8	8	8

Table 3. HFE systems of size $n = 17$ over $GF(3)$ with quadratic degree 6 and the linear degree 3^{r_1}

r_1	0	1	2	3	4	5	6	7	8	9	10	11	12	13	14	15	16
deg-reg	4	4	4	4	4	4	4	4	4	4	4	4	4	4	4	4	4
at step	3	3	3	3	3	3	3	3	3	3	3	3	3	3	3	3	3
deg-size	4	4	4	4	4	4	4	5	5	5	5	5	5	5	5	5	5
at step	8	3	7	8	5	6	6	7	6	6	6	6	6	6	6	6	6
deg-time	4	4	4	4	4	4	4	5	5	5	5	5	5	5	5	5	5
at step	3	3	3	4	5	5	6	7	6	6	6	6	6	6	6	6	6

Table 4. IPHFE system of size $n = 17$ over $GF(3)$ with v internal perturbation variables. The linear degree is 9 and the quadratic degree is 6.

v	0	1	2	3	4	5	6	7	8	9	10	11	12
deg-reg	4	5	6	6	7	7	7	7	7	7	7	7	7
at step	3	4	5	5	6	6	6	6	6	6	6	6	6
deg-size	4	5	6	6	7	7	7	7	7	7	7	7	7
at step	3	4	5	5	6	6	6	6	6	6	6	6	6
deg-time	4	5	6	6	7	7	7	7	7	7	7	7	7
at step	3	4	5	5	6	6	6	6	6	6	6	6	6

Table 5. Random quadratic system of size n over $GF(2)$

n	5	6	7	8	9	10	11	12	13	14	15	16	17	18	19	20	21	22	23	24	25	26	27	28	29
deg-reg	3	3	3	3	4	4	4	4	4	4	4	5	5	5	5	5	5	5	5	6	6	6	6	6	6
at step	2	2	2	2	3	3	3	3	3	3	3	4	4	4	4	4	4	4	4	5	5	5	5	5	5
deg-size	3	3	3	3	4	4	4	4	4	4	4	5	5	5	5	5	5	5	5	6	6	6	6	6	6
at step	2	2	4	3	3	3	3	3	3	3	4	4	4	4	4	4	4	5	4	5	5	5	5	5	5
deg-time	3	3	3	3	4	4	4	4	4	4	4	5	5	5	5	5	5	5	5	6	6	6	6	6	6
at step	2	2	4	2	3	3	3	3	3	3	4	4	4	4	4	4	4	4	5	5	5	5	5	5	5

From all the experiments, we conclude that it seems that it is indeed true that the differences between the degree of regularity and the solving degree for irreducible systems are small. But again, we like to emphasize that the experiments we have done is relatively small in terms of number of variables, and therefore our experiments, though very systematic but are limited by our computing capacity. Some experiments indicate that the situation may not be true for large n. The reason is due to the experiments listed in Table 3 and illustrated in Figure 5.

In the two cases of Figure 5, the quadratic parts are exactly the same (therefore the degree of regularities remains the same), and only the linear parts are different. But in the second case, the linear part is more complicated. This shows that the linear part has a substantial impact. It increases the solving degree by 1 and not just for the case shown in the figure, but in all cases when the linear part had a degree $\geq 3^7$, see Table 3. Therefore, we believe we need more

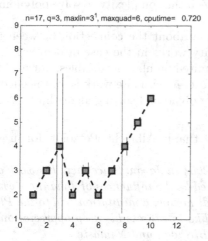
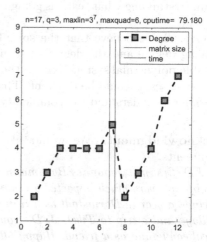

Fig. 5. This is an example of HFE with q=3, n=17, the quadratic degree is at most 6. On the left the linear terms have only degree 1, whereas on the right they are limited by 3^7. This a case where the solving degree can be increased by 1 due to additional linear terms.

experiments with larger n. It will not be a surprise if we find the differences to be bigger than 1 when n is large enough. Therefore we speculate the differences of the two degrees could be dependent on n. For this, we need much more powerful computers to do the experiments, which are now beyond our reach.

3.3   Higher Degree Cases

We performed some examples with higher degree polynomials, in particular, degree 3 polynomials with random coefficients in $GF(2)$ or in $GF(3)$. The overall impression is that in these cases the degree of regularity and the solving degree are the same and occur at the same step. Modifying the equations we have seen examples where the difference was greater than 1. For this case, we need more studies to come to a reasonable conclusion.

4   Conclusion and Discussion

From the experiments, we conclude that indeed for an irreducible quadratic system the difference between the degree of regularity and the solving degree is small. But our experiments are preliminary so far and are limited since they were run on a personal computer with a 64 bit Unix system. The results in Fig. 5 however force us to suspect that maybe the difference between the two degrees can become bigger due to the influence of the linear part. The next step would require to find some way to prove the claim, but for this we need to set up reasonable additional assumptions. This would be a big breakthrough in terms of understanding what really is going on with the complexity to solve polynomial systems.

Overall, we believe that the speculation about the connection between the solving degree and the degree of regularity works in the case of degree 2 irreducible polynomial systems with rather limited number of variables, but for high degree cases, it could be different. Therefore, much more work is still needed to be done to understand the complexity of polynomial solving algorithms.

Acknowledgment. We would like to thank Albrecht Petzoldt for useful comments.

J.D. first met Johannes Buchmann in 2004 in Japan. Since then, his work and his life are very much impacted and inspired by the influences of Johannes, which include a year in Darmstadt as a Humboldt Fellow, a meditation trip to the Plum Village and a trip to Tibet. J. D. would like to express his deep appreciation as a colleague and as a friend. Happy 60th Birthday and Namaste.

References

1. Bardet, M., Faugère, J.-C., Salvy, B., Yang, B.-Y.: Asymptotic expansion of the degree of regularity for semi-regular systems of equations. In: Gianni, P. (ed.) MEGA 2005, Sardinia, Italy (2005)
2. Courtois, N.T., Klimov, A., Patarin, J., Shamir, A.: Efficient algorithms for solving overdefined systems of multivariate polynomial equations. In: Preneel, B. (ed.) EUROCRYPT 2000. LNCS, vol. 1807, pp. 392–407. Springer, Heidelberg (2000), http://www.minrank.org/xlfull.pdf
3. Diem, C.: The XL-algorithm and a conjecture from commutative algebra. In: Lee, P.J. (ed.) ASIACRYPT 2004. LNCS, vol. 3329, pp. 323–337. Springer, Heidelberg (2004)
4. Ding, J.: Inverting the square systems is exponential. Cryptology ePrint Archive, Report 2011/275 (2011), http://eprint.iacr.org/
5. Ding, J., Buchmann, J., Mohamed, M.S.E., Mohamed, W.S.A.E., Weinmann, R.-P.: Mutant XL. In: Talk at the First International Conference on Symbolic Computation and Cryptography (SCC 2008), Beijing (2008)
6. Ding, J., Gower, J., Schmidt, D.: Multivariate Public-Key Cryptosystems. In: Advances in Information Security. Springer (2006) ISBN 0-387-32229-9
7. Ding, J., Hodges, T.J.: Inverting hfe systems is quasi-polynomial for all fields. In: Rogaway, P. (ed.) CRYPTO 2011. LNCS, vol. 6841, pp. 724–742. Springer, Heidelberg (2011)
8. Ding, J., Kleinjung, T.: Degree of regularity for HFE−. Journal of Math-for-Industry 4(2012B-3), 97–104 (2012), http://eprint.iacr.org/
9. Ding, J., Yang, B.-Y.: Post-Quantum Cryptography. Springer, Berlin (2009) ISBN: 978-3-540-88701-0, e-ISBN: 978-3-540-88702-7
10. Ding, J., Yang, B.-Y.: Degree of regularity for hfev and hfev-. In: Gaborit, P. (ed.) PQCrypto 2013. LNCS, vol. 7932, pp. 52–66. Springer, Heidelberg (2013)
11. Dubois, V., Gama, N.: The degree of regularity of hfe systems. In: Abe, M. (ed.) ASIACRYPT 2010. LNCS, vol. 6477, pp. 557–576. Springer, Heidelberg (2010)
12. Faugère, J.-C.: A new efficient algorithm for computing Gröbner bases (F_4). Journal of Pure and Applied Algebra 139, 61–88 (1999)

13. Kipnis, A., Patarin, J., Goubin, L.: Unbalanced Oil and Vinegar signature schemes. In: Stern, J. (ed.) EUROCRYPT 1999. LNCS, vol. 1592, pp. 206–222. Springer, Heidelberg (1999)

14. Lazard, D.: Gröbner-bases, Gaussian elimination and resolution of systems of algebraic equations. In: ISSAC 1983 and EUROCAL 1983. LNCS, vol. 162, pp. 146–156. Springer (March 1983)

15. Mayr, E.W., Meyer, A.: The complexity of the word problems for commutative semigroups and polynomial ideals. Adv. in Math. 46(3), 305–329 (1982)

16. Mohamed, M.S.E., Cabarcas, D., Ding, J., Buchmann, J., Bulygin, S.: MXL$_3$: An efficient algorithm for computing Gröbner bases of zero-dimensional ideals. In: Lee, D., Hong, S. (eds.) ICISC 2009. LNCS, vol. 5984, pp. 87–100. Springer, Heidelberg (2010)

17. Mohamed, M.S.E., Mohamed, W.S.A.E., Ding, J., Buchmann, J.: MXL2: Solving polynomial equations over GF(2) using an improved mutant strategy. In J. Buchmann and J. Ding, editors, *PQCrypto*. In: Buchmann, J., Ding, J. (eds.) PQCrypto 2008. LNCS, vol. 5299, pp. 203–215. Springer, Heidelberg (2008)

18. Yang, B.-Y., Chen, J.-M.: Theoretical analysis of XL over small fields. In: Wang, H., Pieprzyk, J., Varadharajan, V. (eds.) ACISP 2004. LNCS, vol. 3108, pp. 277–288. Springer, Heidelberg (2004)

Shorter Compact Representations
in Real Quadratic Fields

Alan K. Silvester[1], Michael J. Jacobson, Jr.[2,*], and Hugh C. Williams[1]

[1] Department of Mathematics and Statistics, University of Calgary
2500 University Drive NW, Calgary, Alberta, Canada T2N 1N4
aksilves@ucalgary.ca, williams@math.ucalgary.ca
[2] Department of Computer Science, University of Calgary
2500 University Drive NW, Calgary, Alberta, Canada T2N 1N4
jacobs@cpsc.ucalgary.ca

Abstract. Compact representations are explicit representations of algebraic numbers with size polynomial in the logarithm of their height. These representations enable much easier manipulations with larger algebraic numbers than would be possible using a standard representation and are necessary, for example, in short certificates for the unit group and ideal class group. In this paper, we present two improvements that can be used together to reduce significantly the sizes of compact representations in real quadratic fields. We provide analytic and numerical evidence demonstrating the performance of our methods, and suggesting that further improvements using obvious extensions are likely not possible.

Keywords: compact representation, real quadratic field, fundamental unit, infrastructure.

1 Introduction

Let α (> 1) be an algebraic integer in the quadratic order $\mathcal{O}$ of discriminant Δ (> 0). If we put $\alpha = (x + y\sqrt{\Delta})/2$, where the coefficients x and y are rational integers, it is often the case that even when the absolute norm of α, $|N(\alpha)|$, is small, the values of x and y can be very large. Consider the case where $|N(\alpha)| = 1$. In this case α is a unit of $\mathcal{O}$ and therefore a power of the fundamental unit η_Δ of $\mathcal{O}$, but we know (see, for example, Chapter 9 of [16]) that the coefficients in η_Δ can be very large, so much so, that even if Δ is only moderately large it is difficult to impossible to write them down, using conventional decimal representation. For example, when we attempt to solve the famous Cattle Problem of Archimedes we encounter an order of discriminant $\Delta = 410286423278424$ and the coefficients in η_Δ contain about 103,200 decimal digits each. Furthermore, if $\Delta = 990676090995853870156271607886$, a number of 30 decimal digits, then each of the coefficients in η_Δ contains more than 2×10^{15} decimal digits,

---

* The second and third authors are supported in part by NSERC of Canada.

M. Fischlin and S. Katzenbeisser (Eds.): Buchmann Festschrift, LNCS 8260, pp. 50–72, 2013.

see [16, pp. 62,285]. As the average paperback novel contains about one million symbols, this means that it would take over two billion such volumes to record only one of the coefficients. Thus, it is necessary to find a much more compact representation for α other than simply recording the decimal representation of each of the coefficients.

Although a technique for doing this was anticipated in work of Lagarias [17,18], it was Cohen in [4, pp 274, 280-282], who first described in print a method that could be applied to this problem. Somewhat before Cohen's idea had appeared, I (Hugh Williams) had been approached by my graduate student, Gilbert Fung, with the question of how the units in a cubic field with negative discriminant could be represented without recourse to the voluminous decimal representation. This discussion led, in 1991, to our writing a paper [9] on this topic, which we submitted to Mathematics of Computation. Unfortunately, the editor, Dan Shanks, did not seem to know what to do with the paper, and it languished in his care for many months, without being accepted or rejected. At this point, I must confess that I had rather lost interest in this idea, but fortunately I was invited to present a paper at a meeting on Computational Algebra and Number Theory to be held at the University of Sydney in November of 1992. During a previous visit to Saarbrücken, I had explained my ideas to Johannes Buchmann, and he suggested that we work jointly on an account of the technique for application to real quadratic fields. I wrote up a preliminary version of the paper and sent it to Johannes for comments and revisions. This was the paper that I presented at the Sydney meeting. Johannes enlisted the aid of one of his graduate students, Christoph Thiel, and they produced a completely revised version of the paper, with Christoph being added as a third author. Ultimately, this revised paper [2] appeared in the proceedings of the Sydney meeting.

This paper essentially elaborated upon the idea of Cohen, but both extended and formalized it. An improved, but much briefer version appears in Buchmann and Vollmer [3, pp.251-256]. In [16, Chapter 12], we presented another variant which allows us to avoid trying to approximate logarithms and produces somewhat better results than those in [3]. The basic idea of all these techniques is to represent the algebraic number α in terms of a power product which satisfies a number of conditions. In doing this one can drastically reduce the number of digits needed to record the coefficients in α. Furthermore, it can be shown how arithmetic operations can be performed on such representations, leading to more efficient calculations than those required for the standard decimal representation. It must be emphasized that in order to produce a compact representation of α, we usually need an approximate value, within about 1, say, of $\log \alpha$.

The purpose of this paper is to provide an adjustment to our previous definition of a compact representation, which allows us to compute it in fewer iterations than those required for the earlier definition. We also provide an analysis which suggests that these new compact representations are quite likely as small as we can expect to achieve with these methods. These analytical results are backed up by various numerical computations.

2   Background on Quadratic Fields

For more details about quadratic fields, the reader is referred to [16], upon which the following material is based.

Let $D \in \mathbb{Z}$ be an integer, not a perfect square, and greater than 1. The elements of the real quadratic field $\mathbb{K} = \mathbb{Q}(\sqrt{D})$ have the form $\alpha = (a + b\sqrt{D})/c$ for integers a, b, and c. The *conjugate* $\overline{\alpha}$ of a $\alpha \in \mathbb{K}$ is given by $\overline{\alpha} = (a - b\sqrt{D})/c$. The quadratic integers of $\mathbb{K}$ have the form $\alpha = a + b\omega$ where

$$ r = \begin{cases} 1 \text{ if } D \not\equiv 1 \pmod 4 , \\ 2 \text{ otherwise} \end{cases} \quad \text{and} \quad \omega = \frac{r - 1 + \sqrt{D}}{r} , \tag{1} $$

The *height* of a quadratic integer measures its size.

Definition 1. *The* height *of a quadratic integer α is $H(\alpha) = \max\{|\alpha|, |\overline{\alpha}|\}$.*

Recalling that $|N(\alpha)| = |\alpha\overline{\alpha}| \geq 1$, we see that $H(\alpha) \geq 1$ and so an element's height cannot be arbitrarily small.

The set of all quadratic integers of $\mathbb{K}$ is called the *maximal order*, denoted $\mathcal{O}_{\mathbb{K}}$. The *field discriminant* $\Delta_{\mathcal{O}_{\mathbb{K}}}$ is the discriminant of the order $\mathcal{O}_{\mathbb{K}}$, and can be explicitly determined as $\Delta_{\mathcal{O}_{\mathbb{K}}} = 4D/r^2$. Suborders $\mathcal{O}_\Delta$ of $\mathcal{O}_{\mathbb{K}}$ have discriminant $\Delta = f^2 \Delta_{\mathcal{O}_{\mathbb{K}}}, f > 1$.

The smallest unit of $\mathcal{O}_\Delta$ greater than 1 is called the *fundamental unit* and is denoted η_Δ. If $\eta \in \mathcal{O}_\Delta^*$ is a unit then $\eta = \pm\eta_\Delta^n$ for some integer n, i.e., the unit group of a quadratic order $\mathcal{O}_\Delta$ is given by $\mathcal{O}_\Delta^* = \langle -1, \eta_\Delta \rangle$. As the size of η_Δ grows exponentially as Δ increases, we often work with a more manageable quantity called the *regulator*, denoted $\mathcal{R} = \log \eta_\Delta$.

2.1   Ideals

The non-zero ideals of $\mathcal{O}_\Delta$ can be represented as

$$ \mathfrak{a} = S \left[\frac{Q}{r}, \frac{P + \sqrt{D}}{r} \right] , \tag{2} $$

where $S, Q, P \in \mathbb{Z}$, $r \in \{1, 2\}$, $r \mid Q$, and $rQ \mid D - P^2$. We will refer to an ideal as "$S[Q, P]$" where it is understood that S, Q, and P satisfy the conditions listed here. Given two ideals in $S[Q, P]$ representation, well-known formulas originally due to Gauss can compute their product in $S[Q, P]$ representation [16, Ch.5].

A *principal* ideal $\mathfrak{a}$ is an $\mathcal{O}_\Delta$-ideal which can be written as $\mathfrak{a} = (\theta)$ for some $\theta \in \mathcal{O}_\Delta$, in other words it has only a single generator. Two ideals are said to be *equivalent* if there exist non-zero $\alpha, \beta \in \mathcal{O}_\Delta$ such that $(\alpha)\mathfrak{a} = (\beta)\mathfrak{b}$ and we denote this by $\mathfrak{a} \sim \mathfrak{b}$. We remark that we will frequently abuse this notation by writing $\mathfrak{a} = (\gamma)\mathfrak{b}$, where it is understood that $(\gamma) = (\beta/\alpha)$ is a fractional $\mathcal{O}_\Delta$-ideal, i.e., there exists a non-zero $\alpha \in \mathcal{O}_\Delta$ such that $\alpha(\gamma) \subseteq \mathcal{O}_\Delta$.

Our algorithms for compact representations rely on arithmetic with principal ideals that are reduced. An $\mathcal{O}_\Delta$-ideal $\mathfrak{a}$ is *primitive* if it cannot be written as an integer multiple of another ideal $\mathfrak{b}$, i.e., if $\mathfrak{a} \neq (m)\mathfrak{b}$ for any $m \in \mathbb{Z}$, where $|m| > 1$. Using the notation of (2), we say an ideal $\mathfrak{a}$ is primitive if $S = 1$, denoted as $\mathfrak{a} = [Q, P]$. The *norm* $N(\mathfrak{a})$ of an $\mathcal{O}_\Delta$-ideal $\mathfrak{a}$ is the index $|\mathcal{O}_\Delta/\mathfrak{a}|$ and when the ideal $\mathfrak{a}$ is written in the form of (2), we have

$$N(\mathfrak{a}) = S^2 Q/r \ . \tag{3}$$

Finally, an $\mathcal{O}_\Delta$-ideal $\mathfrak{a}$ is *reduced* if it is primitive and there does not exist $\alpha \in \mathfrak{a}$, $\alpha \neq 0$, such that both $|\alpha| < N(\mathfrak{a})$ and $|\overline{\alpha}| < N(\mathfrak{a})$. A useful property of reduced $\mathcal{O}_\Delta$-ideals, when written in the form of (2), is that $0 < P < \sqrt{D}$ and $0 < Q < 2\sqrt{D}$ [16, Cor. 5.8.1, p. 101]. That is, if $\mathfrak{a}$ is a reduced ideal, then $N(\mathfrak{a}) < \sqrt{\Delta}$.

A primitive ideal $\mathfrak{a}$ given by $[Q, P]$ can be reduced by expanding the continued fraction of $\alpha = (P + \sqrt{D})/Q$ as described in [16, Ch.5]. If we start the reduction procedure with the reduced ideal $\mathfrak{a} = \mathfrak{a}_1 = \mathcal{O}_\Delta$, we obtain a sequence of reduced principal ideals $\mathfrak{a}_{i+1} = (\theta_i)\mathfrak{a}_1$. Since the coefficients of a reduced ideal are bounded, there are only finitely many, and consequently this sequence must be periodic. Hence, we can find some minimal $p > 0$ such that $\mathfrak{a}_{p+1} = \mathfrak{a}_1$. These ideals can be arranged into a cycle $\mathcal{C} = \{\mathfrak{a}_1, \mathfrak{a}_2, \ldots, \mathfrak{a}_p\}$ with

$$1 = \theta_1 < \theta_2 < \theta_3 < \cdots < \theta_p < \cdots \ .$$

called the *cycle of reduced principal ideals*. A well-known fact—derived, for instance, from [16, (5.33), p. 113]—is that if the fundamental unit $\eta_\Delta = \theta_{p+1}$. It can be shown that $p \approx O(\mathcal{R}) = O(\Delta^{1/2+\epsilon})$.

2.2 Infrastructure

The *infrastructure*, discovered by Daniel Shanks [20], refers to the group-like structure existing within each equivalence class of ideals in $\mathcal{O}_\Delta$. For our purposes, we focus on the principal class, in particular the set of reduced principal ideals $\mathcal{C}$. The arithmetic properties of this set are key to algorithms for computing compact representations.

Let $\mathfrak{a}_1 = [1, \omega]$ be the first ideal in $\mathcal{C}$. The *distance* of $\mathfrak{a}_i = (\theta_i)$ is defined as $\delta_i = \delta(\mathfrak{a}_i) = \log_2 \theta_i \pmod{\mathcal{R}}$. Let $\mathfrak{a}_i$ and $\mathfrak{a}_j$ be two reduced principal ideals in $\mathcal{C}$. Since they are principal, their product $\mathfrak{a} = \mathfrak{a}_i\mathfrak{a}_j = (\theta_i\theta_j)$ will also be principal. However, $\mathfrak{a}$ may no longer be reduced, but is equivalent to some reduced principal ideal $\mathfrak{a}_l \in \mathcal{C}$. Thus,

$$\mathfrak{a}_l = \left(\frac{\theta_k' \theta_i \theta_j}{m} \right)$$

$$\delta_l = \delta(\mathfrak{a}_l) = \log_2 \left(\frac{\theta_k' \theta_i \theta_j}{m} \right) \equiv \delta_i + \delta_j + \log_2 \frac{\theta_k'}{m} \pmod{\mathcal{R}} \ . \tag{4}$$

We denote by $\mathfrak{a}_i \star \mathfrak{a}_j$ the computation of the reduced ideal equivalent to the product ideal $\mathfrak{a}_i\mathfrak{a}_j$ and refer to this process as a *giant step*. The key observation is that $\mathcal{C}$ is almost a group under this operation — only associativity fails, because instead of having $\delta_l = \delta_i + \delta_j$ in (4), we are stuck with the additional error term $\log_2(\theta'_k/m)$ and so δ_l is only close to $\delta_i + \delta_j$. However, this error term can be bounded in absolute value, say by μ. This bound depends on the particular reduction algorithm selected, but for the method described above, it can be shown [16, p. 175] that $\mu < O(\log \Delta)$ which is quite small compared to $\delta_i, \delta_j \approx O(\mathcal{R})$.

In the rest of this paper, we will refer to one application of the continued fraction algorithm to the ideal $\mathfrak{a}_i$ as a *(forward) baby step*, denoted $\mathfrak{a}_{i+1} = \rho(\mathfrak{a}_i)$. Although we will not derive formulas here, given a reduced principal ideal $\mathfrak{a}_i \in \mathcal{C}$, we can also compute the *backward baby step* $\mathfrak{a}_{i-1} = \rho^{-1}(\mathfrak{a}_i)$ [16, §3.4, p. 64].

2.3　Approximating Distances

While performing computations in the infrastructure, we need to keep track of distances while maintaining accurate approximations in the face of round-off and truncation errors. The method of (f, p) representations [14], adapted from ideas of Hühnlein and Paulus [12], was devised to provide provable bounds on the round-off and truncation errors accumulated during computations. This idea was later refined by the authors of [15] and we will use the method of w-near (f, p) representations, as described in [16, Ch. 11, p. 265].

Let $p \in \mathbb{N}$ and $f \in \mathbb{R}$ be such that $1 \le f < 2^p$. If $\mathfrak{a}$ is a primitive $\mathcal{O}_\Delta$-ideal, then an (f, p) *representation* of $\mathfrak{a}$ is a triple $(\mathfrak{b}, d, k)$ where $\mathfrak{b} \sim \mathfrak{a}$, $d \in \mathbb{N}$ with $2^p < d \le 2^{p+1}$, and $k \in \mathbb{Z}$. In addition, there exists $\theta \in \mathbb{K}$ such that $\mathfrak{b} = (\theta)\mathfrak{a}$ with

$$\left| \frac{\theta}{2^{k-p}d} - 1 \right| < \frac{f}{2^p} .$$

In essence, an (f, p) representation stores both an approximation to the *relative generator* θ and an approximation of its distance, both with precision p. The parameter f is a measure of the approximation error, though it is rarely if ever explicitly computed. If $\mathfrak{b}$ is a reduced $\mathcal{O}_\Delta$-ideal, then $(\mathfrak{b}, d, k)$ is a *reduced (f, p) representation* of $\mathfrak{a}$.

A w-*near* (f, p) representation is a a reduced (f, p) representation $(\mathfrak{b}, d, k)$ of an $\mathcal{O}_\Delta$-ideal $\mathfrak{a}$ with the following two additional conditions:

1. $k < w$ for some $w \in \mathbb{Z}^+$ and
2. if $\rho(\mathfrak{b}) = (\psi)\mathfrak{b}$ then there exist integers d' and k' such that $k' \ge w$, $2^p < d' \le 2^{p+1}$ and

$$\left| \frac{\psi\theta}{2^{k'-p}d'} - 1 \right| < \frac{f}{2^p} .$$

Such representations have the useful property that $\theta \approx 2^w$ and $k \approx w$. Since this property will be used repeatedly in later material, particularly with respect to compact representations, we will state it more formally.

Lemma 1 ([16, Lem. 11.3, p. 270]). *Let $(\mathfrak{b}, d, k)$ be a w-near (f, p) representation of some $\mathcal{O}_\Delta$-ideal $\mathfrak{a}$ with $p > 4$ and $f < 2^{p-4}$. If θ and ψ are defined as above, then*

$$\frac{15N(\mathfrak{b})}{16\sqrt{\Delta}} < \frac{15}{16\psi} < \frac{\theta}{2^w} < \frac{17}{16} \quad and \quad -\log_2 \frac{34\psi}{15} < k - w < 0 .$$

2.4 Algorithms

We define here the basic algorithms we require for performing various computations with (f, p) representations. The majority of these algorithms will not be explicitly presented, rather references to the appropriate sections of [16] will be given.

1. Given $\mu, \nu \in \mathcal{O}_\Delta$, Algorithm IMULT [16, Alg. 12.2, p. 286] computes $\lambda = \mu\nu$.
2. Given an (f, p) representation $(\mathfrak{b}, d, k)$ of $\mathfrak{a}$, we can determine a w-near representation $(\mathfrak{c}, g, h)$ of $\mathfrak{a}$ along with the corresponding relative generator using EWNEAR [16, Alg. 12.1, pp. 286 and 457] provided that $k < w$. In order to implement our improved algorithms, we require a version that also works in the case $k > w$. A modified version of EWNEAR that takes care of this is presented in Appendix A.
3. If $(\mathfrak{a}[x], d_x, k_x)$ and $(\mathfrak{a}[y], d_y, k_y)$ are respectively, x- and y-near (f', p) and (f'', p) representations of $\mathfrak{a} = (1)$, we can employ EADDXY [16, Alg. 12.3, p. 286] to produce an $x + y$-near (f, p) representation $(\mathfrak{a}[x + y], d, k)$ of $\mathfrak{a}$ where $f = 13/4 + f' + f'' + f'f''/2^p$, as well as a relative generator $\lambda \in \mathcal{O}$ such that

$$\mathfrak{a}[x + y] = \left(\frac{\lambda}{N(\mathfrak{a}[x])N(\mathfrak{a}[y])} \right) \mathfrak{a}[x]\mathfrak{a}[y] .$$

4. If $(\mathfrak{a}[x], d', k')$ is an x-near (f', p) representation of the $\mathcal{O}_\Delta$-ideal $\mathfrak{a} = (1)$, algorithm ETRIPLEX (described in Appendix B) computes a $3x$-near (f, p) representation $(\mathfrak{a}[3x], d, k)$ of $\mathfrak{a}$ with $f = 13/4 + 3f' + 3f'^2/2^p + f'^3/2^{2p}$ and

$$\lambda = \frac{a + b\sqrt{D}}{r} \text{ such that } \mathfrak{a}[3x] = \left(\frac{\lambda\theta^3}{N(\mathfrak{a}[x])^3} \right) \mathfrak{a},$$

where $\mathfrak{a}[x] = (\theta)\mathfrak{a}$.

3 Compact Representations

In this section, we describe how to compute a compact representation. Our presentation follows that of [16]; for a more detailed description, see [16, §§12.2–3, pp. 290–304].

The algorithm AX [16, Alg. 11.6, pp. 279–80] computes a reduced principal ideal $\mathfrak{a}$ at distance approximately x from $\mathfrak{a}_1 = (1)$. At the heart of AX is a square-and-multiply routine that uses the binary expansion of x to make a series of giant steps in the infrastructure. For each bit in the binary expansion, we compute the giant step $\mathfrak{a}_j \star \mathfrak{a}_j$—the squaring step—which results in an ideal with roughly double the distance from where we started. If the current bit is 1, then we also adjust the resulting ideal via ρ to correct the distance—the multiplying step.

At each stage of AX, suppose we were to keep track of the relative generator that appears. For the giant steps we would have μ_j such that $\mathfrak{a}'_{j+1} = (\mu_j/N(\mathfrak{a}_j)^2)\mathfrak{a}_j^2$ from EADDXY, and for the adjustment steps we would have ν_j such that $\mathfrak{a}_{j+1} = (\nu_j)\mathfrak{a}'_{j+1}$ from EWNEAR. Then $\mathfrak{a}_{j+1} = (\lambda_j/L_{j+1}^2)\mathfrak{a}_j^2$ with $\lambda_j = \mu_j\nu_j$ (computed with IMULT) and $L_{j+1} = N(\mathfrak{a}_j) < \sqrt{\Delta}$ Note that the ν_j values will be small compared to the μ_j.

At the end of AX, we will not only have an ideal $\mathfrak{a}_n = \mathfrak{a}[x] = (\theta)$ at distance approximately x, but also a list of quadratic integers $\{\lambda_1, \lambda_2, \ldots, \lambda_n\}$, and a list of ideal norms $\{L_1, L_2, \ldots, L_n\}$. At this point it should be clear that if we combine the relative generators λ_j and ideal norms L_j by an appropriate combination of multiplications, divisions, and exponentiations, we will get the generator θ. This leads to the definition of a compact representation.

Definition 2. *For any θ such that $(\theta) = \mathfrak{a}[x] \in \mathcal{O}_\Delta$, a compact representation of θ is*

$$\theta = \prod_{i=0}^{l} \left(\frac{\lambda_i}{L_i^2}\right)^{2^{l-i}}$$

where the following properties are satisfied:

1. *$l = O(\log\log\theta)$ for large θ.*
2. *$\lambda_i \in \mathcal{O}_\Delta$ and L_i is an integer $(0 \le i \le l)$.*
3. *$0 < L_i \le \Delta^{1/2}$ and $H(\lambda_i) = O(\Delta)$ $(0 \le i \le l)$.*
4. *$\pi_j \in \mathcal{O}_\Delta$, $L_{j+1} = |N(\pi_j)|$,*

$$\pi_j = \prod_{i=0}^{j} \left(\frac{\lambda_i}{L_i^2}\right)^{2^{j-i}} \; ,$$

π_j generates a reduced ideal $\mathfrak{b}_j$, where $\mathfrak{b}_0 = \mathfrak{a}[1]$, and

$$L_{i+1}^2 \mathfrak{b}_{i+1} = \lambda_{i+1}\mathfrak{b}_i^2 \quad (0 \le i \le l-1) \, .$$

We remark that this definition is slightly different from that given in [16]. Notice that upon substituting $d_i := L_{i+1}$, $\lambda := d_l$, $L_0 = N((1)) = 1$ and shifting the denominators, we get the same presentation as in [16, (12.8), p. 290].

Returning to the Cattle Problem, a compact representation of $\eta_{410286423278424}$ requires only 1,212 bits, whereas writing out its coefficients explicitly would require 206,400 bits. In general, in order to write down θ using standard decimal representation, we require $O(\log_2\theta)$ bits. However, using a compact representation, we require only $O((\log_2\log_2\theta)\log_2\Delta)$ bits to express θ.

4   Reducing the Size of the Terms

The overall size of a compact representation is determined by two factors: the size of the individual terms and the total number of terms. In this section, we describe a method to reduce the size of the terms.

Consider the sequence of s_i values computed as AX executes, corresponding to the intermediate results produced by applying a square-and-multiply process according to the binary representation of x. Let $x = \sum_{i=0}^{l} 2^{l-i} b_i$ be such a representation and set $s_0 = b_0 (= 1)$. As we progress through AX computing giant steps, ideally we wish to compute

$$\mathfrak{a}[s_{i+1}]' = \mathfrak{a}[s_i]^2 .$$

However because of the way giant steps in the infrastructure work, when we compute $\mathfrak{a}[s_i]^2$ we actually "fall short" of this ideal, computing instead

$$\mathfrak{a}[s_{i+1}]' = (\mu_i)\mathfrak{a}[s_i]^2$$

for a correction factor μ_i corresponding to the error term in (4). We then take the ideal $\mathfrak{a}[s_{i+1}]'$ and, depending on the value of b_i, either set $\mathfrak{a}[s_{i+1}] = \mathfrak{a}[s_{i+1}]'$ or compute a baby-step $\mathfrak{a}[s_{i+1}] = \rho(\mathfrak{a}[s_{i+1}]') = (\nu_i)\mathfrak{a}[s_{i+1}]'$ so that

$$\mathfrak{a}[s_{i+1}] = (\lambda_i)\mathfrak{a}[s_i]^2 .$$

As we also mentioned in Section 3, the μ_i values constitute the bulk of the λ_i terms that we wish to store as a compact representation. With some careful reasoning [16, pp. 445–6], one can show that when using EADDXY as in [16, Alg. 12.3, p. 286] we have

$$O(\Delta^{1/4}) < \mu_i < O(\Delta^{3/4}) . \qquad (5)$$

In other words, while the relative generator μ_i is bounded and cannot become too large, it also cannot become very small.

In the following, we will describe a method to adjust the s_i values to exploit the short-fall we experience and so reduce the upper bound in (5). If we increase the s_i values at each step, we will compute ideals $\mathfrak{a}[s_{i+1}]'$ further along the infrastructure than we want. As before, we still experience a short-fall, but will be closer to our goal of $\mathfrak{a}[s_{i+1}]$ than before. By using a larger and backwards EWNEAR step, we use the relative generator ν_i to cancel out, in a sense, a substantial portion of μ_i.

Let $h \in \mathbb{Z}^+$ and let n be the largest integer such that $x \geq (2^n - 1)h$. Set $y = x + (2^n - 1)h$ and compute the binary representation of $y = \sum_{i=0}^{l} 2^{l-i} b_i$. We iterate the while-loop over $0 \leq i < l - n$ as usual ($s_{i+1} = 2s_i + b_i$), and use $s_{i+1} = 2s_i + b_i - h$ for $l - n \leq i < l$. This yields

$$s_i = \begin{cases} \sum_{j=0}^{i} 2^{i-j} b_j & \text{for } 0 \leq i \leq l-n \\ \sum_{j=0}^{i} 2^{i-j} b_j - (2^{n-l+i} - 1)h & \text{for } l-n < i \leq l. \end{cases}$$

Note that $s_l = y - (2^n - 1)h = x$; thus at the end of the algorithm, we have $\mathfrak{a}[s_l] = \mathfrak{a}[x]$ as desired. Furthermore, we clearly have $s_i > 0$ for $0 \leq i \leq l - n$. now, if $s_i \leq 0$ for some i such that $l - n < i < l$, then

$$s_{i+1} = 2s_i + b_i - h \leq 0$$

because $h \geq 1$. By induction, we get $x = s_l \leq 0$, a contradiction. Thus, $s_i > 0$ for all i such that $0 \leq i \leq l$.

All that remains is to determine an appropriate value for h and from that, determine how much the height of λ_i can be reduced.

Recalling (5), we see that $h = \lceil (1/4) \log_2 \Delta \rceil$ is a good choice. In order to determine how much λ_i is reduced, we must compute a revised bound for $H(\lambda_i)$. As the algorithm executes, it finds a series of reduced principal $\mathcal{O}_\Delta$-ideals $\mathfrak{a}[s_i] = \mathfrak{a}_i = (\pi_i)\mathfrak{a}_1$. By Lemma 1, we can conclude that for an s_{i-1}-near (f, p)-representation of $\mathfrak{a}[s_{i-1}]$ and an s_i-near (f, p)-representation of $\mathfrak{a}[s_i]$ where $l - n < i \leq l$,

$$\frac{15 L_i}{16\sqrt{\Delta}} 2^{s_{i-1}} < \pi_{i-1} < \frac{17}{16} 2^{s_{i-1}} \quad \text{and} \quad \frac{15 L_{i+1}}{16\sqrt{\Delta}} 2^{s_i} < \pi_i < \frac{17}{16} 2^{s_i} . \tag{6}$$

From the definition of a compact representation, we also know that

$$\pi_i = \left(\frac{\lambda_i}{L_i^2} \right) \pi_{i-1}^2 \quad \Longrightarrow \quad \lambda_i = \frac{L_i^2 \pi_i}{\pi_{i-1}^2}, \tag{7}$$

and so combining (6) and (7), we get

$$\lambda_i < L_i^2 \left(\frac{17}{16} 2^{s_i} \right) \left(\frac{16\sqrt{\Delta}}{15 L_i} 2^{-s_{i-1}} \right)^2 = \frac{16 \cdot 17}{15^2} 2^{s_i - 2 s_{i-1}} \Delta .$$

Since $s_i - 2 s_{i-1} = b_i - h$ and $b_i \in \{0, 1\}$ we have

$$0 < \lambda_i < \frac{5}{2} 2^{-\lceil (1/4) \log_2 \Delta \rceil} \Delta \leq \frac{5}{2} \Delta^{-1/4} \Delta = \frac{5}{2} \Delta^{3/4}.$$

We can also show that $|\overline{\lambda_i}| < 5/2 \Delta^{3/4}$ by using the reasoning of [16, p.289]. Hence, our modified algorithm reduces the height of λ_i for $l - n < i \leq l$ from $O(\Delta)$ to $O(\Delta^{3/4})$. The λ_i values for $0 \leq i \leq l - n$ will have height $O(\Delta)$, but there are only a small number of these as $l - n < 2 + \log_2 h$. Thus, they will have little impact on the amount of space needed to record θ.

We refer to a compact representation computed using the ideas above as an h-compact representation.

Theorem 1. *Let $\theta \in \mathcal{O}_\Delta$ such that $\mathfrak{a}[x] = (\theta)$ for some $x \in \mathbb{Z}^+$. The number of bits in an h-compact representation of θ is $O((\log_2 \log_2 \theta) \log_2 \Delta^{3/4})$.*

Proof. From the preceding discussion, we know $H(\lambda_i) < (5/2)\Delta^{3/4}$. As $l = \lceil \log_2 x \rceil$ and $2^x < (16\sqrt{\Delta}/15)\theta$, we also have $l = O(\log_2 \log_2 \theta)$. Thus, we require

$$O(l \log_2 \Delta^{3/4}) = O((\log_2 \log_2 \theta) \log_2 \Delta^{3/4})$$

bits to express θ as an h-compact representation. $\square$

Although our improvement does not change the asymptotic running time, the improvement to the O-constant does yield a significant improvement in practice. Returning to our running example, an h-compact representation of the fundamental unit $\eta_{410286423278424}$ uses only 974 bits, a substantial size reduction of 19.6% as compared to the standard compact representation.

5 Reducing the Number of Terms

In the following, we describe a method to reduce the number of terms in a compact representation, in an effort to further reduce the overall size. Recall that for each step of the algorithm we compute an ideal $(\mathfrak{a}[s_{i+1}])$ at double the distance of the ideal we are currently at $(\mathfrak{a}[s_i])$. In order to store fewer terms, we have to progress further from ideal to ideal, for example, by computing an ideal at triple the distance we are at currently. In other words, instead of computing the binary expansion of x and applying a square-and-multiply routine, we could compute a ternary expansion and use a cube-and-multiply routine, using ETRIPLEX in place of EADDXY. We refer to a compact representation produced in this manner as a 3-compact representation.

To see that this method computes a correct compact representation, note that it produces a series of reduced principal $\mathcal{O}_\Delta$-ideals $\mathfrak{a}[s_i] = \mathfrak{b}_i = (\pi_i)\mathfrak{a}_1$ $(\mathfrak{a}_1 = (1))$ where

$$\left| \frac{2^p \pi_i}{2^{k_i} d_i} - 1 \right| < \frac{f}{2^p} .$$

Moreover, $\pi_i \in \mathcal{O}_\Delta$, $|N(\pi_i)| = N(\mathfrak{a}[s_i]) = N(\mathfrak{b}_i) = L_{i+1}$ and if p is sufficiently large, we can appeal to a result analogous to Theorem 11.9 of [16, p. 280] to ensure that $f < 2^{p-4}$. If we set $\lambda_l = (m_i + n_i\sqrt{D})/r$, then

$$\pi_{i+1} = \left(\frac{\lambda_{i+1}}{L_{i+1}^3} \right) \pi_i^3 \tag{8}$$

where $\pi_0 = \lambda_0$. If we define $L_0 = 1$, then we get

$$\pi_j = \prod_{i=0}^{j} \left(\frac{\lambda_i}{L_i^3} \right)^{3^{j-i}}$$

for $j = 0, 1, \ldots, l$. When $j = l$, we have $s_l = x$, $\mathfrak{a}[x] = \mathfrak{b}_l = (\pi_l)$, and hence $\mathfrak{a}[x] = (\theta)$ where

$$\theta = \prod_{i=0}^{l} \left(\frac{\lambda_i}{L_i^3} \right)^{3^{j-i}} .$$

One simple improvement on this idea that will slightly reduce the sizes of the terms is to use a signed ternary representation of x. Instead of digits $0, 1, 2$, in the ternary representation of x, we use digits $-1, 0, 1$, thereby reducing the average size of terms obtained when $b_i \neq 0$.

Notice that we can also combine the ideas behind the h-compact and 3-compact representations to reduce both the sizes and number of terms. Working through the details of adding "$-h$" to the 3-compact representation, we find we must let n be the largest integer such that $x \geq ((3^n - 1)/2)h$ and set $y = x + ((3^n - 1)/2)h$. Furthermore, using ETRIPLEX, we compute

$$\mathfrak{a}[s_{i+1}]' = (\mu_i)\mathfrak{a}[s_i]^3 = ((\mu_i')\mathfrak{a}[s_i]) \left((\mu_i'')\mathfrak{a}[s_i]^2 \right)$$

for each iteration of the main while-loop. Thus, we have $O(\Delta^{1/4}) < \mu_i', \mu_i'' < O(\Delta^{3/4})$, and since $\mu_i = \mu_i'\mu_i''$, we see $O(\Delta^{1/2}) < \mu_i < O(\Delta^{3/2})$. So our choice of h needs to be increased to $h = \lceil (1/2) \log_2 \Delta \rceil$.

Let $y = \sum_{i=0}^{l} 3^{l-i} b_i$. We now derive bounds on the heights of the λ_i defined above. From (8) we have $\lambda_i = (L_i^3 \pi_i)/\pi_{i-1}^3$, which, when combined with (6) gives

$$\lambda_i < L_i^3 \left(\frac{17}{16} 2^{s_i} \right) \left(\frac{16\sqrt{\Delta}}{15L_i} 2^{-s_{i-1}} \right)^3 < \frac{16^2 \cdot 17}{15^3} 2^{s_i - 3s_{i-1}} \Delta^{3/2} .$$

Since $s_i - 3s_i = b_i - h$ and $b_i \in \{0, \pm 1\}$ we have

$$\lambda_i < \frac{16^2 \cdot 17 \cdot 2}{15^3} \cdot 2^{-h} \Delta^{3/2} < \frac{11}{4} \Delta .$$

Now, since $\overline{\lambda}_i = (L_i^3 \overline{\pi}_i)/\overline{\pi}_{i-1}^3$ and $|\pi_i \overline{\pi}_i| = L_{i+1}$, we find

$$|\overline{\lambda}_i| = \left| \frac{L_i^3 \overline{\pi}_i}{\overline{\pi}_{i-1}^3} \right| = \frac{L_i^3 (L_{i+1}/\pi_i)}{(L_i/\pi_{i-1})^3} = \frac{L_i^3 L_{i+1} \pi_{i-1}^3}{L_i^3 \pi_i} = \frac{L_{i+1} \pi_{i-1}^3}{\pi_i}$$

and thus

$$|\overline{\lambda}_i| < L_{i+1} \left(\frac{17}{16} 2^{s_{i-1}} \right)^3 \left(\frac{16\sqrt{\Delta}}{15L_{i+1}} 2^{-s_i} \right) \le \frac{17^3 \cdot 2}{15 \cdot 16^2} \cdot 2^h \Delta^{1/2} < \frac{11}{4} \Delta . \qquad (9)$$

since $3s_{i-1} - s_i = h - b_i$ and $b_i \in \{0, \pm 1\}$. Thus, we find for a signed $3h$-compact representation that

$$H(\lambda_i) < \frac{11}{4} \Delta \qquad (10)$$

for $l - n < i \le l$. Considering λ_0, we see

$$\frac{15L_1}{16\sqrt{\Delta}} < \lambda_0 < \frac{17}{16} 2^{s_0} = \frac{17}{8} ,$$

as $s_0 = 1$, and so (10) holds for $i = 0$ as well. For $0 < i \le l - n$ we have $H(\lambda_i) \in O(\Delta^{3/2})$, but only for $l - n < 2 + \log_3 h$ terms.

We can now state the definition of a signed $3h$-compact representation. An algorithm to compute such representations is presented in Appendix C.

Definition 3. *For any θ such that $(\theta) = \mathfrak{a}[x] \in \mathcal{O}$, a signed $3h$-compact representation of θ is*

$$\theta = \prod_{i=0}^{l} \left(\frac{\lambda_i}{L_i^3} \right)^{b^{l-i}}$$

where the following properties are satisfied:

1. $l = \lceil \log_3 \log_2 \theta \rceil$.
2. $\lambda_i \in \mathcal{O}_\Delta$ and L_i is an integer $(0 \le i \le l)$.

3. $0 < L_i \leq \Delta^{1/2}$ and $H(\lambda_i) = O(\Delta)$ $(0 \leq i \leq l)$.
4. $\pi_j \in \mathcal{O}_\Delta$, $L_{j+1} = |N(\pi_j)|$,

$$\pi_j = \prod_{i=0}^{j} \left(\frac{\lambda_i}{L_i^3} \right)^{3^{j-i}},$$

π_j generates a reduced ideal $\mathfrak{b}_j$, where $\mathfrak{b}_0 = \mathfrak{a}[1]$ and

$$L_{i+1}^3 \mathfrak{b}_{i+1} = \lambda_{i+1} \mathfrak{b}_i^3 \quad (0 \leq i \leq l - 1).$$

Theorem 2. Let $\theta \in \mathcal{O}_\Delta$ such that $\mathfrak{a}[x] = (\theta)$ for some $x \in \mathbb{Z}^+$. The number of bits in a signed $3h$-compact representation of θ is $O((\log_3 \log_2 \theta) \log_2 \Delta)$.

The proof is analogous to that of Theorem 1.

Going back to our running example, we find that a signed $3h$-compact representation of the fundamental unit $\eta_{410286423278424}$ requires 843 bits. Compared to the compact and h-compact representations respectively, the signed $3h$-compact representation saves us 30.7% and 13.7%. Again, note that the asymptotic size is not changed, but having $\log_3$ terms instead of $\log_2$, combined with the size reduction in terms, further improves the O-constant and the size in practice.

6   Using Larger Bases

Can we extend this idea further? What about a 4-compact, 5-compact, or higher representation?

For the time being, we will occupy ourselves with only signed 4- and 5-compact representations. We can compute a signed quaternary representation of an integer x using digit set $\{-1, 0, 1, 2\}$ or $\{-2, -1, 0, 1\}$. For the signed quinary representation we use $\{-2, -1, 0, 1, 2\}$. When using base 4, we require an algorithm which computes an ideal $\mathfrak{a}[4x]$ from an ideal $\mathfrak{a}[x]$ using w-near (f, p)-representations. An analogous algorithm is also required for base 5.

As with base 3, we consider signed $4h$-compact representations as follows. As in the base-3 case, we need to increase h by a further factor of $(1/4) \log_2 \Delta$ to $h = \lceil (3/4) \log_2 \Delta \rceil$. We also must compute the maximal n such that

$$\frac{x}{(4^n - 1)/3} \geq h,$$

and put $y - x + ((4^n - 1)/3)h$. We find that for most of the λ_i in the resulting algorithm

$$H(\lambda_i) < \frac{45}{8} \Delta^{5/4}$$

for a signed $4h$-compact representation. Furthermore, for $\theta \in \mathcal{O}_\Delta$ such that $\mathfrak{a}[x] = (\theta)$ $(x \in \mathbb{Z}^+)$, the total number of bits required to express θ as a signed $4h$-compact representation is $O((\log_4 \log_2 \theta) \log_2 \Delta^{5/4})$.

The signed $4h$-compact representation of $\eta_{410286423278424}$ using digits $b_i \in \{-1, 0, 1, 2\}$ requires only 832 bits to store. Compared to the signed $3h$-compact representation, this represents an additional savings of 1.3%. The signed $4h$-compact representation using $b_i \in \{-2, -1, 0, 1\}$ requires 843 bits. In general, it seems hard to predict *a priori* which signed base will produce a shorter signed compact representation.

If we set $h = \lceil \log_2 \Delta \rceil$, compute the maximal n such that

$$\frac{x}{(5^n - 1)/4} \geq h,$$

and put $y = x + ((5^n - 1)/4)h$, we can compute a signed $5h$-compact representation. Looking at the heights of the λ_i, we see that for most of the λ_i

$$H(\lambda_i) < \frac{47}{4}\Delta^{3/2}$$

and $O((\log_5 \log_2 \theta) \log_2 \Delta^{3/2})$ bits are needed to store the total representation. The signed $5h$-compact representation of $\eta_{410286423278424}$ requires 875 bits, which is larger than the storage needed for the signed $4h$-compact representations.

At this point, we find the first indications that pursuing this idea to higher powers (i.e., 6-compact and higher representations) may not result in further memory savings. Unfortunately, the increase in size of the individual terms of the compact representations begins to dominate the savings from a decreased overall number of terms. In the remainder of this section, we look at an analytical argument to justify this claim. In the following section, we will present some calculations that confirm, numerically at least, that this analysis is valid.

To determine the overall expected size S_x of the signed base-x h-compact representation in bits, we multiply the base-2 logarithm of the $H(\lambda_i)$ bounds by the corresponding number of terms l. For most of these λ_i ($i \geq 2 + \log_x h$) the $H(\lambda_i)$ bounds are given by some constant B_x multiplied by $\Delta^{(x+1)/4}$. Expanding and converting the logarithms to base 2, we see that

$$S_x = \log_2\left(B_x \Delta^{(x+1)/4}\right) \cdot \log_x \log_2 \theta$$
$$= \left(\frac{\log_2 B_x}{\log_2 x}\right) \log_2 \log_2 \theta + \left(\frac{x+1}{4 \log_2 x}\right) \log_2 \Delta \log_2 \log_2 \theta . \qquad (11)$$

The B_x values are given by

$$B_x = \max\left\{\frac{16^{x-1} \cdot 17}{15^x}, \frac{17^x}{15 \cdot 16^{x-1}}\right\} 2^{\lfloor x/2 \rfloor} = \frac{17}{15} \max\left\{\frac{16}{15}, \frac{17}{16}\right\}^{x-1} 2^{\lfloor x/2 \rfloor}$$

as $17/15$, $16/15$, and $17/16$ are all greater than 1. Thus,

$$B_x < \frac{17}{15}\left(\frac{16}{15}\right)^{x-1} 2^{\lfloor x/2 \rfloor} < \frac{17}{15}\left(\frac{16}{15}\right)^{x-1} 2^{x-1} = \frac{17}{15}\left(\frac{32}{15}\right)^{x-1}$$

and $\log_2 B_x$ is of size $O(x)$. Asymptotically then, the $(x+1)/(4 \log_2 x)$ coefficient will dominate this expression as the discriminant increases. Looking at Figure 1, we see this coefficient has a minimum between $x = 3$ and $x = 4$.

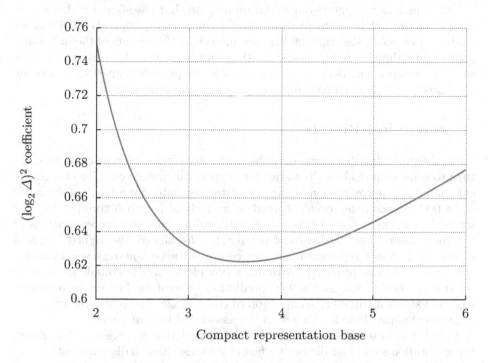

Fig. 1. Plot of $(x+1)/(4\log_2 x)$ coefficients

In this paper, we are most interested in computing compact representations where $\theta = \eta_\Delta$, the fundamental unit. As such, we can write

$$\log_x \log_2 \theta = \log_x \log_2 \eta_\Delta = \log_x \mathcal{R},$$

where $x \in \{2, 3, \ldots, 6\}$. Recall that we can loosely bound the regulator by $\sqrt{\Delta}$ and, after substitution, we are left with

$$l < \log_x \sqrt{\Delta} \tag{12}$$

as an upper bound on the number of terms in our various compact representations. Specializing (11) using (12), we find

$$S_x = \left(\frac{\log_2 B_x}{2\log_2 x} \right) \log_2 \Delta + \left(\frac{x+1}{8\log_2 x} \right) (\log_2 \Delta)^2 .$$

Again, asymptotically, the $(x+1)/(8\log_2 x)$ coefficient will dominate this expression as the discriminant increases and the minimum still occurs between $x = 3$ and $x = 4$. In fact, if we compare the two functions S_3 and S_4, we find that $S_4 < S_3$ for discriminants greater than $10^{16.5}$. In other words, for discriminants larger than about 16 decimal digits, the signed $4h$-compact representation is the most efficient one.

This conclusion supports our initial impression that base-5 and higher representations are not likely to produce shorter compact representations. From an analytic viewpoint, the trade-off between increasing the heights of the individual terms and gaining a representation with a fewer number of terms is no longer working in our favor. Because of this, we will not provide numerical results for the base-5 or higher compact representations in the next section.

7 Numerical Results

Since the preceding discussion only shows the savings in one particular case, we turn to some empirical results to further support our memory-saving claims. We calculated an approximation of the associated regulator for a random sampling of 28,000 discriminants evenly spread from decimal length 5 through 18, and for each discriminant used this to compute various compact representations of the fundamental unit. For each of the regular, h-, signed $3h$-, signed $4h$-, and signed $5h$-compact representations, we computed a best-fit regression line for the data, as well as provided distribution box plots,[1] a 95% confidence interval for our regression line, and a 95% prediction interval for further data points. Figure 2 shows a summary comparison of the average representation length for each discriminant length, along with the associated best-fit curves.

In the previous section, our analysis concluded that the signed $4h$-compact representation should be the most efficient for large enough discriminants. However, the numerical results that follow seems to show that the signed $4h$-compact representation is more efficient all the time. We initially speculated that this discrepancy is caused by our conservative bound on $H(\lambda_i)$. However, this only provides a piece of the answer. In our analysis, we assumed that each digit in the base-3 representation is a 1 and each digit in the base-4 representation is a 2 (or -2). Turning to a probabilistic argument, for a randomly selected number we would expect the proportions of the digits in its representation to approximately be equal. For example, for a signed base-3 representation, we would expect to see 0, 1, and -1 each roughly 33% of the time. If we take this into account, can we derive the following bound on $H(\lambda_i)$ for the signed base-3 case:

$$H(\lambda_i) = \max\left\{\frac{16^2 \cdot 17}{15^3}, \frac{17^3}{15 \cdot 16^2}\right\}\left(\frac{2^{-1}}{3} + \frac{2^0}{3} + \frac{2^1}{3}\right)\Delta < \frac{43}{20}\Delta \, .$$

Similarly for the signed base-4 case, we have

$$H(\lambda_i) = \max\left\{\frac{16^3 \cdot 17}{15^4}, \frac{17^4}{15 \cdot 16^3}\right\}\left(\frac{2^{-1}}{4} + \frac{2^0}{4} + \frac{2^1}{4} + \frac{2^2}{4}\right)\Delta^{5/4} < \frac{13}{5}\Delta^{5/4} \, .$$

Using these probabilistic bounds in S_3 and S_4, we find that S_4 is now strictly less than S_3 for discriminants larger than roughly 10^{14}.

We extended these empirical results further by using the series of discriminants from [10, Tbl. 7.8, p. 101], as well as the regulator approximations given there, to

---

[1] For each box plot, potential outliers have been marked with a "×" symbol.

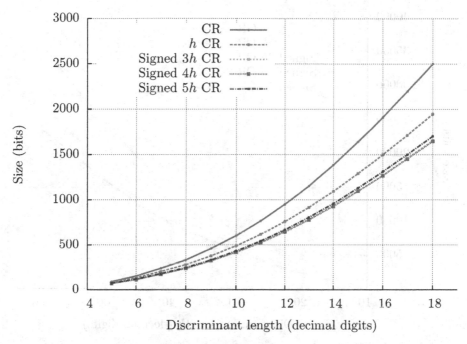

Fig. 2. Comparison of the sizes of some various compact representations for random discriminants (1,000 discriminant values for each length from 5–18)

produce the same variety of compact representations as above for the associated fundamental units. A comparison of the sizes of these representations is shown in Figure 3.

We must be cautious with these extended results. Because of the limited sampling, it is difficult to make a definitive claim on the relative efficiencies of the various h-compact representations at this point. For the majority of these larger discriminants, the signed $4h$-compact representation is the most efficient. However, for a given discriminant, the signed $3h$-compact representation may be just as, or even slightly more, efficient. With further measurements, we expect to find that the signed $4h$-compact representation is most efficient on average.

8   Future Directions

In this paper, we presented two substantial improvements that can be used together to reduce the sizes of a compact representations for certain quadratic integers. The first was noticing that the size of the individual compact representation terms could be reduced by a substantial factor. The second refinement was to notice that the overall number of terms could be reduced by computing larger giant steps on each iteration of the algorithm using bases larger than two. Asymptotically, the signed $4h$-compact representation results in the most

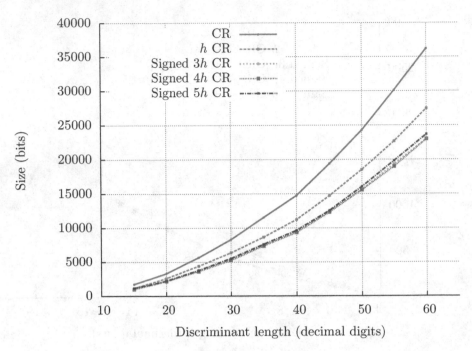

Fig. 3. Comparison of the sizes of some various compact representations for the series of discriminants presented in [10, Tbl. 7.8, p. 101]

efficient balance between larger individual compact representation terms and a reduced overall number of terms. Numerical testing supports this conclusion. In the large-discriminant tests we performed, we found overall memory savings of around 37% as compared to the standard compact representation.

There are other types of number representations we did not investigate which may lead to further memory savings for compact representations. For example, we could consider using a *non-adjacent form* (*NAF*) representation [19], a signed base-2 representation for which the average density of non-zero digits among all NAFs of a given length is approximately $1/3$ [11, Thm. 3.29] as opposed to $1/2$ for the regular binary representation. A *width-w NAF* (*w*NAF) uses odd digits less than 2^{w-1} in absolute value and has average density of non-zero digits $1/(w+1)$, so that NAF is a *w*NAF with $w = 2$. In terms of an overall length the NAF and *w*NAF representation of an integer is at most one more than the length of its binary representation.

Another example is the *double-base* representation of the integer x, given by

$$x = \sum_{i,j} b_{i,j} 2^i 3^j ,$$

where $b_{i,j} \in \{0, 1\}$. These representations only require $O(\log n / \log \log n)$ digits to store and near-canonic representations can be computed via a number of methods [1,5,6,8]. The advantage to this representation over a standard binary or ternary representation is that we require fewer terms. It could be quite beneficial if this numeric representation could be applied to create a double-base compact representation. First, by reducing the overall number of terms of the representation, we would reduce the overall storage requirements by moving from a compact to a 3-compact to a 4-compact representation. Furthermore, by restricting ourselves to the bases 2 and 3, we have the potential of avoiding the per-term expansion we encountered due to the increasing bound on $H(\lambda_i)$.

References

1. Avanzi, R., Dimitrov, V., Doche, C., Sica, F.: Extending scalar multiplication using double bases. In: Lai, X., Chen, K. (eds.) ASIACRYPT 2006. LNCS, vol. 4284, pp. 130–144. Springer, Heidelberg (2006)
2. Buchmann, J., Thiel, C., Williams, H.C.: Short representation of quadratic integers, Mathematics and its Applications, vol. 325, pp. 159–185. Kluwer Academic Publishers, Amsterdam (1995)
3. Buchmann, J., Vollmer, U.: Binary Quadratic Forms, Algorithms and Computation in Mathematics, vol. 20. Springer (2007)
4. Cohen, H.: A Course in Computational Algebraic Number Theory, Graduate Texts in Mathematics, 4th edn., vol. 138. Springer, New York (2000)
5. Dimitrov, V., Imbert, L., Mishra, P.K.: Efficient and secure elliptic curve point multiplication using double-base chains. In: Roy, B. (ed.) ASIACRYPT 2005. LNCS, vol. 3788, pp. 59–78. Springer, Heidelberg (2005)
6. Dimitrov, V.S., Jullien, G.A., Miller, W.C.: An algorithm for modular exponentiation. Information Processing Letters 66, 155–159 (1998)
7. Dixon, V., Jacobson Jr., M.J., Scheidler, R.: Improved exponentiation and key agreement in the infrastructure of a real quadratic field. In: Hevia, A., Neven, G. (eds.) LatinCrypt 2012. LNCS, vol. 7533, pp. 214–233. Springer, Heidelberg (2012)
8. Doche, C., Imbert, L.: Extended double-base number system with applications to elliptic curve cryptography. In: Barua, R., Lange, T. (eds.) INDOCRYPT 2006. LNCS, vol. 4329, pp. 335–348. Springer, Heidelberg (2006)
9. Fung, G.W., Williams, H.C.: Compact representation of the fundamental unit in a complex cubic field (1991) (unpublished manuscript)
10. de Haan, R.: A fast, rigorous technique for verifying the regulator of a real quadratic field. Master's thesis, University of Amsterdam (2004)
11. Hankerson, D., Menezes, A., Vanstone, S.: Guide to Elliptic Curve Cryptography. Springer, New York (2004)
12. Hühnlein, D., Paulus, S.: On the implementation of cryptosystems based on real quadratic number fields (extended abstract). In: Stinson, D.R., Tavares, S. (eds.) SAC 2000. LNCS, vol. 2012, pp. 288–302. Springer, Heidelberg (2001)
13. Imbert, L., Jacobson Jr., M.J., Schmidt, A.: Fast ideal cubing in imaginary quadratic number and function fields. Advances in Mathematics of Communications 4(2), 237–260 (2010)

14. Jacobson Jr., M.J., Scheidler, R., Williams, H.C.: The efficiency and security of a real quadratic field based key exchange protocol. In: Alster, K., Urbanowicz, J., Williams, H.C. (eds.) Public-Key Cryptography and Computational Number Theory, September 11-15 (2000); Walter de Gruyter GmbH & Co., Warsaw (2001)
15. Jacobson Jr., M.J., Scheidler, R., Williams, H.C.: An improved real quadratic field based key exchange procedure. J. Cryptology 19, 211–239 (2006)
16. Jacobson Jr., M.J., Williams, H.C.: Solving the Pell Equation. CMS Books in Mathematics. Springer (2009)
17. Lagarias, J.C.: Succinct certificates for the solvability of binary quadratic diophantine equations (extended abstract). In: Proc. 20th IEEE Symp. on Foundations of Computer Science, pp. 47–54 (1979)
18. Lagarias, J.C.: Succinct certificates for the solvability of binary quadratic diophantine equations. Tech. Rep. Technical Memorandum 81-11216-54, Bell Labs, 28 (1981)
19. Reitwiesner, G.W.: Binary arithmetic. Advances in Computers 1, 231–308 (1960)
20. Shanks, D.: The infrastructure of a real quadratic field and its applications. In: Proc. 1972 Number Theory Conference, University of Colorado, Boulder, pp. 217–224 (1972)
21. Silvester, A.K.: Improving regulator verification and compact representations in real quadratic fields. Ph.D. thesis, University of Calgary, Calgary, Alberta (2012)

A   EWNEAR

We present a version of EWNEAR [16, Alg. 12.1, pp. 286 and 457] that also works when $k > w$. As the new version merely adds some key values which allow the determination of a relative generator, the proof of correctness of EWNEAR will remain unchanged.

Algorithm 3: EWNEAR

Input: $(\mathfrak{b}, d, k)$, w, p, where $(\mathfrak{b}, d, k)$ is a reduced (f, p) representation of some $\mathcal{O}$-ideal $\mathfrak{a}$. Here $\mathfrak{b}[Q/r, (P+\sqrt{D})/r]$, where $P + \lfloor\sqrt{D}\rfloor \geq Q, 0 \leq \lfloor\sqrt{D}\rfloor - P \leq Q$.

Output: $(\mathfrak{c}, g, h)$ a w-near $(f + 9/8, p)$ representation of $\mathfrak{a}$ and a, b, where $\varkappa = (a + b\sqrt{D})/Q$ and $\mathfrak{c} = \varkappa\mathfrak{b}$.

1: **case 1:** $k < w$
    2: Put $B_{-2} = 1, B_{-1} = 0$.
    3: Find $s \in \mathbb{Z}^{\geq 0}$ such that $2^s Q \geq 2^{p+4}$. Put $Q_0 = Q, P_0 = P, M = \lceil 2^{p+s-k+w}Q_0/d\rceil, Q_{-1} = (D - P^2)/Q, T_{-2} = -2^s P_0 + \lfloor 2^s\sqrt{D}\rfloor, T_{-1} = 2^s Q_0, i = 1$.
    4: **while** $T_{i-2} \leq M$ **do**
        5: $q_{i-1} = \lfloor(P_{i-1} + \lfloor\sqrt{D}\rfloor)/Q_{i-1}\rfloor$
        6: $P_i = q_{i-1}Q_{i-1} - P_{i-1}$
        7: $Q_i = Q_{i-2} - q_{i-1}(P_i - P_{i-1})$
        8: $T_{i-1} = q_{i-1}T_{i-2} + T_{i-3}$
        9: $B_{i-1} = q_{i-1}B_{i-2} + B_{i-3}$
      10: $i \leftarrow i + 1$
  11: **end while**
  12: Put $e_{i-1} = \lceil 2^{p-s+3}T_{i-3}/Q_0\rceil$

13: **if** $de_{i-1} \leq 2^{2p-k+w+3}$ **then**

  14: Put $\mathfrak{c} = [Q_{i-2}/r, (P_{i-2} + \sqrt{D})/r]$, $e = e_{i-1}$,
    $a = (T_{i-3} - \lfloor 2^s\sqrt{D} \rfloor)/2^s$, $b = B_{i-3}$.

15: **else**

  16: Put $\mathfrak{c} = [Q_{i-3}/r, (P_{i-3} + \sqrt{D})/r]$, $e = \lceil 2^{p-s+3}T_{i-4}/Q_0 \rceil$,
    $a = (T_{i-4} - \lfloor 2^s\sqrt{D} \rfloor)/2^s$, $b = B_{i-4}$.

17: **end if**

18: Find t such that

$$2^t < \frac{ed}{2^{2p+3}} \leq 2^{t+1}.$$

19: Put

$$g = \left\lceil \frac{ed}{2^{p+t+3}} \right\rceil, \quad h = k + t.$$

20: **end case**

21: **case 2:** $k > w$

  22: Put $B^*_{-2} = 1$, $B^*_{-1} = 0$.

  23: Put $s = p + 4$, $Q^*_0 = Q$, $P^*_0 = P$, $M^* = d2^{k-w+4}$, $Q^*_1 = (D - P^2)/Q$,
    $T^*_{-2} = 2^s Q^*_0$, $T^*_{-1} = 2^s P^*_0 + \lfloor 2^s\sqrt{D} \rfloor$, and $i = 1$.

  24: **while** $T^*_{i-2} < Q^*_i M^*$ **do**

    25: $q^*_i = \lfloor (P^*_{i-1} + \lfloor\sqrt{D}\rfloor)/Q^*_i \rfloor$

    26: $P^*_i = q^*_i Q^*_i - P^*_{i-1}$

    27: $Q^*_{i+1} = Q^*_{i-1} - q^*_i(P^*_i - P^*_{i-1})$

    28: $T^*_{i-1} = q^*_i T^*_{i-2} + T^*_{i-3}$

    29: $B^*_{i-1} = q^*_i B^*_{i-2} + B^*_{i-3}$

    30: $i \leftarrow i + 1$

  31: **end while**

  32: Put $q^*_i = \lfloor (P^*_{i-1} + \lfloor\sqrt{D}\rfloor)/Q^*_i \rfloor$, $P^*_i = q^*_i Q^*_i - P^*_{i-1}$,
    $e = \lceil T^*_{i-2}/2Q^*_i \rceil$, $e' = \lceil T^*_{i-3}/2Q^*_{i-1} \rceil$, $j = 3$.

  33: **while** $e' \geq d2^{k-w+3}$ **do**

    34: $e \leftarrow e'$

    35: $e' \leftarrow \lceil T^*_{i-2-j}/2Q^*_{i-j} \rceil$

    36: $j \leftarrow j + 1$

  37: **end while**

  38: Find t (t') such that

$$2^{t-1} \leq \frac{e}{8d} < 2^t. \quad \left(2^{t'-1} \leq \frac{e'}{8d} < 2^{t'}.\right)$$

  39: Put $\mathfrak{c} = [Q^*_{i-j+3}/r, (P^*_{i-j+3} + \sqrt{D})/r]$, $g = \lceil 2^{p+3+t}d/e \rceil$, $h = k - t$,
    $a = (T^*_{i-2} - B^*_{i-2}\lfloor 2^s\sqrt{\Delta} \rfloor)/2^s$, $b = B^*_{i-j+2}$.

40: **end case**

B   ETRIPLEX

In order to implement a triple-and-add algorithm to use compact representations, a precision analysis for cubing an (f, p)-representation is required. To this end, we include the following theorem.

Theorem 4 ([16, Thm. 11.2, p. 268]). *Let $(\mathfrak{b}, d', k')$ be an (f', p) representation of an $\mathcal{O}_\Delta$-ideal $\mathfrak{a}$. If $d'^3 \leq 2^{3p+1}$, put $d = \lceil d'^3/2^{2p} \rceil$ and $k = 3k'$. If $2^{3p+1} < d'^3 \leq 2^{3p+2}$, put $d = \lceil d'^3/2^{2p+1} \rceil$ and $k = 3k' + 1$. If $d'^3 > 2^{3p+2}$, put $d = \lceil d'^3/2^{2p+2} \rceil$ and $k = 3k' + 2$. Then $(\mathfrak{b}^3, d, k)$ is an (f, p) representation of the product ideal $\mathfrak{a}^3$, where $f = 1 + 3f' + 3f'^2/2^p + f'^3/2^{2p}$.*

Proof. Let $\mathfrak{b} = \theta \mathfrak{a}$ for $\theta \in \mathbb{K}$. By the definition of d in the theorem, it is easy to see that $2^p < d \leq 2^{p+1}$. From the definition of an (f, p) representation, we know

$$\left| \frac{2^{p-k'} \theta}{d'} - 1 \right| < \frac{f'}{2^p},$$

and rearranging this inequality gives

$$\frac{d'}{2^p} \left(1 - \frac{f'}{2^p} \right) < \frac{\theta}{2^{k'}} < \frac{d'}{2^p} \left(1 + \frac{f'}{2^p} \right).$$

As $2^p < d' \leq 2^{p+1}$ and $f'/2^p < 1/16$, we have

$$\frac{d'}{2^p} \left(1 - \frac{f'}{2^p} \right) > 1 \cdot \left(1 - \frac{1}{16} \right) > 0 \quad \text{and} \quad \frac{d'}{2^p} \left(1 + \frac{f'}{2^p} \right) < 2 \cdot \left(1 + \frac{1}{16} \right) < 4,$$

and thus

$$\left(1 - \frac{f'}{2^p} \right)^3 < \frac{2^{3(p-k')} \theta^3}{d'^3} < \left(1 + \frac{f'}{2^p} \right)^3.$$

If we set $f^* = 3f' + 3f'^2/2^p + f'^3/2^{2p}$ then

$$1 - \frac{f^*}{2^p} = 1 - \frac{3f'}{2^p} - \frac{3f'^2}{2^{2p}} - \frac{f'^3}{2^{3p}} < 1 - \frac{3f'}{2^p} + \frac{3f'^2}{2^{2p}} - \frac{f'^3}{2^{3p}} = \left(1 - \frac{f'}{2^p} \right)^3$$

and

$$\left(1 + \frac{f'}{2^p} \right)^3 = 1 + \frac{3f'}{2^p} + \frac{3f'^2}{2^{2p}} + \frac{f'^3}{2^{3p}} = 1 + \frac{3f' + 3f'^2/2^p + f'^3/2^{2p}}{2^p} = 1 + \frac{f^*}{2^p}.$$

Hence

$$1 - \frac{f^*}{2^p} < \frac{2^{3p-3k'} \theta^3}{d'^3} < 1 + \frac{f^*}{2^p}. \tag{13}$$

Now suppose that $d'^3 \leq 2^{3p+1}$. Since $d = d'^3/2^{2p} + \epsilon$ for $0 \leq \epsilon < 1$, (13) becomes

$$1 - \frac{f^*}{2^p} < \frac{2^{p-k} \theta^3}{d - \epsilon} < 1 + \frac{f^*}{2^p}$$

and as $d - \epsilon = d(1 - \epsilon/d)$,

$$\left(1 - \frac{\epsilon}{d}\right)\left(1 - \frac{f^*}{2^p}\right) < \frac{2^{p-k}\theta^3}{d} < \left(1 - \frac{\epsilon}{d}\right)\left(1 + \frac{f^*}{2^p}\right).$$

Looking at the right-hand side of this inequality, $(1 - \epsilon/d) < 1$ so

$$\left(1 - \frac{\epsilon}{d}\right)\left(1 + \frac{f^*}{2^p}\right) < 1 + \frac{f^*}{2^p} < 1 + \frac{1}{2^p} + \frac{f^*}{2^p} = 1 + \frac{f}{2^p};$$

considering the left-hand side, $\epsilon < 1$ and $2^p < d$ so $2^p \epsilon < d$. Rearranging this inequality gives $1 - 1/2^p < 1 - \epsilon/d$ and thus

$$1 - \frac{f}{2^p} = 1 - \frac{1}{2^p} - \frac{f^*}{2^p} < \left(1 - \frac{1}{2^p}\right)\left(1 - \frac{f^*}{2^p}\right) < \left(1 - \frac{\epsilon}{d}\right)\left(1 - \frac{f^*}{2^p}\right).$$

It follows that

$$\left|\frac{2^{p-k}\theta^3}{d} - 1\right| < \frac{f}{2^p}$$

and $(\mathfrak{b}^3, d, k)$ is an (f, p) representation of $\mathfrak{a}^3$, where $\mathfrak{b}^3 = \theta^3 \mathfrak{a}^3$. The theorem follows by applying similar arguments when $2^{3p+1} < d'^3 \leq 2^{3p+2}$ and when $d'^3 > 2^{3p+2}$. $\qquad\square$

In practice, ideal cubing could be accomplished by a square and a multiplication. Another option is to use the dedicated ideal cubing algorithm NUCUBE [13, Alg. 4] described by Imbert, Jacobson, and Schmidt. An extended version of this algorithm for w-near representations is presented in [7]. Finally, ETRIPLEX is obtained by extending the algorithm further to also produce the corresponding relative generator. For a complete description, see [21, §§ 5.3].

C   Algorithm to Compute Signed $3h$-Compact Representations

We present here an algorithm to compute a signed $3h$-compact representation, based on the ideas of Section 5.

Algorithm 5: 3HCRAX

Input: x, p, where $x \in \mathbb{Z}^+$ and $2^p > 11.2x \max\{16 \log_2 x\}$.

Output: $(\mathfrak{a}[x], d, k)$, (m_i, n_i), and L_i, where $(\mathfrak{a}[x], d, k)$ is an x-near (f, p) representation of $\mathfrak{a} = (1)$ with $f < 2^{p-4}$, (m_i, n_i) are pairs of integers, and $L_i \in \mathbb{Z}^+$ for $i = 0, 1, \ldots, l$ where l is such that $x = \sum_{j=0}^{l} 3^{l-j} b_j$ and $b_0 \neq 0$, $b_j \in \{0, 1, 2\}$.

1: Put $h = \lceil (1/2) \log_2 \Delta \rceil$ and compute the maximal n such that

$$\frac{x}{(3^n - 1)/2} \geq h,$$

and put $y = x + ((3^n - 1)/2)h$.

2: Compute the signed ternary representation of y with

$$y = \sum_{i=0}^{l} 3^{l-i} b_i \text{ and } b_0 \neq 0, b_i \in \{-1, 0, 1\} \quad (1 \leq i \leq l).$$

3: Put

$$Q = r, \quad P = r \left\lfloor \frac{\lfloor \sqrt{\Delta} \rfloor - r + 1}{r} \right\rfloor + r - 1, \quad (\mathfrak{b}, d, k) = ([Q, P], 2^p + 1, 0),$$

$s = b_0$, $L_0 = 1$, and $i = 0$.

4: Put $((\mathfrak{b}_0, d_0, k_0), m_0, n_0) = \text{EWNEAR}((\mathfrak{b}, d, k), s, p)$.

5: **while** $i < l - n$ **do**

   6: Put $L_{i+1} = N(\mathfrak{b}_i)$ and

$$((\mathfrak{b}_{i+1}, d_{i+1}, k_{i+1}), m_{i+1}, n_{i+1}) = \text{ETRIPLEX}((\mathfrak{b}_i, d_i, k_i), s, p).$$

   7: Set $s \leftarrow 3s + b_{i+1}$.

   8: **if** $b_{i+1} \neq 0$ **then**

      9: Put $N = N(\mathfrak{b}_{i+1})$ and set

$$((\mathfrak{b}_{i+1}, d_{i+1}, k_{i+1}), m'_{i+1}, n'_{i+1}) \leftarrow \text{EWNEAR}((\mathfrak{b}_{i+1}, d_{i+1}, k_{i+1}), s, p).$$

    10: Set $(m_{i+1}, n_{i+1}) \leftarrow \text{IMULT}(m_{i+1}, n_{i+1}, m'_{i+1}, n'_{i+1}, N)$.

  11: **end if**

  12: Set $i \leftarrow i + 1$.

13: **end while**

14: **while** $i < l$ **do**

  15: Put $L_{i+1} = N(\mathfrak{b}_i)$ and

$$((\mathfrak{b}_{i+1}, d_{i+1}, k_{i+1}), m_{i+1}, n_{i+1}) = \text{ETRIPLEX}((\mathfrak{b}_i, d_i, k_i), s, p).$$

  16: Set $s \leftarrow 3s + b_{i+1} - h$.

  17: Put $N = N(\mathfrak{b}_{i+1})$ and set

$$((\mathfrak{b}_{i+1}, d_{i+1}, k_{i+1}), m'_{i+1}, n'_{i+1}) \leftarrow \text{EWNEAR}((\mathfrak{b}_{i+1}, d_{i+1}, k_{i+1}), s, p).$$

  18: Set $(m_{i+1}, n_{i+1}) \leftarrow \text{IMULT}(m_{i+1}, n_{i+1}, m'_{i+1}, n'_{i+1}, N)$.

  19: Set $i \leftarrow i + 1$.

20: **end while**

21: Put $L_{l+1} = N(\mathfrak{b}_l)$ and $(\mathfrak{a}[x], d, k) = (\mathfrak{b}_l, d_l, k_l)$.

Factoring Integers by CVP Algorithms

Claus Peter Schnorr

Fachbereich Informatik und Mathematik,
Goethe-Universität Frankfurt, PSF 111932,
60054 Frankfurt am Main, Germany
schnorr@cs.uni-frankfurt.de

Abstract. We use pruned enumeration algorithms to find lattice vectors close to a specific target vector for the prime number lattice. These algorithms generate multiplicative prime number relations modulo N that factorize a given integer N. The algorithm NEW ENUM performs the stages of exhaustive enumeration of close lattice vectors in order of decreasing success rate. For example an integer $N \approx 10^{14}$ can be factored by about 90 prime number relations modulo N for the 90 smallest primes. Our randomized algorithm generated for example 139 such relations in 15 minutes. This algorithm can be further optimized. The optimization for larger integers N is still open.

Keywords: Factoring integers, enumeration of close lattice vectors, the prime number lattice.

1 Introduction and Surview

The algorithm ENUM of [SE94, SH95] locally performs stages in order of decreasing success rate and often finds short / close vectors much faster than previous **SVP** and **CVP** algorithms of KANNAN [Ka87] and FINCKE, POHST [FP85] that disregard the success rate of stages. The NEW ENUM algorithm for **SVP** / **CVP** presented in section 3 performs all stages in order of decreasing success rate, stages with high success rate are done first. This greatly reduces the number of stages that precede the finding of a shortest / closest lattice vector.

Section 4 summarizes results on time bounds of NEW ENUM for **SVP** / **CVP** for a basis $\mathbf{B} = [\mathbf{b}_1, ..., \mathbf{b}_n]$ that satisfies GSA (meaning that the local reduction strength of the reduced basis is "uniform" for all 2-dimensional basis blocks). Prop. 1 shows that NEW ENUM finds under "linear" pruning a shortest lattice vector $\mathbf{b}$ under the volume heuristics in polynomial time (without proving that $\mathbf{b}$ is shortest) if the *relative density* $rd(\mathcal{L})$ of $\mathcal{L}$ satisfies $rd(\mathcal{L}) \leq \left(\sqrt{\frac{e\pi}{2n}} \frac{\lambda_1}{\|\mathbf{b}_1\|}\right)^{1/2}$. Here λ_1 is the first successive minimum of $\mathcal{L}$ and $rd(\mathcal{L})$ is defined by $\lambda_1 = rd(\mathcal{L}) \gamma_n^{1/2} (\det \mathcal{L})^{1/n}$ and the Hermite constant γ_n. Theorem 1 upper bounds the **SVP**-time of NEW ENUM without the volume heuristics. Cor. 1 translates Theorem 1 from **SVP** to **CVP** and shows that the **CVP** for $\mathcal{L}$ and the target vector $\mathbf{t} \in \text{span}(\mathcal{L})$ is solved in time $2^{O(n)}$ and linear space if $rd(\mathcal{L}) = O(n^{-1/2})$, $\|\mathcal{L} - \mathbf{t}\| = O(\lambda_1)$ and $\mathbf{b}_1$ is a nearly shortest vector of $\mathcal{L}$. Cor. 3 shows under

M. Fischlin and S. Katzenbeisser (Eds.): Buchmann Festschrift, LNCS 8260, pp. 73–93, 2013.

the volume heuristics that, given a target vector $\mathbf{t}$, a closest lattice vector can be found, without proving optimal closeness, in polynomial time if $rd(\mathcal{L}) \leq \left(\sqrt{\frac{e\pi}{2n}} \frac{\lambda_1}{\|\mathbf{b}_1\|}\right)^{1/2}$, $\|\mathcal{L} - \mathbf{t}\| = O(\lambda_1)$ and if the found closest vector behaves randomly (CA). The latter situation comes close to the one occurring in factoring large integers.

Sections 5 and 6 study factoring integers N from **CVP** solutions for the prime number lattice and a target vector $\mathbf{N}$ that represents N. These **CVP** solutions provide smooth integers $u, v, |u - vN|$ that factorize over the smallest n primes, however the smoothness of $|u - vN|$, required for the factorization of N, only holds under conditions. It helps to use a prime number lattice of relative density $rd(\mathcal{L}) = o(n^{-1/4})$. A main problem is to prune the enumeration of lattice vectors close to $\mathbf{N}$ toward small values v and $|u - vN|$. In each round this algorithm randomly scales the lattice basis such that distinct pairs (u, v) are found in case of success. We explain as example the factorization of some $N \approx 10^{14}$ using the $n = 90$ smallest primes and clever pruning of NEW ENUM within 10 minutes time. This algorithm can be further optimized, in particular for large n, N.

2   Lattices

Let $\mathbf{B} = [\mathbf{b}_1, ..., \mathbf{b}_n] \in \mathbb{R}^{m \times n}$ be a basis matrix consisting of n linearly independent column vectors $\mathbf{b}_1, ..., \mathbf{b}_n \in \mathbb{R}^m$. They generate the lattice $\mathcal{L}(\mathbf{B}) = \{\mathbf{Bx} \mid \mathbf{x} \in \mathbb{Z}^n\}$ consisting of all integer linear combinations of $\mathbf{b}_1, ..., \mathbf{b}_n$, the *dimension* of $\mathcal{L}$ is n. The *determinant* of $\mathcal{L}$ is $\det \mathcal{L} = (\det \mathbf{B}^t \mathbf{B})^{1/2}$ for any basis matrix $\mathbf{B}$ and the transpose $\mathbf{B}^t$ of $\mathbf{B}$. The *length* of $\mathbf{b} \in \mathbb{R}^m$ is $\|\mathbf{b}\| = (\mathbf{b}^t \mathbf{b})^{1/2}$.

Let $\lambda_1, \ldots, \lambda_n$ denote the successive minima of $\mathcal{L}$ and $\lambda_1 = \lambda_1(\mathcal{L})$ is the length of the shortest nonzero vector of $\mathcal{L}$. The HERMITE constant γ_n is the minimal γ such that $\lambda_1^2 \leq \gamma (\det \mathcal{L})^{2/n}$ holds for all lattices of dimension n.

Let $\mathbf{B} = \mathbf{QR} \in \mathbb{R}^{m \times n}$, $R = [r_{i,j}]_{1 \leq i,j \leq n} \in \mathbb{R}^{n \times n}$ the unique **QR**-factorization: $\mathbf{Q} \in \mathbb{R}^{m \times n}$ is isometric (with pairwise orthogonal column vectors of length 1) and $\mathbf{R} \in \mathbb{R}^{n \times n}$ is upper-triangular with positive diagonal entries $r_{i,i}$. The **QR**-factorization also provides the Gram-Schmidt coefficients $\mu_{j,i} = r_{i,j}/r_{i,i}$ which are rational for integer matrices $\mathbf{B}$. The orthogonal projection $\mathbf{b}_i^*$ of $\mathbf{b}_i$ in $\mathrm{span}(\mathbf{b}_1, ..., \mathbf{b}_{i-1})^\perp$ has length $r_{i,i} = \|\mathbf{b}_i^*\|$.

LLL-*Bases*. A basis $\mathbf{B} = \mathbf{QR}$ is LLL-*reduced* or an LLL-*basis* for $\delta \in (\frac{1}{4}, 1]$ if

1. $|r_{i,j}|/r_{i,i} \leq \frac{1}{2}$ for all $j > i$,   **2.** $\delta r_{i,i}^2 \leq r_{i,i+1}^2 + r_{i+1,i+1}^2$ for $i = 1, ..., n-1$.

Obviously, LLL-bases satisfy $r_{i,i}^2 \leq \alpha \, r_{i+1,i+1}^2$ for $\alpha := 1/(\delta - \frac{1}{4})$. [LLL82] introduced LLL-bases focusing on $\delta = 3/4$ and $\alpha = 2$. A famous result of [LLL82] shows that LLL-bases for $\delta < 1$ can be computed in polynomial time and that they nicely approximate the successive minima:

3. $\alpha^{-i+1} \leq \|\mathbf{b}_i\|^2 \lambda_i^{-2} \leq \alpha^{n-1}$ for $i = 1, ..., n$,   **4.** $\|\mathbf{b}_1\|^2 \leq \alpha^{\frac{n-1}{2}} (\det \mathcal{L})^{2/n}$.

A basis $\mathbf{B} = \mathbf{QR} \in \mathbb{R}^{m \times n}$ is an HKZ-*basis* (HERMITE, KORKINE, ZOLOTAREFF) if $|r_{i,j}|/r_{i,i} \leq \frac{1}{2}$ for all $j > i$,   and if each diagonal entry $r_{i,i}$ of $\mathbf{R} = [r_{i,j}] \in$

$\mathbb{R}^{n \times n}$ is minimal under all transforms of $\mathbf{B}$ to $\mathbf{BT}$, $\mathbf{T} \in GL_n(\mathbb{Z})$ that preserve $\mathbf{b}_1, ..., \mathbf{b}_{i-1}$.

A basis $\mathbf{B} = \mathbf{QR} \in \mathbb{R}^{m \times n}$. $\mathbf{R} = [r_{i,j}]$ is a BKZ-*basis* for block size k, i.e., a BKZ-k basis if the matrices $[r_{i,j}]_{h \leq i,j < h+k} \in \mathbb{R}^{k \times k}$ form HKZ-bases for $h = 1, ..., n - k + 1$, see [SE94].

A famous problem is the shortest vector problem (**SVP**): Given a basis of $\mathcal{L}$ find a shortest nonzero vector of $\mathcal{L}$, i.e., a vector of length λ_1.

Closest vector problem (**CVP**): Given a basis of $\mathcal{L}$ and a target $\mathbf{t} \in \text{span}(\mathcal{L})$ find a closest vector $\mathbf{b}' \in \mathcal{L}$ such that $\|\mathbf{t} - \mathbf{b}'\| = \|\mathbf{t} - \mathcal{L}\| =_{def} \min\{ \|\mathbf{t} - \mathbf{b}\| \mid \mathbf{b} \in \mathcal{L} \}$.

The efficiency of our algorithms depends on the lattice invariant $rd(\mathcal{L}) := \lambda_1 \gamma_n^{-1/2} (\det \mathcal{L})^{-1/n}$ which we call the *relative density* of $\mathcal{L}$. Note that $rd(\mathcal{L}) = \lambda_1(\mathcal{L}) / \max \lambda_1(\mathcal{L}')$ holds for the maximum of $\lambda_1(\mathcal{L}')$ over all lattices $\mathcal{L}'$ of $\dim \mathcal{L} = \dim \mathcal{L}'$ and $\det \mathcal{L} = \det \mathcal{L}'$.

Clearly $0 < rd(\mathcal{L}) \leq 1$ holds for all $\mathcal{L}$, and $rd(\mathcal{L}) = 1$ if and only if $\mathcal{L}$ has maximal density. Lattices of maximal density and γ_n are known for $n = 1, ..., 8$ and $n = 24$.

3 A Novel Enumeration of Short Lattice Vectors

We first outline the novel **SVP**-algorithm based on the success rate of stages. NEW ENUM improves the algorithm ENUM of [SE94, SH95]. We recall ENUM and present NEW ENUM as a modification that essentially performs all stages of ENUM in decreasing order of success rates. Previous **SVP**-algorithms solve **SVP** by a full exhaustive search, disregard the success rate of stages, and prove to have found a shortest nonzero lattice vector. Our novel **SVP**-algorithm NEW ENUM finds a shortest lattice vector $\mathbf{b}$ rather fast by performing the stages in order of decreasing success rate.

Let $\mathbf{B} = [\mathbf{b}_1, ..., \mathbf{b}_n] = \mathbf{QR} \in \mathbb{Z}^{m \times n}$, $\mathbf{R} = [r_{i,j}]_{1 \leq i,j \leq n} \in \mathbb{R}^{n \times n}$ be the given basis of $\mathcal{L} = \mathcal{L}(\mathbf{B})$. Let $\pi_t : \text{span}(\mathbf{b}_1, ..., \mathbf{b}_n) \to \text{span}(\mathbf{b}_1, ..., \mathbf{b}_{t-1})^\perp = \text{span}(\mathbf{b}_t^*, ..., \mathbf{b}_n^*)$ for $t = 1, ..., n$ denote the orthogonal projections and let $\mathcal{L}_t = \mathcal{L}(\mathbf{b}_1, ..., \mathbf{b}_{t-1})$.

The Success Rate of Stages. The vector $\mathbf{b} = \sum_{i=t}^n u_i \mathbf{b}_i \in \mathcal{L}$ and $A \geq \lambda_1^2$ are given at stage $(u_t, ..., u_n)$ of ENUM [SH95]. That stage calls the substages $(u_{t-1}, ..., u_n)$ such that $\|\pi_{t-1}(\sum_{i=t-1}^n u_i \mathbf{b}_i)\|^2 \leq A$. Note that $\|\sum_{i=1}^n u_i \mathbf{b}_i\|^2 = \|\zeta_t + \sum_{i=1}^{t-1} u_i \mathbf{b}_i\|^2 + \|\pi_t(\mathbf{b})\|^2$, where $\zeta_t := \mathbf{b} - \pi_t(\mathbf{b}) \in \text{span}\, \mathcal{L}_t$ is $\mathbf{b}$'s orthogonal projection in $\text{span}\, \mathcal{L}_t$. Stage $(u_t, ..., u_n)$ and its substages exhaustively enumerate the intersection $\mathcal{B}_{t-1}(\zeta_t, \rho_t) \cap \mathcal{L}_t$ for the sphere $\mathcal{B}_{t-1}(\zeta_t, \rho_t) \subset \text{span}\, \mathcal{L}_t$ with radius $\rho_t := (A - \|\pi_t(\mathbf{b})\|^2)^{1/2}$ and center ζ_t.

The GAUSSIAN volume heuristics estimates $|\mathcal{B}_{t-1}(\zeta_t, \rho_t) \cap \mathcal{L}_t|$ for $t > 1$ to

$$\beta_t =_{def} \text{vol}\, \mathcal{B}_{t-1}(\zeta_t, \rho_t) / \det \mathcal{L}_t.$$

Here $\mathrm{vol}\,\mathcal{B}_{t-1}(\zeta_t, \rho_t) = V_{t-1}\rho_t^{t-1}$, $V_{t-1} = \pi^{\frac{t-1}{2}}/(\frac{t-1}{2})! \approx (\frac{2e\pi}{t-1})^{\frac{t-1}{2}}/\sqrt{\pi(t-1)}$ is the volume of the unit sphere of dimension $t-1$, and $\det \mathcal{L}_t = r_{1,1}\cdots r_{t-1,t-1}$. If $\zeta_t \bmod \mathcal{L}_t$ is uniformly distributed the expected size of this intersection satisfies $\mathrm{E}_{\zeta_t}[\,\#(|\mathcal{B}_{t-1}(\zeta_t, \rho_t) \cap \mathcal{L}_t)\,] = \beta_t$. This holds because $1/\det \mathcal{L}_t$ is the number of lattice points of $\mathcal{L}_t$ per volume in span $\mathcal{L}_t$.

The success rate β_t has been used in [SH95] to speed up ENUM by cutting stages of very small success rate. NEW ENUM proceeds differently, it first performs all stages with $\beta_t \geq 2^{-s}t$ and collects during this process the stages with $\beta_t < 2^{-s}t$ in the list L. Thereafter NEW ENUM performs the stages of L with $\beta_t \geq 2^{-s-1}t$. The test $\beta_t \geq 2^{-s}t$ gives priority to stages of small t, stages of large t require a higher success rate. The analysis in section 4 is independent of the factor t in $\beta_t < 2^{-s}t$.

We will use that $A := \frac{n}{4}(\det \mathbf{B}^t\mathbf{B})^{1/n} > \lambda_1^2$ holds for $n \geq 10$ since $\gamma_n < \frac{n}{4}$ for $n \geq 10$.

Optimal Value of A. If λ_1 is known it is best to set the input A to $A = \lambda_1^2$.

Outline of New Enum

INPUT BKZ-basis $\mathbf{B} = \mathbf{QR} \in \mathbb{Z}^{m \times n}$, $\mathbf{R} = [r_{i,j}] \in \mathbb{R}^{n \times n}$ for block size 32,
OUTPUT a sequence of $\mathbf{b} \in \mathcal{L}(\mathbf{B})$ of decreasing length terminating with $\|\mathbf{b}\| = \lambda_1$.
1. $s := 10$, $L := \emptyset$, $A := \frac{n}{4}(\det \mathbf{B}^t\mathbf{B})^{1/n}$ (we call s the *level*)
2. *Perform via algorithm* ENUM *of [SE94, SH95], all stages with* $\beta_t \geq 2^{-s}t$:
    Upon entry of stage $(u_t, ..., u_n)$ compute β_t. If $\beta_t < 2^{-s}t$ store information about $(u_t, ..., u_n)$ in the list L of *delayed stages*. Otherwise perform stage $(u_t, ..., u_n)$ on level s, and as soon as some $\mathbf{b} \in \mathcal{L} - \mathbf{0}$ of length $\|\mathbf{b}\|^2 \leq A$ has been found, give out $\mathbf{b}$ and set $A := \|\mathbf{b}\|^2 - 1$.
3. $s := s + 1$, IF $L \neq \emptyset$ THEN GO TO 2 (*to perform all stages*
    $(u_t, ..., u_n)$ *of L with* $\beta_t \geq 2^{-s}t$.)
ELSE *terminate* .

Running in Linear Space. If instead of storing the list L we restart NEW ENUM in step 3 on the level $s+1$ then NEW ENUM runs in linear space and its running time increases at most by a factor n.

Practical Optimization. NEW ENUM computes $\mathbf{R}$, $\beta_t, V_t, \rho_t, c_t$ in floating point and $\mathbf{b}$, $\|\mathbf{b}\|^2$ in exact arithmetic. The final output $\mathbf{b}$ has length $\|\mathbf{b}\| = \lambda_1$, but this is only known when the more expensive final search does not find a vector shorter than $\mathbf{b}$.

Reason of Efficiency. For short vectors $\mathbf{b} = \sum_{i=1}^n u_i\mathbf{b}_i \in \mathcal{L}$ the stages $(u_t, ..., u_n)$ have large success rate β_t. If $\mathbf{b}$ is short then so are the projections $\pi_t(\mathbf{b})$. (On average $\|\pi_t(\mathbf{b})\|^2 \approx \frac{n-t+1}{n}\|\mathbf{b}\|^2$ holds for a random $\mathbf{b} \in_R \mathcal{B}_n(\mathbf{0}, \lambda)$ of length λ.) Therefore $\rho_t^2 = A - \|\pi_t(\mathbf{b})\|^2$ and β_t are large. New Enum tends to output very short lattice vectors $\mathbf{b}$ first.

Consider the case $A = \lambda_1^2$. Prior to finding the shortest lattice vector $\mathbf{b}' = \sum_{i=1}^n u_i'\mathbf{b}_i$ NEW ENUM essentially performs only stages $(u_t, ..., u_n)$ of success rate $\beta_t = V_{t-1}\rho_t^{t-1}/\det \mathcal{L}_t$ where on average $\rho_t^2 = \lambda_1^2 - \|\pi_t(\mathbf{b}')\|^2 \approx \frac{t-1}{n}\lambda_1^2$ since on average $\|\pi_t(\mathbf{b}')\|^2 \approx \frac{n-t+1}{n}\lambda_1^2$. While ENUM calls nearly all stages $(u_t, ..., u_n)$ of $\beta_t > 0$ NEW ENUM only calls about a $(\frac{n-t+1}{n})^{\frac{n-t+1}{2}}$ fraction of them prior to finding $\mathbf{b}'$ and delays the rest to be performed later than $(u'_t, ..., u'_n)$.

NEW ENUM is particularly fast for small λ_1. The size of its search space is proportional to λ_1^n, and is by Prop. 1 heuristically polynomial if $rd(\mathcal{L}) = o(n^{-1/4})$. Having found $\mathbf{b}'$ NEW ENUM proves $\|\mathbf{b}'\| = \lambda_1$ in exponential time by a complete exhaustive enumeration.

Notation. We use the following function $c_t : \mathbb{Z}^{n-t+1} \to \mathbb{R}$:

$$c_t(u_t, ..., u_n) = \|\pi_t(\textstyle\sum_{i=t}^{n} u_i \mathbf{b}_i)\|^2 = \textstyle\sum_{i=t}^{n}(\textstyle\sum_{j=i}^{n} u_j r_{i,j})^2.$$

Clearly $c_t(u_t, ..., u_n) = (\textstyle\sum_{j=t}^{n} u_j r_{t,j})^2 + c_{t+1}(u_{t+1}, ..., u_n).$

Given $u_{t+1}, ..., u_n$ ENUM tries the $u_t \in \mathbb{Z}$ close to $-y_t := -\sum_{i=t+1}^{n} u_i r_{t,i}/r_{t,t}$ in order of increasing distance $|u_t + y_t|$, recursively as $u_t := \lceil -y_t \rceil$, $u_t := \text{next}(u_t, -y_t)$:

$$\lceil -y_t \rfloor, \lceil -y_t \rfloor - \sigma_t, \lceil -y_t \rfloor + \sigma_t, \lceil -y_t \rfloor - 2\sigma_t, \lceil -y_t \rfloor + 2\sigma_t, \cdots$$

for $\sigma_t := sign(\lceil -y_t \rfloor + y_t) \in \{\pm 1\}$, $sign(0) := 1$, where $\lceil r \rfloor =_{def} \lceil r - 0.5 \rceil$ denotes the nearest integer to $r \in \mathbb{R}$. The iteration $u_t := next(u_t, -y_t)$ increases or preserves $|u_t + y_t|$ and $c_t(u_t, ..., u_n)$, decreases or preserves ρ_t and β_t so that ENUM performs the stages $(u_t, ..., u_n)$ for fixed $u_{t+1}, ..., u_n$ in order of increasing $c_t(u_t, ..., u_n)$ and decreasing success rate β_t. Note that $next(u_t, -y_t) = next_{\sigma_t, \nu_t}(u_t, -y_t)$ is a simple function of the number ν_t of iterations of next and the initial sign σ_t. The center $\zeta_t = \mathbf{b} - \pi_t(\mathbf{b}) = \sum_{i=t}^{n} u_i(\mathbf{b}_i - \pi_t(\mathbf{b}_i)) \in span(\mathcal{L}_t)$ changes continuously within NEW ENUM.

Algorithm Enum [SH95]

INPUT BKZ-basis $\mathbf{B} = \mathbf{QR} \in \mathbb{Z}^{m \times n}$, $\mathbf{R} = [r_{i,j}] \in \mathbb{R}^{n \times n}$ for block size 20,
OUTPUT $\mathbf{b} \in \mathcal{L}(\mathbf{B})$ such that $\mathbf{b} \neq \mathbf{0}$ has minimal length.

1. FOR $i = 1, ..., n$ DO $c_i := u_i := y_i := 0$

    $u_1 := 1$, $t := t_{max} := 1$, $\bar{c}_1 := c_1 := \|\mathbf{b}_1\|^2$. $(c_t = c_t(u_t, ..., u_n)$
    always holds for the current t, $\bar{c}_1$ is the current minimum of c_1)

2. WHILE $t \leq n$ #*perform stage* $(u_t, ..., u_n)$:
    $c_t := c_{t+1} + (u_t + y_t)^2 r_{t,t}^2$
    IF $c_t < \bar{c}_1$
        THEN IF $t = 1$ THEN $\bar{c}_1 := c_1$, $\mathbf{b} := \sum_{i=1}^{n} u_i \mathbf{b}_i$
                ELSE $t := t - 1$, $y_t := \sum_{i=t+1}^{t_{max}} u_i r_{t,i}/r_{t,t}$, $u_t := \lceil -y_t \rfloor$
        ELSE [$t := t + 1$, $t_{max} := \max(t, t_{max})$
                IF $t = t_{max}$ THEN $u_t := u_t + 1$ ELSE $u_t := next(u_t, -y_t)$].

3. output $\mathbf{b}$

New Enum for SVP

INPUT BKZ-basis $\mathbf{B} = \mathbf{QR} \in \mathbb{Z}^{m \times n}$, $\mathbf{R} = [r_{i,j}] \in \mathbb{R}^{n \times n}$ for block length 32,

OUTPUT a sequence of $\mathbf{b} \in \mathcal{L}(\mathbf{B})$ such that $\|\mathbf{b}\|$ decreases to λ_1.

1. $L := \emptyset$, $t := t_{max} := 1, s := 10$, FOR $i = 1, ..., n$ DO $c_i := u_i := y_i := 0$, $u_1 := 1$,
   $c_1 := r_{1,1}^2$, $A := \frac{n}{4} (\det \mathbf{B}^t \mathbf{B})^{1/n}$ ($c_t = c_t(u_t, ..., u_n)$ *always holds for the current t*)

2. WHILE $t \le n$ #*perform stage* $(u_t, ..., u_n)$:
   $c_t := c_{t+1} + (u_t + y_t)^2 r_{t,t}^2$,
   IF $c_t > A$ THEN GO TO 2.1,
   $\rho_t := (A - c_t)^{1/2}$, $\beta_t := V_{t-1}\rho_t^{t-1}/(r_{1,1} \cdots r_{t-1,t-1})$,
   IF $t = 1$ THEN [$\mathbf{b} := \sum_{i=1}^n u_i \mathbf{b}_i$,
            IF $\|\mathbf{b}\|^2 < A$ THEN output $\mathbf{b}$, $A := \|\mathbf{b}\|^2 - 1$, GO TO 2],
   IF $\beta_t \ge 2^{-s}t$ THEN [$t := t - 1$, $y_t := \sum_{i=t+1}^{t_{max}} u_i r_{t,i}/r_{t,t}$, $u_t := \lceil -y_t \rceil$,
            $\sigma_t := sign(u_t + y_t)$, $\nu_t := 1$, GO TO 2]
            ELSE store $(u_t, ...u_n, y_t, c_t, \sigma_t, \nu_t)$ in L.

2.1. $t := t + 1$, $t_{max} := \max(t, t_{max})$,
       IF $t = t_{max}$ THEN $u_t := u_t + 1$, $\nu_t := 1$, $y_t := 0$
       ELSE $u_t := \text{next}_{\sigma_t, \nu_t}(u_t, -y_t)$, $\nu_t := \nu_t + 1$.

3. $s := s + 1$, perform all delayed stages $(u_t, ..., u_n, y_t, c_t, \sigma_t, \nu_t)$ of L on level
   s and delete them. Delay new stages with $\beta_{t'} < 2^{-s}t'$, $t' \le t$ and store
   $(u_{t'}, ..., \nu_{t'})$ in L.

4. IF $L \ne \emptyset$ THEN GO TO 3 ELSE terminate.

Performing in step 3 a delayed stage $(u_t, ..., u_n, y_t, c_t, \sigma_t, \nu_t)$ means to restart the algorithm in step 2 with that information. The recursion initiated by this restart does not perform any stages $(u_{t''}, ..., u_n)$ with $t'' > t$. These stages have already been performed. Therefore, within step 2.1 the running t-value t' must be restricted not to surpass by the t-value at the restart.

Pruned New Enum for CVP. Given a target vector $\mathbf{t} = \sum_{i=1}^n \tau_i \mathbf{b}_i \in \text{span}(\mathcal{L}) \subset \mathbb{R}^m$ we minimize $\|\mathbf{t} - \mathbf{b}\|$ for $\mathbf{b} \in \mathcal{L}(\mathbf{B})$. [Ba86] solves $\|\mathbf{t} - \mathbf{b}\|^2 \le \frac{1}{4}\sum_{i=1}^n r_{i,i}^2$ in polynomial time for an LLL-basis $\mathbf{B} = \mathbf{QR}$, $\mathbf{R} = [r_{i,j}]$.

Adaption of NEW ENUM *to* **CVP**. We adapt NEW ENUM to solve $\|\mathbf{t} - \mathbf{b}\|^2 < \ddot{A}$. Initially we set $\ddot{A} := 0.01 + \frac{1}{4}\sum_{i=1}^n r_{i,i}^2$ so that $\|\mathbf{t} - \mathcal{L}\|^2 < \ddot{A}$. Having found some $\mathbf{b} \in \mathcal{L}$ such that $\|\mathbf{t} - \mathbf{b}\|^2 < \ddot{A}$ NEW ENUM gives out $\mathbf{b}$ and decreases $\ddot{A}$ to $\|\mathbf{t} - \mathbf{b}\|^2$.

Optimal Value of $\ddot{A}$. If the distance $\|\mathbf{t} - \mathcal{L}\|$ or a close upper bound of it is known then we initially choose $\ddot{A}$ to be that close upper bound. This prunes away many irrelevant stages.

At stage $(u_t, ..., u_n)$ NEW ENUM searches to extend the current $\mathbf{b} = \sum_{i=t}^n u_i \mathbf{b}_i \in \mathcal{L}$ to some $\mathbf{b}' = \sum_{i=1}^n u_i \mathbf{b}_i \in \mathcal{L}$ such that $\|\mathbf{t} - \mathbf{b}'\|^2 < \ddot{A}$. The expected number of such $\mathbf{b}'$ is for random $\mathbf{t}$:

$$\ddot{\beta}_t = V_{t-1}\ddot{\rho}_t^{t-1}/\det \mathcal{L}(\mathbf{b}_1, ..., \mathbf{b}_{t-1}) \quad \text{for} \quad \ddot{\rho}_t := (\ddot{A} - \|\pi_t(\mathbf{t} - \mathbf{b})\|^2)^{1/2}.$$

Previously, stage $(u_{t+1}, ..., u_n)$ determines u_t to yield the next integer minimum of

$$c_t(\tau_t - u_t, ..., \tau_n - u_n) := \|\pi_t(\mathbf{t} - \mathbf{b})\|^2$$
$$= (\textstyle\sum_{i=t}^n (\tau_i - u_i)r_{t,i})^2 + c_{t+1}(\tau_{t+1} - u_{t+1}, ..., \tau_n - u_n).$$

Given $u_{t+1}, ..., u_n$, $\|\pi_t(\mathbf{t}-\mathbf{b})\|^2$ is minimal for $u_t = \lceil -\tau_t - \sum_{i=t+1}^n (\tau_i - u_i)r_{t,i}/r_{t,t} \rfloor$.
NEW ENUM solves **CVP** for $(\mathcal{L}, \mathbf{t})$ by solving **CVP** for $(\pi_t(\mathcal{L}), \pi_t(\mathbf{t}))$ for $t = n, ..., 1$.

New Enum for CVP

INPUT BKZ-basis $\mathbf{B} = \mathbf{QR} \in \mathbb{Z}^{m \times n}$ for block size 32, $\mathbf{R} = [r_{i,j}] \in \mathbb{R}^{n \times n}$,

    $\mathbf{t} = \sum_{i=1}^n \tau_i \mathbf{b}_i \in \mathrm{span}(\mathcal{L})$, $\tau_1, ..., \tau_n \in \mathbb{Q}^n$, $\ddot{A} \in \mathbb{Q}$ such that $\|\mathbf{t} - \mathcal{L}(\mathbf{B})\|^2 < \ddot{A}$.

OUTPUT A sequence of $\mathbf{b} = \sum_{i=1}^n u_i \mathbf{b}_i \in \mathcal{L}(\mathbf{B})$ such that $\|\mathbf{t} - \mathbf{b}\|$ decreases to $\|\mathbf{t} - \mathcal{L}\|$.

1. $s := 10$, $t := n$, $L := \emptyset$, $y_n := \tau_n$, $u_n := \lceil y_n \rfloor$, $\ddot{c}_{n+1} := 0$, (We call s the level)
    ($\ddot{c}_t = c_t(\tau_t - u_t, ..., \tau_n - u_n)$ always holds for the current $t, u_t, ..., u_n$)

2. WHILE $t \leq n$ #perform stage $(u_t, ..., u_n)$:
    $\ddot{c}_t := \ddot{c}_{t+1} + (u_t - y_t)^2 r_{t,t}^2$,
    IF $\ddot{c}_t \geq \ddot{A}$ THEN GO TO 2.1,
    $\ddot{\rho}_t := (\ddot{A} - \ddot{c}_t)^{1/2}$, $\ddot{\beta}_t := V_{t-1}\ddot{\rho}_t^{t-1}/(r_{1,1} \cdots r_{t-1,t-1})$,
    IF $t = 1$ THEN [output $\mathbf{b} := \sum_{i=1}^n u_i \mathbf{b}_i$, $\ddot{A} := \|\mathbf{t} - \mathbf{b}\|^2$, GO TO 1]
    IF $\ddot{\beta}_t \geq 2^{-s}t$ THEN [$t := t - 1$, $y_t := \tau_t + \sum_{i=t+1}^n (\tau_i - u_i)r_{t,i}/r_{t,t}$,
                $u_t := \lceil y_t \rfloor$, $\sigma_t := sign(u_t + y_t)$, $\nu_t := 1$, GO TO 2]
           ELSE store $(u_t, ..., u_n, y_t, \ddot{c}_t, \sigma_t, \nu_t)$ in L,

2.1. $t := t + 1$, $u_t := next_{\sigma_t, \nu_t}(u_t, y_t)$, $\nu_t := \nu_t + 1$.

3. $s := s + 1$, perform all delayed stages $(u_t, ..., u_n, y_t, \ddot{c}_t, \sigma_t, \nu_t)$ of L on level s and delete them from L. Delay all new stages with $\ddot{\beta}_{t'} < 2^{-s}t'$, $t' \leq t$ and store $(u_{t'}, ..., u_n, y_{t'}, \ddot{c}_{t'}, \sigma_{t'}, \nu_{t'})$ in L.

4. IF $L \neq \emptyset$ THEN GO TO 3 ELSE *terminate*.

---

4 Performance of Pruned New Enum for SVP and CVP

We present results that are proven in the full version of the paper [S10]. Proposition 1 bounds, for a different type of pruning, the time to find an **SVP**-solution $\mathbf{b}'$ without proving $\lambda_1 = \|\mathbf{b}'\|$. Finding an unproved shortest vector $\mathbf{b}'$ is easier than proving $\|\mathbf{b}'\| = \lambda_1$. NEW ENUM finds an unproved shortest lattice vector $\mathbf{b}'$ in polynomial time under the following assumptions:

- the given lattice basis $\mathbf{B} = [\mathbf{b}_1, ..., \mathbf{b}_n]$ and the relative density $rd(\mathcal{L})$ of $\mathcal{L}(\mathbf{B})$ satisfy

$$rd(\mathcal{L}) \leq \left(\sqrt{\tfrac{e\pi}{2n}} \tfrac{\lambda_1}{\|\mathbf{b}_1\|}\right)^{\frac{1}{2}},$$ i.e., both $\mathbf{b}_1$ and $rd(\mathcal{L})$ are sufficiently small.

SA: NEW ENUM finds a shortest lattice vector $\mathbf{b}'$ of $\mathcal{L}$ such that $\|\pi_t(\mathbf{b}')\|^2 \lesssim \frac{n-t+1}{n}\lambda_1^2$ for all t.

- the vol. heur. is close: $\mathcal{M}_t^\rho := |\mathcal{B}_{n-t+1}(\mathbf{0}, \rho_t) \cap \pi_t(\mathcal{L})| \approx \frac{V_{n-t+1}\rho_t^{n-t+1}}{\det \pi_t(\mathcal{L})}$ for $\rho_t^2 = \frac{n-t+1}{n}\lambda_1^2$.

GSA: The basis $\mathbf{B} = \mathbf{QR} = \mathbf{Q}[r_{i,j}]$ satisfies $r_{i,i}^2/r_{i-1,i-1}^2 = q$ for $i = 2, ..., n$ for some $q < 0$

Remarks. **1.** If GSA holds with $q \geq 1$ the basis $\mathbf{B}$ satisfies $\|\mathbf{b}_i\| \leq \frac{1}{2}\sqrt{i+3}\,\lambda_i$ for all i and $\|\mathbf{b}_1\| = \lambda_1$. Therefore, $q < 1$ unless $\|\mathbf{b}_1\| = \lambda_1$. GSA means that the reduction of the basis is "locally uniform". It is easier to work with the idealized property that all $r_{i,i}/r_{i-1,i-1}$ are equal. [BL05] studies "nearly equality". B. LANGE [L13] shows that GSA can be replaced by the weaker property that the reduction potential of $\mathbf{B}$ is sufficiently small. GSA has been used in [S03, NS06, GN08, S07, N10] and in the security analysis of NTRU in [H07, HHHW09].

2. The assumption SA is supported by a fact proven in the full paper of [GNR10]:
$$\Pr[\, \|\pi_t(\mathbf{b}')\|^2 \leq \tfrac{n-t+1}{n}\lambda_1^2 \text{ for } t = 1, ..., n\,] = \tfrac{1}{n}$$
holds for random $\mathbf{b}' \in \mathcal{B}_n(\mathbf{0}, \lambda_1)$. The probability $1/n$ increases by iterating the search for a shortest lattice vector by statistical independent trials via permuted bases.

3. Failings of the volume heuristics. For the lattice $\mathbb{Z}^n$ we have for any $a = \Theta(1)$ and $n \geq n_0(a)$:
$$\#\{\mathbf{x} \in \mathbb{Z}^n \mid \|\mathbf{x}\|^2 \leq an\} \geq \left(2e\sqrt{n/a}\right)^{\sqrt{an}} = n^{\Theta(\sqrt{n})},$$
whereas the volume heuristics estimates this cardinality to $O(1)$ for $a \leq \frac{1}{2e\pi}$, also see Figure 1 of [MO90]. [GN08] reports that extensive experiments on high density random lattices show only negligible errors of the volume heuristics. The situation for low density lattices as $\mathcal{L} = \mathbb{Z}^n$ and small radius $\rho_t \ll \sqrt{n}\,\lambda_1$ is less clear.

4. *A trade-off* between $\|\mathbf{b}_1\|/\lambda_1$ and $rd(\mathcal{L})$ under GSA. B. LANGE observed that
$$\|\mathbf{b}_1\|/\lambda_1 = \|\mathbf{b}_1\|/(rd(\mathcal{L})\gamma_n^{1/2}\det(\mathcal{L})^{\frac{1}{n}}) = q^{\frac{1-n}{4}}/(rd(\mathcal{L})\gamma_n^{1/2}).$$
Therefore $rd(\mathcal{L})\,\gamma_n^{1/2}\,\|\mathbf{b}_1\|/\lambda_1 \leq 1$ implies under GSA that $q \geq 1$ and thus $\|\mathbf{b}_1\| = \lambda_1$. Hence the trade-off implies $rd(\mathcal{L})\gamma_n^{1/2}\,\|\mathbf{b}_1\|/\lambda_1 > 1$ unless $\|\mathbf{b}_1\| = \lambda_1$. Moreover, solving SVP with approximation factor $1/(rd(\mathcal{L})\,\gamma_n^{1/2})$ and a basis satisfying GSA already solves SVP exactly.

Also this trade-off implies $n^{\frac{1}{2}+b}rd(\mathcal{L}) > \|\mathbf{b}_1\|/(\det \mathcal{L})^{\frac{1}{n}} = q^{\frac{1-n}{4}} > 1$ for $q < 1$ and $n \geq n_0$ due to $\gamma_n < \frac{n}{e\pi}$ [KL78]. This shows that the time bound of Theorem 1 is at best exponential $2^{O(n)}$.

All our time bounds must be multiplied by the work load per stage, a modest polynomial factor covering the steps performed at stage $(u_t, ..., u_n)$ of NEW ENUM before going to a subsequent stage.

Proposition 1. *Let the given lattice basis* $\mathbf{B} \in \mathbb{Z}^{m \times n}$ *satisfy* $rd(\mathcal{L}) \leq \left(\sqrt{\frac{e\pi}{2n}}\frac{\lambda_1}{\|\mathbf{b}_1\|}\right)^{\frac{1}{2}}$ *and GSA. If a shortest lattice vector* $\mathbf{b}'$ *satisfies SA then* ENUM *and* NEW ENUM, *pruned to stages satisfying* $c(u_t, ..., u_n) \leq \frac{n-t+1}{n}\lambda_1^2$, *find such* $\mathbf{b}'$ *under the volume heuristics in polynomial time.*

Proof. (included for completeness) For simplicity we assume that λ_1 is known. If we prune all stages $(u_t, ..., u_n)$ of NEW ENUM that satisfy $c(u_t, ..., u_n) > \frac{n-t+1}{n}\lambda_1^2$ this does not cut off any shortest lattice vector $\mathbf{b}'$ satisfyng SA. The volume heuristics and SA show for $\rho_t^2 = \frac{n-t+1}{n}\lambda_1^2$:

$$\mathcal{M}_t^\rho := \#\mathcal{B}_{n-t+1}(\mathbf{0}, \rho_t) \cap \pi_t(\mathcal{L}) \le (\sqrt{\tfrac{n-t+1}{n}}\,\lambda_1)^{n-t+1} V_{n-t+1}/(r_{t,t}\cdots r_{n,n})$$
$$< (\tfrac{\sqrt{2e\pi}\lambda_1}{\sqrt{n}})^{n-t+1}/(r_{t,t}\cdots r_{n,n}).$$

We used Stirling's approximation of $(n - t + 1)!$ in approximating V_{n-t+1}. (attention: the volume heuristics underestimates $\#\mathcal{B}_{n-t+1}(\mathbf{0}, \rho_t) \cap \pi_t(\mathcal{L})$ for the center $\mathbf{0}$ and small radius ρ_t.)

Moreover GSA and $\|\mathbf{b}_i^*\| = \|\mathbf{b}_1\| q^{\frac{i-1}{2}}$ yield

$$(r_{t,t}\cdots r_{n,n}) = \det \pi_t(\mathcal{L}) = \|\mathbf{b}_1\|^{n-t+1} q^{\sum_{i=t-1}^{n-1} i/2}.$$

We see from $q^{\frac{n-1}{2}} = \frac{(\det \mathcal{L})^{\frac{2}{n}}}{\|\mathbf{b}_1\|^2} = \frac{\lambda_1^2}{\gamma_n rd(\mathcal{L})^2} \frac{1}{\|\mathbf{b}_1\|^2}$ and $\gamma_n \le \frac{n}{e\pi}$ for $n \ge n_0$ [KL78] that

$$\mathcal{M}_t^\rho \le (\tfrac{\lambda_1}{\|\mathbf{b}_1\|}\sqrt{\tfrac{2e\pi}{n}})^{n-t+1} (\sqrt{\tfrac{n}{e\pi}}\, rd(\mathcal{L})\tfrac{\|\mathbf{b}_1\|}{\lambda_1})^{n-\frac{(t-1)(t-2)}{n-1}}.$$

Evaluating this upper bound at $rd(\mathcal{L}) = (\sqrt{\tfrac{e\pi}{2n}}\tfrac{\lambda_1}{\|\mathbf{b}_1\|})^{\frac{1}{2}}$ yields

$$\mathcal{M}_t^\rho \le (\sqrt{\tfrac{n}{2e\pi}}\tfrac{\|\mathbf{b}_1\|}{\lambda_1})^{-n+t-1} (\sqrt{\tfrac{n}{2e\pi}}\tfrac{\|\mathbf{b}_1\|}{\lambda_1})^{+\frac{n}{2}-\frac{1}{2}\frac{(t-1)(t-2)}{n-1}}.$$

- The latter upper bound on $\mathcal{M}_t^\rho$ has for $t \le n$ the maximum 1 at $t = n$. This proves that $\max_{1 \le t \le n} \mathcal{M}_t^\rho \le 1$ and thus proves Proposition 1. $\qquad\square$

Finding a vector $\mathbf{b} \in \mathcal{L}(\mathbf{B})$ such that $0 < \|\mathbf{b}\|/\lambda_1 \le \sqrt{\tfrac{e\pi}{2n}}/rd(\mathcal{L}(\mathbf{B}))$ can still be hard in worst case.

The translation of Prop. 1 in Cor. 1, Cor. 2 from **SVP** to **CVP** for a random target $\mathbf{t}$ gets rid of the volume heuristics. The volume heuristics provably holds on the average for the **CVP** of minimizing $\|\mathbf{b} - \mathbf{t}\|$ for $\mathbf{b} \in \mathcal{L}(\mathbf{B})$ given a random $\mathbf{t} \in \mathrm{span}(\mathcal{L})$. On the average $\|\mathbf{t} - \mathcal{L}\|$ may be large for random $\mathbf{t}$.

Theorem 1. *Given a lattice basis* $\mathbf{B} \in \mathbb{Z}^{m \times n}$ *satisfying GSA and* $\|\mathbf{b}_1\| \le \sqrt{e\pi}\, n^b \lambda_1$ *for some* $b \ge 0$, NEW ENUM *solves* **SVP** *and proves to have found a solution in time* $2^{O(n)}(n^{\frac{1}{2}+b} rd(\mathcal{L}))^{\frac{n+1+o(1)}{4}}$.

Recall from remark 4 that $n^{\frac{1}{2}+b} rd(\mathcal{L}) \ge 1$ holds under GSA for $\|\mathbf{b}_1\| \le \sqrt{e\pi}\, n^b \lambda_1$ or else $\|\mathbf{b}_1\| = \lambda_1$. Interestingly Cor. 1, the translation of Theorem 1 from **SVP** to **CVP**, shows that the corresponding **CVP**-algorithm solves many important **CVP**-problems in simple exponential time $2^{O(n)}$ and linear space.

[HS07] proves the time bound $n^{n/2+o(n)}$ for solving **CVP** by KANNAN's **CVP**-algorithm [Ka87]. Minimizing $\|\mathbf{b}\|$ for $\mathbf{b} \in \mathcal{L} - \{\mathbf{0}\}$ and minimizing $\|\mathbf{t} - \mathbf{b}\|$ for $\mathbf{b} \in \mathcal{L}$ require nearly the same work if $\|\mathbf{t} - \mathcal{L}\| \approx \lambda_1$. In fact the proof of Theorem 1 yields:

Corollary 1. [S10] *Given a basis* $\mathbf{B} = [\mathbf{b}_1, ..., \mathbf{b}_n]$ *satisfying GSA,* $\|\mathbf{b}_1\| \le \sqrt{e\pi}\, n^b \lambda_1$ *with* $b \ge 0$ *and* $\mathbf{t} \in \mathrm{span}(\mathcal{L})$ *with* $\|\mathcal{L} - \mathbf{t}\| \le \lambda_1$, NEW ENUM *solves this* **CVP** *in time* $2^{O(n)}(n^{\frac{1}{2}+b} rd(\mathcal{L}))^{\frac{n}{4}}$.

Corollary 1 proves under GSA, $rd(\mathcal{L}) = O(n^{-\frac{1}{2}-b})$ and $\|\mathcal{L} - \mathbf{t}\| \le \lambda_1$ the CVP time bound $2^{O(n)}$ even using linear space (by iterating NEW ENUM for

$s = 1, ..., O(n)$ without storing delayed stages). Moreover it proves under GSA and $\|\mathbf{b}_1\| = O(\lambda_1)$ and $\|\mathcal{L} - \mathbf{t}\| \leq \lambda_1$ the time bound $n^{\frac{n+o(n)}{n}}$ However subexponential time remains unprovable due remark 4 of section 4.

CA translates the assumption SA from **SVP** to **CVP**:

CA: $\|\pi_t(\mathbf{t} - \ddot{\mathbf{b}})\|^2 \lesssim \frac{n-t+1}{n} \|\mathbf{t} - \mathcal{L}\|^2$ holds for $t = 1, ..., n$ and NEW ENUM's **CVP**-solution $\ddot{\mathbf{b}}$.

CA holds with probability $1/n$ for random $\ddot{\mathbf{b}} \in \text{span}(\mathcal{L})$ that is statistically independent of the given basis of $\mathcal{L}$ [GNR10]. B. LANGE [La13] proves that this probability $1/n$ increases to 1 for the increased bound $\|\pi_t(\mathbf{t} - \ddot{\mathbf{b}})\|^2 \lesssim \frac{n-t+1}{n} \|\mathbf{t} - \mathcal{L}\|^2(1 + 1/\sqrt{n})$.

Corollary 2. [S10] *Given a basis* $\mathbf{B} = [\mathbf{b}_1, ..., \mathbf{b}_n] \in \mathbb{Z}^{m \times n}$ *of* $\mathcal{L}$ *that satisfies* GSA, $\|\mathbf{b}_1\| = O(\lambda_1)$ *and* $rd(\mathcal{L}) \leq \left(\sqrt{\frac{e\pi}{2n}} \frac{\lambda_1}{\|\mathbf{b}_1\|}\right)^{\frac{1}{2}}$. *Let the target vector* $\mathbf{t}$ *and its closest lattice vector* $\ddot{\mathbf{b}}$ *found by* NEW ENUM *satisfy* CA *then* NEW ENUM *finds* $\ddot{\mathbf{b}}$ *for random* $\mathbf{t}$ *in average time* $n^{O(1)}\mathbf{E}_\mathbf{t}[(\|\mathbf{t} - \mathcal{L}\|/\lambda_1)^n]$.

Cor. 1 and Cor. 2 do not use the questionable volume heuristics. Cor. 2 eliminates the volume heuristics by randomizing the target vector $\mathbf{t}$. Without this randomization Cor. 3 proves a polynomial time bound under the volume heuristics if in addition $\|\mathbf{t} - \mathcal{L}\| \lesssim \lambda_1$ holds. It remains to analyze the error of the polynomial time bound of Cor. 3 resulting from the volume heuristics. Prop. 1 translates into

Corollary 3. *Let a basis* $\mathbf{B} = [\mathbf{b}_1, ..., \mathbf{b}_n] \in \mathbb{Z}^{m \times n}$ *of* $\mathcal{L}$ *be given satisfying* GSA, $\|\mathbf{b}_1\| = O(\lambda_1)$ *and* $rd(\mathcal{L}) \leq \left(\sqrt{\frac{e\pi}{2n}} \frac{\lambda_1}{\|\mathbf{b}_1\|}\right)^{\frac{1}{2}}$. *If the target vector* $\mathbf{t}$ *and its closest lattice vector* $\ddot{\mathbf{b}}$ *found by* NEW ENUM *satisfies* CA *and* $\|\mathbf{t} - \mathcal{L}\| \lesssim \lambda_1$ NEW ENUM *finds* $\ddot{\mathbf{b}}$ *under the volume heuristics in polynomial time.*

B. LANGE [La13] shows that GSA for $\mathbf{B}$ can be replaced by a less rigid condition, namely that the "reduction potential" $\prod_{\ell_i \geq 1} \ell_i$ for $\ell_i = \|\mathbf{b}_i^*\|/(\det \mathcal{L})^{1/n}$ of the basis $\mathbf{B}$ is sufficiently small. This is important since even well-reduced bases of the prime number lattice $\mathcal{L}(\mathbf{B}_{n,c})$ do not well approximate GSA.

5   Factoring via CVP Solutions for the Prime Number Lattice

Let N be a positive integer that is not a prime power. Let $p_1 < \cdots < p_n$ denote the smallest n primes. Let the prime factors of N be larger than p_n. A classical method factors N via $n + O(1)$ modular equations $\prod_{i=1}^n p_i^{e_i} = \pm \prod_{i=1}^n p_i^{e_i'}$ mod N. We construct such modular equations from **CVP** solutions for the prime number lattice $\mathcal{L}(\mathbf{B}_{n,c})$ with basis $\mathbf{B}_{n,c} = [\mathbf{b}_1, ..., \mathbf{b}_n] \in \mathbb{R}^{(n+1) \times n}$ and target vector $\mathbf{N} \in \mathbb{R}^{n+1}$ for some constant $c > 0$:

$$\mathbf{B}_{n,c} = \begin{bmatrix} \sqrt{\ln p_1} & 0 & 0 \\ 0 & \ddots & 0 \\ 0 & 0 & \sqrt{\ln p_n} \\ N^c \ln p_1 & \cdots & N^c \ln p_n \end{bmatrix}, \quad \mathbf{N} = \begin{bmatrix} 0 \\ \vdots \\ 0 \\ N^c \ln N \end{bmatrix}, \quad (5.1)$$

$$\left(\det \mathcal{L}(\mathbf{B}_{n,c}) \right)^2 = \left(\prod_{i=1}^n \ln p_i \right) (1 + N^{2c} \sum_{i=1}^n \ln p_i),$$

$$\left(\det \mathcal{L}(\mathbf{B}_{n,c}) \right)^{2/n} = \ln p_n \cdot (1 \pm o(1)) \cdot N^{2c/n}$$

as the prime number theorem implies $\prod_{i-1}^n \ln p_i^{1/n} / \ln p_n = 1 - o(1)$ for $n \to \infty$.

We identify the vector $\mathbf{b} = \sum_{i=1}^n e_i \mathbf{b}_i \in \mathcal{L}(\mathbf{B}_{n,c})$ with the pair (u, v) of relative prime integers

$$u = \prod_{e_i > 0} p_i^{e_i}, \quad v = \prod_{e_i < 0} p_i^{-e_i} \in \mathbb{N}.$$

Clearly $e_1, ..., e_n \in \{0, \pm 1\}$ if and only if uv is square-free. Let $\hat{z} = N^c \ln \frac{u}{vN}$ denote the last coordinate of $\mathbf{b} - \mathbf{N}$.

Outline of the Factoring Method. We compute vectors $\mathbf{b} = \sum_{i=1}^n e_i \mathbf{b}_i \in \mathcal{L}(\mathbf{B}_{n,c})$ close to $\mathbf{N}$ such that $|u - vN| \leq p_n^{O(1)}$ factorizes as $|u - vN| = \prod_{i=1}^n p_i^{e_i'}$. This yields a non-trivial relation

$$u = \prod_{e_i > 0} p_i^{e_i} = \pm \prod_{i=1}^n p_i^{e_i'} \mod N. \quad (5.2)$$

We write $n + 1$ such relations with $p_0 = -1$ as $\prod_{i=0}^n p_i^{e_{i,j} - e_{i,j}'} = 1 \mod N$ for $j = 1, ..., n + 1$. Any solution $t_1, ..., t_{n+1} \in \{0, 1\}$ of the equations

$$\sum_{j=1}^{n+1} t_j (e_{i,j} - e_{i,j}') = 0 \mod 2 \quad \text{for } i = 0, ..., n \quad (5.3)$$

solves $X^2 = 1 \mod N$ by $X = \prod_{i=0}^n p_i^{\frac{1}{2} \sum_{j=1}^{n+1} t_j (e_{i,j} - e_{i,j}')} \mod N$. In case that $X \neq \pm 1 \mod N$

this yields two non-trivial factors $\gcd(X \pm 1, N) \notin \{1, N\}$ of N.

The linear system of equations (5.3) can be solved within $O(n^3)$ bit operations. We neglect this minor part of the work load of factoring N. This reduces factoring N to finding about n vectors $\mathbf{b} \in \mathcal{L}(\mathbf{B}_{n,c})$ for which $|u - vN|$ factorizes over $p_1, ..., p_n$.

Lemma 1. 1. *Let the uv of $\mathbf{b} \in \mathcal{L}(\mathbf{B}_{n,c})$ be square-free, $|\hat{z}|^2 \leq \|\mathbf{b} - \mathbf{N}\|^2 / n$ and $|u - vN| = o(vN)$ then $\|\mathbf{b} - \mathbf{N}\|^2 \leq (1 + 1/n) \ln(v^2 N) + o(1)$.*

2. *If $\|\mathbf{b} - \mathbf{N}\|^2 = O(\ln N)$ and $v \leq N^{c-1}(\ln N)^k$ then $|u - vN| = O((\ln N)^{k+1/2})$.*

Proof. 1. A factor $p_i^{e_i}$ of uv contributes $e_i \ln p_i$ to $\ln uv$ and $e_i^2 \ln p_i$ to $\|\mathbf{b} - \mathbf{N}\|^2$. As $e_i = e_i^2$ we

have $$\|\mathbf{b} - \mathbf{N}\|^2 = \ln uv + N^{2c} |\ln \frac{u}{vN}|^2 = \ln uv + |\hat{z}|^2,$$

where $\ln uv$ is the contribution of the $\sqrt{\ln p_i}$ entries of $\mathbf{B}_{n,c}$. We have $u = vN \pm |u - vN| = vN \pm o(vN) = vN(1 + \pm o(1))$, hence $\ln uv \leq \ln(v^2 N(1 + o(1))) \leq \ln v^2 N + o(1)$. This proves **1**.

We conclude that most likely $\|\mathcal{L}(\mathbf{B}_{n,c}) - \mathbf{N}\|^2 = O(\ln N)$.

2. Applying $\|\mathbf{b} - \mathbf{N}\|^2 = O(\ln N)$ to the last coordinate $\hat{z} = N^c \ln(\frac{u}{vN})$ of $\mathbf{b} - \mathbf{N}$ we see that

$$N^{-c}|\hat{z}| = \left| \ln \tfrac{u}{vN} \right| = \left| \ln(1 + \tfrac{u-vN}{vN}) \right| = O(N^{-c}(\ln N)^{1/2}) = o(1)$$

due to $c > 0$. As $\ln(1 + x) = x - \frac{x^2}{2} + \frac{x^3}{3} \cdots$ this proves for $x = \frac{u-vN}{vN} = O((N^{-c}(\ln N)^{1/2})$, and thus $|u - vN| \leq O(vN^{1-c}(\ln N)^{1/2}) \leq O((\ln N)^{k+1/2})$ due to $v \leq N^{c-1}(\ln N)^{k+1/2}$. This proves **2**.

We conclude for $c > 1$, $p_n > (\ln N)^{k+1/2}$ that most likely any $\mathbf{b}$ close to $\mathbf{N}$ having a small v yields a relation (5.2). $\square$

An integer is called *y-smooth*, if it has no prime factor larger than y. Let $\Psi(X, y)$ denote the number of integers in $[1, X]$ that are y-smooth. DICKMAN [1930] has shown for constant $z \geq 1$ that

$$\Psi(X.y) \sim X\rho(z) \text{ as } X \to \infty \text{ where } X = y^z. \tag{5.4}$$

$\rho(z)$ is known as Dickman's de Bruijn ρ-function, see [G08] for a recent surview. [G08] proves

$$\rho(z) = 1 - \ln z \text{ for } 1 \leq z \leq 2$$

$$\rho(z) = \left(\tfrac{e+o(1)}{z \ln z}\right)^z = 1/z^{z+o(z)} \text{ for } z \to \infty \tag{5.5}$$

HILDEBRAND [H84] proved for $X = y^z$ under the Riemann Hypothesis that

$$\Psi(X, y) = X\rho(z)\left(1 + O\left(\tfrac{\ln(z+1)}{\ln y}\right)\right) \tag{5.6}$$

uniformly holds for $1 \leq z \leq y^{1/2-\varepsilon}$, $y \geq 2$ and any fixed $\varepsilon > 0$ and

Let $\Phi(N, p_n, \sigma)$ denote the number of triples $(u, v, |u - vN|) \in \mathbb{N}^3$ that are p_n-smooth and bounded as $v, |u - vN| \leq p_n^\sigma$. We conclude from (5.6) that

$$\Phi(N, p_n, \sigma) = O\left(2p_n^{2\sigma} \rho\left(\tfrac{\ln(Np_n^\sigma)}{\ln p_n}\right)\rho^2(\sigma)\right) \tag{5.7}$$

uniformly holds for $\frac{\ln N}{\ln p_n} + \sigma \leq p_n^{1/2-\varepsilon}$ if the p_n-smoothness events of $u, v, |u - vN|$ are nearly statistically independent. We will apply (5.7) in a range where $\frac{\ln N}{\ln p_n} + \sigma \ll p_n^{1/2}$ and we will neglect the $O(1)$-factor of (5.7).

Proof of (5.7). There are $2p_n^{2\sigma}$ pairs of integers u, v such that $0 < v, |u - vN| \leq p_n^\sigma$. Clearly $u \leq Np_n^\sigma + p_n^\sigma \leq p_n^z$ holds for $z = \frac{\ln(N+1)}{\ln p_n} + \sigma$. Then (5.6) for $X = (N + 1)p_n^\sigma$ and $y = p_n$ shows that the fraction of u that are p_n-smooth is $\rho(z)\left(1 + O\left(\tfrac{\ln(z+1)}{\ln p_n}\right)\right)$ if $\frac{\ln N}{\ln p_n} + \sigma \leq p_n^{1/2-\varepsilon}$.

Moreover (5.6) for $X = p_n^\sigma$, $y = p_n$, $z = \sigma$ shows that the fraction of $0 < v \le p_n^\sigma$ that are p_n-smooth is $\rho(\sigma)\big(1 + O\big(\frac{\ln(\sigma+1)}{\ln p_n}\big)\big)$ if $\sigma \le p_n^{1/2-\varepsilon}$. Therefore the statistical independence of the p_n-smoothness events of $u, v, |u-vN|$ implies (5,7) if $\ln(z+1) = O(\ln p_n)$ holds in both cases. The latter holds due to $\frac{\ln N}{\ln p_n} + \sigma \le p_n^{1/2-\varepsilon}$.

Example Factoring. Let $N = 100000980001501 \approx 10^{14}$ and $n = 90, p_{90} = 463$. (5.7) shows that there are $\Theta(6.4 \cdot 10^5)$ relations (5.2) such that $v, |u-vN| \le 463^3$ are p_n-smooth. M. Charlet has constructed several hundreds such relations (5.2) for the above N. For this N the following program is particular efficient for $N^c = 10^{10}$, $c \approx 5/7$ and pruned to stages with success rate $\ddot{\beta}_t \ge 2^{-14}$. For the first time this recommends to use $c < 1$ as well as relatively small prime bases and to use extreme pruning.

A Program for Finding Relations (5.2) Efficiently. Initially the given basis $\mathbf{B}_{n,c}$ gets strongly BKZ-reduced with block size 32 and the target vector $\mathbf{N}$ is shifted modulo lattice vectors into the ground mesh of the reduced basis. The initial value $\ddot{A}$, the upper bound on $\|\mathbf{N} - \mathcal{L}(\mathbf{B}_{n,c})\|^2$ is set to $\frac{11}{5\,4} \sum_{i=1}^n r_{i,i}^2$.

LOOP. In each round the vectors of the reduced basis of $\mathcal{L}(\mathbf{B}_{n,c})$ and the shifted $\mathbf{N}$ are randomly scaled as follows. For $i = 1, ..., n$ with probabiliy $1/2$ all i-th coordinates of the basis vectors and the shifted target vector are multiplied by 2. (This nearly excludes the "scaled" primes p_i to appear as factors of uv in relations (5.2) resulting from **CVP**-solutions.) The scaled basis gets slightly reduced by BKZ-reduction of block size 20. Then NEW ENUM for **CVP** is called to search for lattice vectors that are close to the shifted target vector $\mathbf{N}$. NEW ENUM always decreases $\ddot{A}$ to the square distance to $\mathbf{N}$ of the closest found lattice vector. But whenever a relation (5.2) has been found NEW ENUM stops further decreasing $\ddot{A}$ for this round. Whenever a new closer lattice vector is found it is checked whether it yields a relation (5.2). The scaling per rround makes sure that the algorithm produces distinct relations (5.2).

Previously B. Lange had implemented NEW ENUM for **SVP** and by hand picking various prime bases $\mathbf{B}_{n,c}$ for various c and subsets of the smallest 125 primes he found 45 relations (5.2) for the above N. M. Charlet has adjusted that program from **SVP** to **CVP**. He generated several hundred of relations (5.2). We report on his experiments.

Performance. The algorithm found in one run of 15 minutes and 350 rounds 139 relations. On average it found a relation every 6.5 seconds. This amounts to a factoring time of 10 minutes for $N \approx 10^{14}$. The majority of rounds does not find a relation. But 74 of the 139 found relations have been found in rounds with multiple successes. Here are the first 30 of the 139 example relations, they mostly satisfy $|u - vN| \le p_{90}^3$.

	u	v	$\lvert u - vN \rvert$
1.	$19 \cdot 29^2 \cdot 31 \cdot 73 \cdot 109 \cdot 139 \cdot 211 \cdot 259$	415	$22 \cdot 11 \cdot 37 \cdot 437$
2.	$29 \cdot 37 \cdot 83 \cdot 139 \cdot 191 \cdot 269 \cdot 307 \cdot 443$	865	$2 \cdot 11 \cdot 239 \cdot 383$
3.	$2 \cdot 3 \cdot 17^2 \cdot 103 \cdot 263 \cdot 317 \cdot 379 \cdot 443$	25	$13 \cdot 173$
4.	$2 \cdot 5 \cdot 47 \cdot 83 \cdot 157 \cdot 179 \cdot 307 \cdot 331 \cdot 421$	469	$19 \cdot 43 \cdot 373$
5.	$7^2 \cdot 13 \cdot 41 \cdot 43 \cdot 107 \cdot 109 \cdot 113 \cdot 131 \cdot 409 \cdot 461$	365571	$2^4 \cdot 5 \cdot 11^2 \cdot 197 \cdot 433$
6.	$2 \cdot 7 \cdot 13 \cdot 31 \cdot 107 \cdot 127 \cdot 149 \cdot 179 \cdot 383 \cdot 397 \cdot 439$	1364937	$3 \cdot 5 \cdot 11 \cdot 61 \cdot 337 \cdot 419$
7.	$43 \cdot 131 \cdot 139 \cdot 193 \cdot 307 \cdot 353 \cdot 401 \cdot 439$	28829	$2 \cdot 3^2 \cdot 5^2 \cdot 13 \cdot 41 \cdot 109$
8.	$19 \cdot 31 \cdot 53 \cdot 61 \cdot 67 \cdot 131 \cdot 163 \cdot 241 \cdot 313$	2055	$2^2 \cdot 59 \cdot 71 \cdot 89$
9.	$13^2 \cdot 17 \cdot 101 \cdot 137 \cdot 199 \cdot 229 \cdot 277 \cdot 331$	1661	$2^6 \cdot 3 \cdot 19 \cdot 233$
10.	$19 \cdot 101 \cdot 107 \cdot 127 \cdot 131 \cdot 179 \cdot 191 \cdot 211 \cdot 370$	93398	$3^3 \cdot 13 \cdot 29 \cdot 109 \cdot 167$
11,	$2 \cdot 7^2 \cdot 31 \cdot 37 \cdot 163 \cdot 181 \cdot 257 \cdot 281 \cdot 433$	1037	$3^2 \cdot 251 \cdot 421$
12.	$2^2 \cdot 7 \cdot 17 \cdot 101 \cdot 131 \cdot 157 \cdot 167 \cdot 239 \cdot 347 \cdot 349$	47793	$5 \cdot 13 \cdot 29 \cdot 83 \cdot 151$
13.	$3 \cdot 79 \cdot 83 \cdot 131 \cdot 167 \cdot 263 \cdot 331^2$	124	$7^2 \cdot 13 \cdot 139$
14.	$11 \cdot 13 \cdot 151 \cdot 181 \cdot 251 \cdot 271 \cdot 419 \cdot 439$	489	$2^4 \cdot 139 \cdot 199$
15.	$2^2 \cdot 11 \cdot 19 \cdot 41 \cdot 79 \cdot 199 \cdot 281 \cdot 331 \cdot 421$	211	$5 \cdot 149 \cdot 397$
16.	$2 \cdot 3 \cdot 5 \cdot 7 \cdot 11 \cdot 59 \cdot 73 \cdot 167 \cdot 173 \cdot 257 \cdot 421$	311	$53 \cdot 457$
17.	$2 \cdot 3 \cdot 13 \cdot 29 \cdot 67 \cdot 103 \cdot 109 \cdot 193 \cdot 331$	25	$7 \cdot 73^2$
18.	$3^2 \cdot 17 \cdot 47 \cdot 53^2 \cdot 67 \cdot 113 \cdot 179^2$	49	$2^3 \cdot 5 \cdot 331$
19.	$11 \cdot 47 \cdot 53 \cdot 107 \cdot 179 \cdot 241 \cdot 271 \cdot 283$	97	$2^6 \cdot 7 \cdot 13 \cdot 17$
20.	$2 \cdot 11 \cdot 47 \cdot 53 \cdot 131 \cdot 149 \cdot 151 \cdot 241 \cdot 353 \cdot 439$	60323	$3 \cdot 7 \cdot 41 \cdot 89 \cdot 347$
21.	$2 \cdot 11 \cdot 59 \cdot 67 \cdot 197 \cdot 263 \cdot 337 \cdot 461$	7	$3 \cdot 5^2 \cdot 179$
22.	$53 \cdot 59 \cdot 79^2 \cdot 137 \cdot 149 \cdot 223 \cdot 421$	374	$7 \cdot 151 \cdot 347$
23.	$11 \cdot 17 \cdot 53 \cdot 73 \cdot 97 \cdot 137 \cdot 191 \cdot 223 \cdot 307 \cdot 337$	423677	$2^2 \cdot 3 \cdot 7^2 \cdot 13 \cdot 19 \cdot 193$
24.	$7^2 \cdot 53 \cdot 79 \cdot 137 \cdot 149 \cdot 223 \cdot 271 \cdot 367 \cdot 443$	411477	$2 \cdot 5 \cdot 211 \cdot 263 \cdot 283$
25.	$5 \cdot 7 \cdot 59 \cdot 73^2 \cdot 89 \cdot 131 \cdot 173 \cdot 419$	93	$2^2 \cdot 107 \cdot 179$
26.	$7 \cdot 17 \cdot 47 \cdot 53 \cdot 67 \cdot 131 \cdot 179 \cdot 179 \cdot 199 \cdot 307 \cdot 401$	114091	$2^2 \cdot 3^2 \cdot 11^2 \cdot 13 \cdot 29^2$
27.	$2 \cdot 3^2 \cdot 23 \cdot 47 \cdot 67 \cdot 139 \cdot 239 \cdot 331 \cdot 383 \cdot 419$	23005	$17 \cdot 53 \cdot 71 \cdot 173$
28.	$2 \cdot 3^3 \cdot 5 \cdot 11 \cdot 31 \cdot 47 \cdot 73 \cdot 191 \cdot 293 \cdot 397$	67	$281 \cdot 283$
29.	$2 \cdot 5 \cdot 47 \cdot 83 \cdot 157 \cdot 179 \cdot 307 \cdot 331 \cdot 421$	469	$19 \cdot 43 \cdot 373$
30.	$7 \cdot 41 \cdot 103 \cdot 127 \cdot 173 \cdot 197 \cdot 251 \cdot 337 \cdot 419$	45347	$2^4 \cdot 3 \cdot 19^2 \cdot 283$

Most example relations have square factors. Relations 18-26 have all been found in one round of 10 seconds. For most examples v and $\lvert u - vN \rvert$ are of about the same order.

In the above list of relations (5.2) the $u, v, \lvert u - vN \rvert$ have between 12 - 23 distinct prime factors ≤ 463. Since each such prime has been nearly eliminated with probability $1/2$ the number of possible remaining relations per round is $\Theta(6.4 \cdot 10^5)$, by (5.7), divided by 2^{12} to 2^{23}. Therefore only 83 out of the 350 rounds found a relation. This also indicates that by decreasing the probability $1/2$ of "scaling" the primes we can speed up the finding of relations.

Future Optimizations. 1. Restrict the scaling of the first $n/2$ coordinates of the reduced basis vectors of $\mathcal{L}(\mathbf{B}_{n,c})$, e.g. scale them only with probability $\leq 1/4$. Then the first $n/2$ primes can more freely occur as factors of uv.
2. We can increase c to $c+0.2$ by multiplying the last coordinates of the vectors of the BKZ-basis of $\mathcal{L}(\mathbf{B}_{n,c})$ by $N^{0.2}$, then again BKZ-reduce the basis. Such an iterative increase of c gives access to larger c-values at low costs for the BKZ-reduction. Theorem 3 shows that there are exponentially many relations (5.2) for $c > 1$.
3. For large N, N^c it is crucial to prune NEW ENUM towards small v, and small $|u - vN|$ because large $|u - vN|$ are rarely p_n-smooth. We can eliminate large v by pruning all stages $(u_t, ..., u_n)$ for which the v of $\mathbf{b} = \sum_{i=t}^{n} u_i \mathbf{b}_i$ is "to large". If $v \leq N^{c-1}(\ln N)^\alpha$ holds for $\mathbf{b} \in \mathcal{L}(\mathbf{B}_{n,c})$ then $|u - vN| \leq (\ln N)^{\alpha+1/2}$ holds by Lemma 1. part **2**, and thus $\mathbf{b}$ most likely provides a relation (5.2).
4. Before calling in a round NEW ENUM for **CVP** it makes sense to first approximate a closest lattice vector to $\mathbf{N}$ by Babai's nearest plane method [Ba86]. The LLL-version of this method is very fast and even a BKZ-version with block size 20 is very efficient and should already find many relations (5.2). In particular this should work well for large dimension n.

Factoring Large Integers $\mathbf{N} = \Theta(2^{2000})$. We extend the search of relations (5.2) to large, non smooth v. This greatly increases the number of relations (5.2) that are accessible by pruned **CVP** computations. We represent v by modifying the last coordinate of the target vector $\mathbf{N}$ from $N^c \ln(N)$ to $N^c \ln(vN)$ and let $\mathbf{N}_v$ denote the modified target vector. $\mathbf{N}_v$ yields the same last coordinate $\ddot{z} = N^c(\ln(u/v) - \ln N)$ of $\mathbf{b} - \mathbf{N}_v$ as our previous representation of v via $\mathbf{N} = \mathbf{N}_1$ and $\mathbf{b} = \sum_{i=1}^{n} u_i \mathbf{b}_i$ with some negative u_i. Now a stage of NEW ENUM is of the form $(u_t, ..., u_n, v) \in \mathbb{N}_0^{n-t+2}$, $v \geq 1$. NEW ENUM must enumerate all vectors $\mathbf{b} = \sum_{i=1}^{n} u_i \mathbf{b}_i$ together with all v such that $1 \leq v \leq N^{c-1}$ and $|u - vN| \leq p_n$ for $u = \prod_{i=1}^{n} p_i^{u_i}$. Stages are performed in the order of decreasing success rate $\ddot{\beta}_t$. The GAUSSIAN volume heuristics shows that $\ddot{\beta}_t$ is proportional to $(\ddot{A} - ||\pi_t(\mathbf{b} - \mathbf{N}_v)||^2)^{(n-t+1)/2}$. We can direct the enumerated v via clever pruning into any direction $v = \Theta(N^{c-1}2^{-j})$ for $1 \leq 2^j \leq N^{c-1}$. We hope that the pruned enumeration can be done in polynomial time for each j.

Recall that $v \leq N^{c-1}$ and $||\mathbf{b} - \mathbf{N}||^2 = O(\ln N)^{1/2}$ implies by Lemma 1, part 2 that $|u - vN| = O(\ln N)$. Note that this also holds for arbitrary non smooth v and the modified target vector $\mathbf{N}_v$.

Let $\Upsilon(N, n, c)$ denote the number of relations (5.2) corresponding to some u, v such that $|u - vN| \leq p_n$, u is p_n-smooth and $1 \leq v \leq N^{c-1}$. (5.5) shows that

$$\Upsilon(N, n, c) \geq N^{c-1}\left(\frac{e+o(1)}{\bar{z} \ln \bar{z}}\right)^{\bar{z}} 2 p_n \quad \text{for} \quad \bar{z} = \frac{c \ln N}{\ln p_n}$$

provided that the p_n-smoothness of $1 \leq u \leq N^c$ is nearly statistically independent of the event that $1 \leq v \leq N^{c-1}$ and $|u - vN| \leq p_n$.

We get for $2^{2000} \leq N \leq 2 \cdot 2^{2000}$, $n = 2500$, $p_n = 24281$ that
$$\Upsilon(N, n, 3.5) \geq e^{115}, \quad \Upsilon(N, n, 4) \geq e^{200}, \quad \Upsilon(N, n, 5) \geq e^{470},$$
$$\Upsilon(N, n, 6) \geq e^{666}, \quad \Upsilon(N, n, 7) \geq e^{834}, \quad \Upsilon(N, n, 8) \geq e^{985}.$$

If most of these (u, v) get enumerated by NEW ENUM with clever pruning we should find by randomly scaling primes about p_n relations (5.2) and factor N in polynomial time. The pruning into the most important directions $v = \Theta(N^{c-1})$ and the parameters $c, \ddot{A}$ have still to be optimized.

Comparison with [S93]. Our new results show an enormous progress compared to the previous approach of [S93]. [S93] reports on experiments for $N = 2131438662079 \approx 2.1 \cdot 10^{12}$, $N^c = 10^{25}$, $c \approx 2.0278$ and the prime number basis of dimension $n = 125$ with diagonal entries $\ln p_i$ for $i = 1, ..., n$ instead of $\sqrt{\ln p_i}$. The larger diagonal entries $\ln p_i$ require a larger c and this increases the time for the construction of relations (5.2). The latter took 10 hours per found relation on a PC of 1993. ADLEMAN [Ad95] introduced the diagonal elements $\sqrt{\ln p_i}$ of $\mathbf{B}_{n,c}$ in translating the method of [S93] from the $\| \ \|_1$-norm used in [S93] to the square norm.

6   Factoring N via the Prime Number Lattice with $c > 1$

Let $p_n = (\ln N)^{\alpha}$ and $c = (\ln N)^{\beta} > 1$. Theorem 3 shows that there are exponentially many vectors $\mathbf{b} \in \mathcal{L}(\mathbf{B}_{n,c})$ that are sufficiently close to $\mathbf{N}$ and provide a relation (5.2). In fact Lemma 1. part **2.** shows for $c > 1 - \frac{k \ln \ln N}{\ln N}$ that vectors $\mathbf{b}$ close to $\mathbf{N}$ with small v yield small $|u - vN|$ providing a relation (5.2). But an efficient construction of relations (5.2) via **CVP**-solutions requires clever pruning towards small v because complete enumeration can be expensive for large c, see Theorem 4.

The Number of Lattice Vectors $\mathbf{b} \in \mathcal{L}(\mathbf{B}_{n,c})$ Such That $|u - vN| = 1$. We denote

$$M_{n,c} = \left\{ (u, v) \in \mathbb{N}^2 \ \middle| \ \begin{array}{l} |u - vN| = 1, \ N^{c-1}/2 \leq v \leq N^{c-1} \\ u, v \text{ are } p_n\text{-smooth} \end{array} \right\}.$$

The v of $(u, v) \in M_{n,c}$ is quite large for $c \gg 1$ but due to Lemma 1 part **2** we can efficiently prune NEW ENUM for **CVP** towards bounded values v and $|u - vN| = p_n^{O(1)}$. Theorem 2 shows that there are exponentially many relations (5.2) for sufficiently large c, α.

Theorem 2.   *Assuming that the equation $|u - \lceil u/N \rceil N| = 1$ is for random $u = \Theta(N^c)$ nearly statistically independent from the event that $u, \lceil u/N \rceil$ are p_n-smooth then $\#M_{n,c} \geq N^{\varepsilon + o(\varepsilon)}$ holds for $\varepsilon > 0$ if $\alpha > 2\beta + 2 > 2$ and $\frac{\alpha + \alpha\varepsilon - 1 - \beta}{\alpha - 2\beta - 2} < c = (\ln N)^{\beta}$.*

Proof. (5.5) shows for $X = N$, $p_n = (\ln N)^\alpha = N^{1/z}$, $z = \ln N/\alpha \ln \ln N$ that

$$\Psi(N, p_n) = \left(\tfrac{e+o(1)}{z \ln z}\right)^z = z^{-z+o(z)} \quad \text{holds for } z \to \infty.$$

The assumption of the theorem and the lower bound on $\Psi(N^c, p_n)$ yields for $z, N \to \infty$:

$$\#M_{n,c} \geq N^{c-1}(zc - z)^{-zc+z}(zc)^{-zc+o(z)},$$

$$\ln \#M_{n,c} \geq (c-1)\ln N - z\big((c-1)\ln(zc-z) + c\ln zc\big)(1 + o(1)).$$

Here N^{c-1} counts twice the number of v, $\frac{1}{2}N^{c-1} \leq v \leq N^{c-1}$. For every such v there are two $u = vN \pm 1$, and $(zc - z)^{-zc+z+o(z)}$, $(zc)^{-zc+o(z)}$ lower bound the portions of those v and u that are p_n-smooth. We assume again that the p_n-smoothness events for the u and v are nearly statistical independent. Hence we get for $\frac{\alpha+\alpha\varepsilon-1-\beta}{\alpha-2\beta-2} < c = (\ln N)^\beta$ and $\alpha > 2\beta + 2$ that

$$\ln \#M_{n,c} > (c-1)\ln N - \frac{(2c-1)\ln N \ln(zc)}{\alpha \ln \ln N}(1+o(1)) \qquad \text{as } z = \ln N/\alpha \ln \ln N$$

$$> (c-1)\ln N - \frac{(2c-1)\ln N(1+\beta)\ln \ln N}{\alpha \ln \ln N}(1+o(1)) \text{ as } \ln z < \ln \ln N \text{ and } \ln c = \beta \ln \ln N$$

$$= \ln N(-1 + \tfrac{\alpha c - (2c-1)(1+\beta)(1+o(1))}{\alpha}) = \ln N(-1 + \tfrac{c(\alpha-2\beta-2)+1+\beta}{\alpha(1+o(1))}) \geq (\varepsilon - o(\varepsilon))\ln N$$

since $-1 + \frac{c(\alpha-2\beta-2)+1+\beta}{\alpha} > \varepsilon$ holds due to $\frac{\alpha+\alpha\varepsilon-1-\beta}{\alpha-2\beta-2} < c$. Hence $\#M_{n,c} \geq N^{\varepsilon+o(\varepsilon)}$. $\qquad \square$

Theorem 3. *The vector $\mathbf{b} = \sum_{i=1}^n e_i \mathbf{b}_i \in \mathcal{L}(\mathbf{B}_{n,c})$ that is closest to $\mathbf{N}$ yields a non-trivial relation (5.2) provided that $M_{n,c} \neq \emptyset$ and $v \leq O(N^{c-1}\ln N)$ and $p_n > (\ln N)^{3/2}$.*

Proof. Let $\mathbf{b}' = \sum_{i=1}^n e_i' \mathbf{b}_i \in \mathcal{L}(\mathbf{B}_{n,c})$ be the vector corresponding to some $(u', v') \in M_{n,c}$, $u' = \prod_{e_i'>0} p_i^{e_i'}$, $v' = \prod_{e_i'<0} p_i^{-e_i'}$ such that $|u' - v'N| = 1$.

We have $N^{c-1}/2 < v' < N^{c-1}$ and thus $v' = \eta N^{c-1}$ with $\frac{1}{2} < \eta < 1$. Lemma 1 part **1** shows

$$\|\mathbf{b}' - \mathbf{N}\|^2 \leq (2c-1)\ln N + \eta^{-2} + O(1) + |\hat{z}|^2 = O(\ln N).$$

This bound holds a fortiori for the lattice vector $\mathbf{b} \in \mathcal{L}(\mathbf{B}_{n,c})$ that is closest to $\mathbf{N}$. Consider the $(u, v) \in \mathbb{N}^2$ corresponding to $\mathbf{b}$. Lemma 1 part **2** shows that $|u - vN| = O((\ln N)^{1+1/2})$ holds for $c \geq 1$, $k = 1$ due to the small v. Hence $\mathbf{b}$ provides a relation (5.2) since $p_n > (\ln N)^{3/2}$. $\qquad \square$

Theorem 4. *Let $\alpha > 2\beta + 2$ and $\frac{\alpha+\alpha\varepsilon-1-\beta}{\alpha-2\beta-2} < c = (\ln N)^\beta$ for some $\varepsilon > 0$. Let a reduced version of the basis $\mathbf{B}_{n,c}$ of $\mathcal{L}$ be given that satisfies GSA, $\|\mathbf{b}_1\|^2 = 2c\ln N + O(1)$. Then $\|\mathbf{b}_1\|^2 = \lambda_1^2 + O(1)$, $\|\mathcal{L} - \mathbf{N}\| < \lambda_1$ and $rd(\mathcal{L}) = o(n^{-1/4})$. Let $\mathbf{N}_r \in_R \text{span}(\mathcal{L})$ be a random version of $\mathbf{N}$. If NEW ENUM finds a vector $\ddot{\mathbf{b}} \in \mathcal{L}$ closest to $\mathbf{N}_r$ that satisfies CA it finds $\ddot{\mathbf{b}}$ in average time $n^{O(1)}\mathbf{E}_{\mathbf{N}_r}[(\|\mathbf{N}_r - \mathcal{L}\|/\lambda_1)^n]$ and under the volume heuristics for the specific $\mathbf{N}$ of (5.1) in pol. time.*

Proof. We first prove that $\lambda_1^2 = 2c \ln N + O(1)$. $M_{n,c} \neq \emptyset$ holds by Theorem 2, moreover we see from that proof argument that there exist $u = \prod_{i \leq n} p_i^{e_i}$ and $v = \prod_{i \leq n} p_i^{e_i'}$ with $e_i, e_i' \in \{0,1\}$ such that $u = \Theta(N^c)$, $|u-v| \leq 2$, $\gcd(u,v) = 1$. In particular, let

$$\widetilde{M}_{n,c} = \left\{ (u,v) \in \mathbb{N}^2 \;\middle|\; \begin{array}{l} |u-v| = 1, \; N^c/2 \leq v \leq N^c \\ u, v \text{ are } p_n\text{-smooth} \end{array} \right\}.$$

Then $\#\widetilde{M}_{n,c} \geq N^c (zc)^{-2zc+o(z)}$ holds for $z = \ln N/(\alpha \ln \ln N)$, and thus

$$\ln \#\widetilde{M}_{n,c} \geq \ln N\left(\tfrac{\alpha - 2 - 2\beta - o(1)}{\alpha} c\right) = \Theta(c \ln N).$$

Similar to the proof of Lemma 1 part **2.** we have

$\|\sum_{i \leq n}(e_i - e_i')\mathbf{b}_i\|^2 = 2c \ln N + O(1) + N^{2c} \ln(u/v)^2 = 2c \ln N + O(|u-v|)^2$,

where $\ln(u/v) = \ln(1 + \tfrac{u-v}{v}) = \Theta(|u-v|N^{-c})$. Hence $\lambda_1^2 \leq 2c \ln N + O(1)$. On the other hand Lemma 5.3 of [MG02] proves that $\lambda_1^2 > 2c \ln N$ if the prime $p_1 = 2$ is neglected. Hence the claim.

Lemma 1 part **1** shows for $(u,v) \in M_{n,c}$ and $n \geq \ln N$ that

$\|\mathcal{L} - \mathbf{N}\|^2 \leq \|\mathbf{b} - \mathbf{N}\|^2 \leq (2c-1)\ln N + O(1)$, and thus $\|\mathcal{L} - \mathbf{N}\|^2/\lambda_1^2 \leq 1 - 1/2c + o(1)$.

For $c \geq 1$ we minimize $\|\mathbf{b} - \mathbf{N}\|$ for $\mathbf{b} \in \mathcal{L}$ by solving SVP for the lattice with basis $[\mathbf{N}, \mathbf{B}_{n,c}]$. In particular, $\|\mathcal{L}(\mathbf{B}_{n,1}) - \mathbf{N}\| \leq \lambda_1$ and thus $rd(\mathcal{L}(\mathbf{N}, \mathbf{B}_{n,1})) \leq rd(\mathcal{L}(\mathbf{B}_{n,1}))$.

Next we bound $rd(\mathcal{L})$ for $\mathcal{L} = \mathcal{L}(\mathbf{B}_{n,c}))$. For $n = (\ln N)^\alpha/(\alpha \ln \ln N)(1 + o(1))$ we have

$$\gamma_n(\det \mathcal{L})^{\frac{2}{n}} \geq \tfrac{n}{2e\pi}(\ln p_n \pm o(1)) \cdot N^{2c/n} = \tfrac{n}{2e\pi} N^{2c/n}(1 \pm o(1)).$$

Hence $\qquad\qquad rd(\mathcal{L}) = \lambda_1/(\sqrt{\gamma_n}(\det \mathcal{L})^{\frac{1}{n}}) = \left(\tfrac{2e\pi \, 2c \ln N}{n}\right)^{\frac{1}{2}}/N^{c/n}(1 \pm o(1))$.

We have for $\beta < \alpha/2 - 1$ that $c = (\ln N)^\beta < (\ln N)^{\alpha/2 - 1}$ and $\tfrac{c \ln N}{(\ln N)^\alpha} \leq \tfrac{1}{(\ln N)^{\alpha/2}} = o(n^{-1/2})$.

We see from $c/n \leq (\ln N)^{\beta - \alpha} \alpha \ln \ln N \, (1 + o(1))$ and $\beta - \alpha + 1 < -\alpha/2$ that

$$N^{c/n} = N^{\frac{1}{\ln N}(\ln N)^{\beta - \alpha + 1} \alpha \ln \ln N \, (1 + o(1))}$$

$$\leq e^{(\ln N)^{-\alpha/2} \alpha \ln \ln N \,(1 + o(1))} = e^{o(1)} = 1 + o(1).$$

Hence $rd(\mathcal{L}) = O\left(\left(\tfrac{c \ln N \, \alpha \ln \ln N}{(\ln N)^\alpha}\right)^{1/2}\right) = o(n^{-1/4})$ since $2 + 2\beta < \alpha$.

By Corollary 2 NEW ENUM minimizes $\|\mathbf{b} - \mathbf{N_r}\|$ for $\mathbf{b} \in \mathcal{L}$ for random $\mathbf{N_r} \in \mathrm{span}(\mathcal{L})$ under GSA and CA in average time $n^{O(1)} \mathbf{E}_{N_r}[(\|\mathbf{N_r} - \mathcal{L}\|/\lambda_1)^n]$. This proves the first claim.

Now assume the volume heuristics and consider the $\mathbf{N}$ of (5.1). Then following the proof of Prop. 1 and Cor. 3 NEW ENUM finds for the α of $(\ln N)^\alpha = p_n$ and $c = (\ln N)^\beta$ some $\mathbf{b} \in \mathcal{L}(\mathbf{B}_{n,c})$ that minimizes $\|\mathbf{b} - \mathbf{N}\|$ in polynomial time, without proving correctness of the minimization. This proves the claim. However, the volume heuristics can underestimate by far the number of $\mathbf{b} \in \mathcal{L}$ such that $\|\mathbf{b} - \mathbf{N}\| \leq \lambda_1$ if $\mathbf{N}$ is very close to the lattice. □

Without the volume heuristics the time bound of Theorem 5 for the $\mathbf{N}$ of (5.1) increases to $n^{O(1)}(R_\mathcal{L}/\lambda_1)^n$ where $R_\mathcal{L} = \max_{\mathbf{u} \in \mathrm{span}(\mathcal{L})} \|\mathcal{L} - \mathbf{u}\|$ is the covering radius of $\mathcal{L}$. The factor $(R_\mathcal{L}/\lambda_1)^n$ overestimates NEW ENUM's running time because $\|\mathcal{L} - \mathbf{N}\| < \lambda_1$ NEW ENUM enumerates lattice points in a ball of radius $\|\mathcal{L} - \mathbf{N}\| < \lambda_1 \ll R_\mathcal{L} = \max_{\mathbf{u} \in \mathrm{span}(\mathcal{L})} \|\mathcal{L} - \mathbf{u}\|$.

Constructing a nearly shortest vector of $\mathcal{L}(\bar{\mathbf{B}}_{n,c})$ for an extended basis $\bar{\mathbf{B}}_{n,c}$. In order to factor N heuristically in polynomial time via Theorem 5 we may need a very short vector of the prime number lattice. For this we extend the basis $\mathbf{B}_{n,c} = [\mathbf{b}_1, ..., \mathbf{b}_n] \in \mathbb{R}^{(n+1) \times n}$ by a suitable prime $\bar{p}_{n+1}$ and the corresponding vector $\bar{\mathbf{b}}_{n+1} = [0, ...0, \sqrt{\ln(\bar{p}_{n+1})}, N^c \ln(\bar{p}_{n+1})]^t \in \mathbb{R}^{n+2}$ to $\bar{\mathbf{B}}_{n,c} = [\bar{\mathbf{b}}_1, ..., \bar{\mathbf{b}}_n, \bar{\mathbf{b}}_{n+1}] \in \mathbb{R}^{(n+2) \times (n+1)}$ and construct such a short vector of the extended lattice. We construct the prime $\bar{p}_{n+1}$ such that $\bar{p}_{n+1} = \Theta(N^c)$ and $|u - \bar{p}_{n+1}| = O(1)$ holds for a square-free $(\ln N)^\alpha$-smooth integer $u = \prod_{i=1}^n p_i^{e_i}$. From the initial part of the proof of Theorem 6 and that of Lemma 2 we see that $\|\bar{\mathbf{b}}\|^2 = 2c \ln N + O(1) = \lambda_1^2(\mathcal{L}(\bar{\mathbf{B}}_{n,c}) + O(1)$ holds for $\bar{\mathbf{b}} := \sum_{i=1}^n e_i \bar{\mathbf{b}}_i - \bar{\mathbf{b}}_{n+1}$. This construction is efficient, we generate $u = \prod_i p_i = \Theta(N^c)$ at random and test the $\bar{p}$ near to u for primality. If the density of primes near the u is not exceptionally small we find a prime $\bar{p}_{n+1} = u + O(1)$ within $O(c \ln N)$ primality tests on such $\bar{p}$. Therefore $\bar{\mathbf{B}}_{n,c}$ and a nearly shortest vector $\bar{\mathbf{b}}$ of $\mathcal{L}(\bar{\mathbf{B}}_{n,c})$ can be found in probabilistic polynomial time. A single $(\bar{\mathbf{B}}_{n,c}, \bar{\mathbf{b}})$ can be used to solve all **CVP**'s for the factorization of all integers of the order of N.

Corollary 4. *Integers N can be factored under the vol. heuristics, GSA and CA in polynomial time by solving $(\ln N)^\alpha$ **CVP**'s for the prime number lattice of dimension $n < (\ln N)^\alpha$ and $c = (\ln N)^\beta$ such that $0 < \frac{\alpha-\beta-1}{\alpha-2\beta-2} < c$ and $\|\mathbf{b}_1\|^2 = 2c \ln N + O(1)$. This lattice satisfies $rd(\mathcal{L}) = o(n^{-1/4})$.*

Proof. We apply Theorems 3, 4 and 5 to $c = (\ln N)^\beta$. Then the condition $0 < \frac{\alpha-1-\beta}{\alpha-2\beta-2} < c$ required for Theorem 4 holds for arbitrary $0 < \beta < \alpha/2 - 1$ and sufficiently large N. The proof of Theorem 5 shows that $rd(\mathcal{L}) = o(n^{-1/4})$ is clearly smaller than required for Prop. 1 and Cor. 3. Therefore the errors of the volume heuristics should not be extreme. □

Acknowledgment. I am indebted B. Lange and M. Charlet (U. Frankfurt) for implementing NEW ENUM for **SVP** and **CVP** towards the generation of prime number relations for factoring integers N and for pointing out some inconsistencies in early versions of this manuscript.

References

[Ad95]   Adleman, L.A.: Factoring and lattice reduction. Manuscript (1995)

[Ba86]   Babai, L.: On Lovász lattice reduction and the nearest lattice point problem. Combinatorica 6 (1), 1–13 (1986)

[BL05]   Buchmann, J., Ludwig, C.: Practical lattice basis sampling reduction. eprint.iacr.org, TR 072 (2005)

[Ch13]   Charlet, M.: Faktorisierung ganzer Zahlen mit dem NEW ENUM-Gitteralgorithmus. Diplomarbeit, Frankfurt (2013)

[FP85]   Fincke, U., Pohst, M.: Improved methods for calculating vectors of short length in a lattice, including a complexity analysis. Math. of Comput. 44, 463–471 (1985)

[GN08]   Gama, N., Nguyen, P.Q.: Predicting lattice reduction. In: Smart, N.P. (ed.) EUROCRYPT 2008. LNCS, vol. 4965, pp. 31–51. Springer, Heidelberg (2008)

[GNR10]   Gama, N., Nguyen, P.Q., Regev, O.: Lattice enumeration using extreme pruning. In: Gilbert, H. (ed.) EUROCRYPT 2010. LNCS, vol. 6110, pp. 257–278. Springer, Heidelberg (2010)

[G08]   Granville, A.: Smooth numbers: computational number theory and beyond. Algorithmic Number Theory 44, 267–323 (2008)

[H84]   Hildebrand, A.: Integers free of large prime factors and the Riemann hypothesis. Mathematika 31, 258–271 (1984)

[HHHW09]   Hirschhorn, P.S., Hoffstein, J., Howgrave-Graham, N., Whyte, W.: Choosing NTRUEncrypt parameters in light of combined lattice reduction and MITM approaches. In: Abdalla, M., Pointcheval, D., Fouque, P.-A., Vergnaud, D. (eds.) ACNS 2009. LNCS, vol. 5536, pp. 437–455. Springer, Heidelberg (2009)

[H07]   Howgrave-Graham, N.: A hybrid lattice–reduction and meet-in-the-middle attiack against NTRU. In: Menezes, A. (ed.) CRYPTO 2007. LNCS, vol. 4622, pp. 150–169. Springer, Heidelberg (2007)

[Ka87]   Kannan, R.: Minkowski's convex body theorem and integer programming. Math. Oper. Res. 12, 415–440 (1987)

[La13]   Lange, B.: Neue Schranken für SVP-Approximation und SVP-Algorithmen. Dissertation, Frankfurt (2013)

[LLL82]   Lenstra Jr., H.W., Lenstra, A.K., Lovász, L.: Factoring polynomials with rational coefficients. Mathematische Annalen 261, 515–534 (1982)

[L86]   lOVász, L.: An Algorithmic Theory of Numbers, Graphs and Convexity. SIAM (1986)

[MG02]   Micciancio, D., Goldwasser, S.: Complexity of Lattice Problems: A Cryptographic Perspective. Kluwer Academic Publishers, Boston (2002)

[MV09]   Micciancio, D., Voulgaris, P.: Faster exponential time algorithms for the shortest vector problem. ECCC Report No. 65 (2009)

[N10]   Nguyen, P.Q.: Hermite's Constant and Lattice Algorithms. In: Nguyen, P.Q., Vallée, B. (eds.) The LLL Algorithm. Springer (January 2010)

[S87]   Schnorr, C.P.: A hierarchy of polynomial time lattice basis reduction algorithms. Theoret. Comput. Sci. 53, 201–224 (1987)

[S93]   Schnorr, C.-P.: Factoring integers and computing discrete logarithms via Diophantine approximation. In: Davies, D.W. (ed.) EUROCRYPT 1991. LNCS, vol. 547, pp. 281–293. Springer, Heidelberg (1991)

[SE94]   Schnorr, C.P., Euchner, M.: Lattce basis reduction: Improved practical algorithms and solving subset sum problems. Mathematical Programming 66, 181–199 (1994), http://www.mi.informatik.uni-frankfurt.de/

[SH95]   Schnorr, C.-P., Hörner, H.H.: Attacking the Chor–Rivest cryptosystem by improved lattice reduction. In: Guillou, L.C., Quisquater, J.-J. (eds.) EUROCRYPT 1995. LNCS, vol. 921, pp. 1–12. Springer, Heidelberg (1995)

[S03]   Schnorr, C.P.: Lattice reduction by sampling and birthday methods. In: Alt, H., Habib, M. (eds.) STACS 2003. LNCS, vol. 2607, pp. 145–156. Springer, Heidelberg (2003), www.mi.informatik.uni-frankfurt.de

[S07]   Schnorr, C.P.: Progress on LLL and lattice reduction. In: Phong, P.Q., Vallée, B. (eds.) Proceedings LLL+25, Caen, France, June 29-July 1. The LLL Algorithm (2007), www.mi.informatik.uni-frankfurt.de/

[S10]   Schnorr, C.P.: Average Time Fast SVP and CVP Algorithms for Low Density Lattices and the Factorisation of Integers (2010), www.mi.informatik.uni-frankfurt.de

Solving the Elliptic Curve Discrete Logarithm Problem Using Semaev Polynomials, Weil Descent and Gröbner Basis Methods – An Experimental Study

Michael Shantz and Edlyn Teske

University of Waterloo, Canada
{mcshantz,eteske}@uwaterloo.ca

Johannes Buchmann zu Ehren

Abstract. At ASIACRYPT 2012, Petit and Quisquater suggested that there may be a subexponential-time index-calculus type algorithm for the Elliptic Curve Discrete Logarithm Problem (ECDLP) in characteristic two fields. This algorithm uses Semaev polynomials and Weil Descent to create a system of polynomial equations that subsequently is to be solved with Gröbner basis methods. Its analysis is based on heuristic assumptions on the performance of Gröbner basis methods in this particular setting. While the subexponential behaviour would manifest itself only far beyond the cryptographically interesting range, this result, if correct, would still be extremely remarkable. We examined some aspects of the work by Petit and Quisquater experimentally.

1 Introduction

Throughout this paper, let E be an elliptic curve over $\mathbb{F}_{2^n}$,

$$E/\mathbb{F}_{2^n} : y^2 + xy = x^3 + a_2 x^2 + a_6$$

where $a_2, a_6 \in \mathbb{F}_{2^n}{}^*$ and such that the trace $\mathrm{Tr}_{\mathbb{F}_{2^n}/\mathbb{F}_2}(a_2) = 1$. (In particular, if n is odd, then we can set $a_2 = 1$.) For such an elliptic curve, and an integer $m \geq 2$, the m-th *Semaev polynomial* S_m [10] is the unique polynomial in m variables with the following property: $S_m(\overline{x}_1, \ldots, \overline{x}_m) = 0$ for $\overline{x}_1, \ldots, \overline{x}_m \in \overline{\mathbb{F}_{2^n}}$ if and only if there exist $\overline{y}_1, \ldots, \overline{y}_m \in \overline{\mathbb{F}_{2^n}}$ with $(\overline{x}_i, \overline{y}_i) \in E(\overline{\mathbb{F}_{2^n}})$ and such that $\sum_{i=1}^{m}(\overline{x}_i, \overline{y}_i) = \infty \in E(\overline{\mathbb{F}_{2^n}})$. Here, $\sum_{i=1}^{m}(\overline{x}_i, \overline{y}_i)$ denotes the addition of m points on the elliptic curve E, and ∞ denotes the point at infinity. For example, $S_2(x_1, x_2) = x_1 + x_2$ since $(\overline{x}_1, \overline{y}_1) + (\overline{x}_2, \overline{y}_2) = \infty$ in $E(\overline{\mathbb{F}_{2^n}})$ if and only if $\overline{x}_1 = \overline{x}_2$. For a fixed integer m, Semaev polynomials can be used to find relations on the set of points in $E(\mathbb{F}_{2^n})$. This is done by combining Weil descent and Gröbner basis techniques to find certain zeroes of S_m. This yields an index-calculus type algorithm to solve the *elliptic curve discrete logarithm problem (ECDLP)*: given $P \in E(\mathbb{F}_{2^n})$ and $Q \in \langle P \rangle$, find ℓ such that $Q = \ell P$. Variants of this algorithm

M. Fischlin and S. Katzenbeisser (Eds.): Buchmann Festschrift, LNCS 8260, pp. 94–107, 2013.
© Springer-Verlag Berlin Heidelberg 2013

are due to Gaudry [7] and Diem [3]. We will refer to it as *Diem's algorithm* for the remainder of this paper.

This paper is motivated by the works of Faugère, Perret, Petit and Renault [6], and Petit and Quisquater [9] who both emphasize the special form of the multivariate polynomial equations arising in Diem's algorithm, thereby suggesting a significant speed-up in the running time of special-purpose Gröbner basis techniques. While Diem [3] gives a runtime complexity of $\exp(O(n(\log n)^{1/2}))$, Faugère, Perret, Petit and Renault [6] derive complexity bounds for solving $S_m = 0$ based on the so-called *Linearization Method* and suggest a runtime complexity of $O(2^{\omega t})$ with $2.376 \leq \omega \leq 3$ ($\omega = $ the linear algebra constant) and $t \approx n/2$. (Compare this with the running time of the parallelized Pollard Rho method of $O(2^{n/2})$). This running time is under the unproven yet plausible heuristic assumption that a certain set of equations arising in the algorithm is linearly independent. Petit and Quisquater [9] conduct a more aggressive analysis of Diem's algorithm, and claim a running time complexity that is subexponential in the input size. More specifically, under the assumption on bounds on the so-called degree of regularity of a the system of equations that arises in the computation, the ECDLP over $\mathbb{F}_{2^n}$ is said to be solved in time $O(2^{cn^{2/3}\log n})$, where $c = 2\omega/3$. Petit and Quisquater also give estimated running times for specific values of the extension degree n, which indicate that Diem's algorithm outperforms the Pollard rho method for $n > 2000$. On the other hand, both papers [6,9] suggest that for large enough n (and m), the use of the so-called hybrid method (for computing a Gröbner basis) should produce even further speed-up. While these results seem to be no threat for current ECDLP-based cryptographic systems, they do require further study. The purpose of this paper is to do exactly this.

In particular, the contributions of this paper are the following:

- We confirm and extend the Petit-Quisquater experimental data [9] on the degree of regularity, up to $n = 29$. (Petit-Quisquater give data for $n = 11, 17$ only.)
- We suggest the *Delta Method* to achieve speed-up.
- We study the effect of various realizations of the hybrid method.
- Lastly, for the case n even only, we report on experiments with subfield-based factor-bases.

2   Preliminaries

2.1   Semaev's Summation Polynomials

For an integer $m \geq 2$, the m-th Semaev polynomial has been defined in the introduction. Further,

$$S_3(x_1, x_2, x_3) = x_1^2 x_2^2 + x_1^2 x_3^2 + x_2^2 x_3^2 + x_1 x_2 x_3 + a_6$$

([3, Lemma 3.4]), and recursively, for $m \geq 4$:

$$S_m(x_1, \ldots, x_m) = \operatorname{Res}_X(S_{m-k}(x_1, \ldots, x_{m-k-1}, X), S_{k+2}(x_{m-k}, \ldots, x_m, X)),$$

where $1 \leq k \leq m-3$. By definition S_m is symmetric, and S_m has degree 2^{m-2} in each x_i for $m \geq 2$. For Semaev polynomials to solve the ECDLP, we think of m to be small. For example, in the running time analysis by Petit and Quisquater, best running times are achieved with $m \leq 4$ if $n \leq 2000$, and with $m \leq 14$ for $n \leq 100,000$. Petit and Quisquater argue that using a method by Collin's [2], the calculation of S_m can be done in time $O(2^{m(m+1)})$.

2.2 An Index Calculus for the ECDLP

In our description of Diem's algorithm for the ECDLP, using Semaev's summation polynomials, we follow Petit and Quisquater [9].

Input:

- $\mathbb{F}_{2^n} = \mathbb{F}_2[z]/(f(z))$, where $\deg f = n$ and f irreducible/$\mathbb{F}_2$.
  We view $\mathbb{F}_{2^n}$ as a vector space over $\mathbb{F}_2$ of dimension n.
- Elliptic curve $E/\mathbb{F}_{2^n}$, $P \in E(\mathbb{F}_{2^n})$, $Q \in <P>$.

Output:

- The least positive integer ℓ such that $Q = \ell P$.

Algorithm:

- *Find a Factor Basis $\mathcal{F}_V$:*
  - Fix $n' \in [1,n] \cap \mathbb{Z}$.
  - Choose a subspace $V \subseteq \mathbb{F}_{2^n}/\mathbb{F}_2$ of $\dim V = n'$.
  - Set $\mathcal{F}_V = \{(x,y) \in E(\mathbb{F}_{2^n}); x \in V\}$
  (among pairs (x,y), (x,y') with $y \neq y'$, take only those (x,y) with lexicographically smaller y).

- *Compute Relations:*
  - Fix m.
  - Do about $2^{n'}$ times:

    - *REPEAT*
      - Take $a,b \in_R [0, 2^n] \cap \mathbb{Z}$.
      - Set $R = aP + bQ =: (x_R, y_R)$.
      - Look for $(x_1, \ldots, x_m) \in V^m$ with $S_{m+1}(x_1, \ldots, x_m, x_R) = 0$.
      UNTIL such $x_1, \ldots, x_m$ are found.

    - For $j = 1, \ldots, m$, compute $y_1, \ldots, y_m$ such that $R_j := (x_j, y_j) \in \mathcal{F}_V$.
    - Find $e_j \in \{\pm 1\}$ such that $R + \sum_{j=1}^{m} e_j R_j = \infty$.
  - *Result: about $2^{n'}$ Relations:*

$$\sum_{i=1}^{\#\mathcal{F}_V} \epsilon_{ik} R_i + a_k P + b_k Q = \infty$$

  where $\epsilon_{ik} = \{0, 1, -1\}$.

– *Linear Algebra Step:*
  Use sparse matrix Linear Algebra to find a linear dependency among the
  relations.
  Then easily obtain a solution ℓ to the ECDLP $Q = \ell P$.
  Return ℓ.

Notes:

– According to Faugère et al. [6], one should choose m, n' such that $mn' \approx n$.
– If $\dim V = n'$, then Diem [3] shows that $\#\mathcal{F}_V \approx 2^{n'}$.
– If $\dim V = n'$, the probability that $S_{m+1}(x_1, \ldots, x_m, x_R) = 0$ with
  $(x_1, \ldots, x_m) \in V$ is, on average, $\approx \frac{2^{mn'-n}}{m!}$.

Weil Descent. It remains to discuss how to solve $S_{m+1}(x_1, \ldots, x_m, x_R) = 0$
with the added constraint that $x_1, \ldots, x_m \in V$. This is achieved via Weil descent,
followed by Gröbner basis techniques. More specifically, one does the following:

– Choose a basis $\{\Theta_1, \ldots, \Theta_n\}$ of $\mathbb{F}_{2^n}/\mathbb{F}_2$.
– Choose a basis $\{v_1, \ldots, v_{n'}\}$ of $V/\mathbb{F}_2$.
– Introduce mn' new variables x_{ij}, $1 \le i \le m, 1 \le j \le n'$:
  Write

$$x_i = \sum_{j=1}^{n'} x_{ij} v_j, \qquad i = 1, \ldots, m.$$

– Substitute the x_i into S_{m+1}.
  Decompose each v_j, $j = 1, \ldots, n'$, and x_R, into the Θ_s, $s \in \{1, \ldots, n\}$.
  Reduce any $x_{ij}^2 - x_{ij} = 0$.
– Obtain equation:

$$0 = S_{m+1}(x_1, \ldots, x_m, x_R)$$
$$= S_{m+1}(\sum_{j=1}^{n'} x_{1j} v_j, \ldots, \sum_{j=1}^{n'} x_{mj} v_j, x_R)$$
$$= [f]_1 \Theta_1 + \cdots + [f]_n \Theta_n$$

where $[f]_s \subset \mathbb{F}_2[x_{11}, \ldots, x_{mn'}]$ for $s = 1, \ldots, n$.

Thus, and after adding the field equations, one has to solve the set of polynomial
equations in mn' variables

$$\begin{aligned} [f]_s &= 0, & s &= 1, \ldots, n, \\ x_{ij}^2 - x_{ij} &= 0, & i &= 1, \ldots, m; j = 1, \ldots, n', \end{aligned} \qquad (2.1)$$

for which one can use Gröbner basis techniques such as F4 [4] or F5 [5]. See [6]
and [9] for details.

2.3   The Hybrid Method

Reconsider the system (2.1). The hybrid method [1] (see also [8]) works as follows:

– Choose k variables among the x_{ij}, label them $y_1, \ldots, y_k$. There are 2^k possible choices to assign values to the k-tuple $(y_1, \ldots, y_k)$.
– For each such assignment, try to solve the new system in $mn' - k$ variables via a Gröbner basis calculation.
– With probability $\approx \frac{2^{mn'-n}}{m!}$, one of these new systems yields a solution to (2.1).

Using the hybrid method will require doing more Gröbner basis calculations, but since we are fixing k variables each time, the number of variables in the system is reduced. This produces an over-determined system which causes the Gröbner basis algorithms to run much faster.

Fixing k variables could require doing up to 2^k times as many Gröbner basis calculations. Thus we require a speed up of at least 2 each time k goes up by 1. Consequently, setting k too high causes the solution time for a single polynomial system to start increasing, since the decrease in running time for each of the 2^k required Gröbner basis calculations is not sufficient to make up for the increased number of calculations.

Note that setting $k = n$ corresponds to an exhaustive search. This approach is only efficient for very small n.

Faugère et al. [6] observed in experiments that with a suitable choice of k and some tweaking, the hybrid method is faster than solving (2.1) directly. They speculate that the hybrid method gives speedup by a factor m in the exponent.

3   Supporting the Petit-Quisquater Analysis, and More Experimental Evidence

In their Table 3, Petit and Quisquater [9] give running time estimates for Diem's algorithm for the ECDLP in $E(\mathbb{F}_{2^n})$ for various extension degrees $50 \leq n \leq 100,000$. These data illustrate the claimed subexponential behaviour, and suggest that for large enough n ($n \geq 2000$), Diem's algorithm outperforms the Pollard rho method. We reproduce their table in Table 3.1.

The Petit-Quisquater analysis is based on the assumption that for the Semaev polynomial equations (2.1),

$$\text{degree of regularity} = \text{first fall degree} + o(1). \tag{3.1}$$

Here, the degree of regularity D_{reg} is the degree of the largest Macaulay matrix appearing in a Gröbner basis computation with the algorithm F5 (cf. [9]), while the first fall degree $D_{\text{firstfall}}$ is the degree at which a non-trivial degree fall occurs during a Gröbner basis computation (see [9, Definition 2]).

We have the following facts ([9]):

– In general: $D_{\text{reg}} \geq D_{\text{firstfall}}$.

Table 3.1. Petit-Quisquater complexity estimates for the ECDLP in $E(\mathbb{F}_{2^n})$. Here, t_S = time to compute the mth Semaev polynomial, t_R = time to generate $2^{n'}$ relations, t_{LA} = time for the linear algebra step, and $T = \max\{t_S, t_R, t_{LA}\}$.

n	m	n'	t_S	t_R	t_{LA}	T
50	2	25	6	97	57	97
100	2	50	6	137	108	137
160	2	80	6	177	168	177
200	2	100	6	202	209	209
500	3	167	12	393	344	393
1000	3	250	20	664	512	664
2000	4	500	20	965	1013	1013
5000	6	833	42	1926	1682	1926
10000	7	1429	56	3020	2873	3020
20000	9	2222	90	4986	4462	4986
50000	11	4545	132	9030	9110	9110
100000	14	7143	210	14762	14306	14762

- (3.1) is true for many systems analyzed in the context of multivariate cryptosystems.
- $D_{\text{firstfall}} \leq m^2 + 1$.
- The running time to solve (2.1) with $F4$ or $F5$ is $O(n^{\omega D_{\text{reg}}})$. Memory requirements: $O(n^{2D_{\text{reg}}})$.

In their Table 2, Petit and Quisquater support (3.1) with experimental evidence for the Semaev polynomial equations, for $n = 11, 17$, $m = 2, 3$.

3.1 Extending the Petit-Quisquater Data

Mimicking the Petit-Quisquater experiments [9], we reproduced and expanded their Table 2. For this, we did the following; We fixed n, n' and m. We constructed $\mathbb{F}_{2^n} = \mathbb{F}_2[z]/(f(z))$ with f a Conway polynomial of degree n. We chose a random elliptic curve over $\mathbb{F}_{2^n}$ of order twice a prime. We chose the vector space V of dimension n' with basis $\{1, z, \ldots, z^{n'-1}\}$. We picked a random point $R = (x_R, y_R)$ of prime order on the elliptic curve and used Magma on an AMD Opteron Processor 6168 to solve the $(m+1)$-st Semaev polynomial associated with R, that is, to solve $S_{m+1}(x_1, \ldots, x_m, x_R) = 0$. This was done for 20 random curves. We measured the average degrees of regularity, the average time for solving $S_{m+1} = 0$, and the maximum memory requirement for that computation. Selected results are shown in Table 3.2 (for $m = 2$), and in Table 3.3 (for $m = 3$). They agree with the Petit-Quisquater data (given for $n = 11, 17$) and also confirm that the average degrees of regularity are lower than the assumed upper bound $m^2 + 1$. Further, for $10 \leq n \leq 21$ and $m = 3$, we observed $D_{\text{reg}} = 7, 8$ most often, and always $D_{\text{reg}} \leq 9$.

We repeated the computations for 20 random points R on the Koblitz curves $y^2 + xy = x^3 + x^2 + 1$ for various values of n, and $m = 2, 3$. Results are given in Table 3.4.

Table 3.2. Solving $S_3(x_1, x_2, x_R) = 0$. Random curves: $a_6 \in_R \mathbb{F}_{2^n}$. Average degrees of regularity, average running times and maximum memory use.

n	n'	m	$m^2 + 1$	D_{reg}	Time (s)	Max Mem (MB)
11	6	2	5	3.0	0	11
11	5	2	5	2.7	0	11
13	7	2	5	3.0	0	11
13	6	2	5	3.2	0	11
15	8	2	5	3.1	0	11
15	7	2	5	3.2	0	11
17	9	2	5	3.1	0	11
17	8	2	5	2.8	0	11
23	12	2	5	4.0	0	29
23	11	2	5	3.0	0	12
29	15	2	5	4.0	3	97
29	13	2	5	3.0	0	13

Table 3.3. Solving $S_4(x_1, x_2, x_3, x_R) = 0$. Random curves: $a_6 \in_R \mathbb{F}_{2^n}$. Average degrees of regularity, average running times and maximum memory use.

n	n'	m	$m^2 + 1$	D_{reg}	Time (s)	Max Mem (MB)
11	4	3	10	7.0	1	24
11	3	3	10	6.4	0	11
13	4	3	10	7.0	1	23
13	3	3	10	6.0	0	11
15	5	3	10	7.0	15	188
15	3	3	10	6.0	0	11
17	6	3	10	7.2	220	2143
17	3	3	10	6.0	0	11
21	7	3	10	7.0	6910	27235
21	3	3	10	6.0	0	11

4   The Delta Method

So far, we always worked with parameters m and n' such that $mn' \approx n$. In fact, previous work [9,6] used work with $n' = \lceil n/m \rceil$. A closer look at the data in Tables 3.2 and 3.3 however suggests that choosing $n' < n/m$ is favourable: for each value of n, the first row gives the data for $n' = \lceil n/m \rceil$ while the second row gives the data for an optimized $n' < n/m$. We notice lower average degrees of regularity, running times and memory requirements almost throughout. This gives rise to the *Delta method*. Let

$$\Delta := n - mn' > 0.$$

Table 3.4. Solving $S_3(x_1, x_2, x_R) = 0$ and $S_4(x_1, x_2, x_3, x_R) = 0$. Koblitz curves $y^2 + xy = x^3 + x^2 + 1$. Average degrees of regularity, average running times and maximum memory use.

n	n'	m	$m^2 + 1$	D_{reg}	Time (s)	Max Mem (MB)
11	6	2	5	3.0	0	11
11	4	3	10	7.1	1	24
13	7	2	5	3.1	0	11
13	4	3	10	7.0	1	23
15	8	2	5	3.1	0	11
15	5	3	10	7.0	16	189
17	9	2	5	3.0	0	11
17	6	3	10	7.1	211	2139
23	12	2	5	4.0	0	28
29	15	2	5	4.0	2	95

There is a tradeoff to consider when picking the value of n', as a lower n' value will reduce the chances of finding a decomposition of R into m points in the factor base, but will also reduce the factor base and therefore the number of relations needed in Diem's algorithm. The important experimental observation is that lowering n' causes the complexity of the Gröbner basis calculation (using the F4 algorithm) to decrease dramatically.

More precisely, decreasing n' by 1 decreases the chance of a decomposition being found by a factor of 2^m. It also decreases the number of relations needed by a factor of 2, since $2^{n'} + c$ relations are needed to solve the ECDLP (for some small constant c). Thus we require a speedup of 2^{m-1} in Gröbner basis calculation times in order for the Delta method to offer an improvement.

Also note that we expect that for the majority of polynomial systems for which we are trying to calculate a Gröbner basis, the system will have **no** solution. This follows from that an elliptic curve point R can be decomposed into m points of the factor basis if and only if the $m + 1$-st Semaev polynomial associated with R has a solution. Thus the Delta method with $\Delta > 0$ is useful as long as the time required to calculate a Gröbner basis for systems with no solution decreases by a factor of slightly over 2 each time n' is reduced by 1 (and the time required for systems with a solution does not increase by too much). For $n = 26$, we do in fact get this required decrease for $n' > 11$ For $n = 34$, we get the decrease for $n' \geq 14$. Table 4.1 shows some selected average degrees of regularity and Gröbner basis running times (using Magma on an AMD Opteron 6168); to obtain these data, we used the same experiment as in Section 3.1, but separated the data into the cases that the system $S_3(x_1, x_2, x_R) = 0$ had a solution or not. We took averages over 5 calculations in each case. Observe that looking at solvable and unsolvable systems separately, we can see that solvable systems have a degree of regularity between 3 and 5, while unsolvable systems are between 2 and 4.

Table 4.1. Average degree of regularity and Gröbner basis running time, for Semaev polynomial systems with, or without solution

n	m	n'	Degree of Regularity		Gröbner Basis Running Time	
			No Sol	Sol	No Sol	Sol
26	2	13	2.8	4.0	0.39	0.99
26	2	12	2.4	3.0	0.06	0.14
26	2	11	2.4	3.0	0.01	0.03
26	2	10	2.6	4.0	0.01	0.02
34	2	17	2.8	4.0	12.31	55.20
34	2	16	3.2	4.0	1.42	3.14
34	2	15	2.8	3.0	0.21	0.47
34	2	14	2.6	3.0	0.03	0.07
34	2	13	2.0	4.0	0.02	0.04

Obviously and unfortunately, we cannot increase Δ arbitrarily in order to decrease the ECDLP running time. As we continue to increase Δ the Gröbner basis times stop decreasing as rapidly, and the overall running time starts going back up. Our experimental data suggests that for the case $m = 2$, the optimal value of Δ is given by

$$\Delta = \begin{cases} 2\lfloor \frac{n-15}{6} \rfloor & \text{if } n \text{ is even} \\ 2\lfloor \frac{n-15}{6} - 1 \rfloor & \text{if } n \text{ is odd} \end{cases}$$

At present, our best explanation for the remarkable decrease in Gröbner basis calculation time offered by the Delta method is that decreasing n' gives a system with the same number of equations, but in fewer variables. This creates a more over-determined system which can be solved more efficiently by F4. Why the F4 algorithm works so remarkably well needs more investigation. Decreasing n' often results in the polynomial systems having a lower degree of regularity. However, this is not always the case. In fact, there are cases where a decrease in n' results in both faster F4 times and a higher average degree of regularity. Clearly, a more detailed theoretic explanation for the Delta method's success is still needed.

The above optimal Δ values are based on experimental results for $m = 2$ and values of n from 25 to 40. For each n value, we decreased n' from $\lceil \frac{n}{m} \rceil$ until the overall ECDLP running time started going up. For each choice of n and n', we solved Semaev polynomials until we found five systems with a solution. Data for $n = 42$ and $n = 48$ was also generated, and our formula remains valid at these higher n values.

There are also some results available for $m = 3$ and values of n from 10 to 20. Using a value of $m = 3$ is much slower than using $m = 2$, as is expected based on the complexity analysis in [9]. We do however expect that for higher values of n, a choice of $m = 3$ will be preferable. Interestingly, the total ECDLP times for $m = 3$ increase faster than the times for $m = 2$, so our experimental results suggest that using $m = 2$ will always remain the better option. This result is actually to be expected, as the theoretical running time determined by Petit and Quisquater increases at a faster rate for $m = 2$ than it does for $m = 3$ until we reach $n > 200$. The relation gathering stage is in fact faster for $m = 2$ than $m = 3$ until $n > 600$. Note that the theoretical switch to $m = 3$ occurs earlier, around $n = 290$, but that this is due to the complexity of the linear algebra stage, not the relation gathering stage.

5   Experiments with the Hybrid Method

Recall the hybrid method from Section 2.3, in which we fix k variables $y_1, \ldots, y_k \in \mathbb{F}_2$ among the mn' variables and perform a Gröbner basis computation for each of the 2^k possible assignments for $(y_1, \ldots, y_k) \in \mathbb{F}_2^k$ (or until a solution to $S_{m+1} = 0$ has been found). We experimentally determined optimal values of k for various n (as we could not make sensible use of existing Magma code for the hybrid method that allegedly determines such k). We worked with $m = 2$ and $21 \leq n \leq 40$, and used the same set-up for our experiments as before, with the exception that we used a Magma implementation of the hybrid method by Bettale [8] instead of the built-in Magma function GröbnerBasis. For each value of n, we used $n' = \lfloor \frac{n}{m} \rfloor$ (corresponding to $\Delta = 0$ or 1) and incremented k from 0 until the average running time to solve $S_3(x_1, x_2, x_R) = 0$ stopped improving. Optimal values for k are given in Table 5.1.

Table 5.1. Optimal k-values in the hybrid method

n	23	24	25	26	27	28	29	30	31	32	33	34	35	36	37	38	39	40
n'	11	12	12	13	13	14	14	15	15	16	16	17	17	18	18	19	19	20
k	0	1	0	3	1	3	2	3	2	5	3	5	4	5	4	6	5	7

We confirmed some speed-up for optimal k over $k = 0$, but the observed speed-up for the overall ECDLP running time was not always as much as with the Delta method. Specifically, for $n > 28$, an optimal choice of Δ produced better results with the Delta method than the hybrid method with an optimal choice of k.

5.1   Block Hybrid Method versus Standard Hybrid Method

When using the hybrid method, we can choose which of the mn' variables occurring in the final system to fix. Let $X_1, X_2, \ldots, X_m$ be the m variables occurring in the original Semaev polynomial. Let $x_1, x_2, \ldots, x_{mn'}$ be the mn' variables we get after doing the Weil descent, where X_i is a function of the block of variables $x_{(i-1)n'+1}, \ldots, x_{in'}$, for $i = 1, \ldots, m$.

The above results are based on fixing the first k variables, x_1 to x_k. Thus we fix all the variables corresponding to X_i before starting to fix the variables corresponding to X_{i+1}.

Other approaches are certainly possible however. For example, in the block hybrid method, the order in which we fix the variables is to fix the first variables $x_{(i-1)n'+1}$ occurring in each X_i, then to fix the second variables $x_{(i-1)n'+2}$, and so on. So for $n' = 3$, $m = 2$ and $k = 3$ we would fix the variables x_1, x_2, x_4.

We ran the same tests for the block hybrid method as we ran for the standard hybrid method, although only for n from 21 to 28. For optimal k values, the block hybrid method performed worse than the standard hybrid method for every value of n. The optimal block hybrid times were always within a factor of 2 of the optimal standard hybrid times, however. Even for non-optimal k values, the block hybrid method was normally slower. In particular, if we take $k = m$ so that we fix the first variable in each block, then the block hybrid method was slower than the standard hybrid method by a factor of over 10 times.

In conclusion, simply fixing the first k variables is preferable to the block hybrid method.

5.2   Combining the Hybrid and Delta Methods

Since the hybrid and Delta methods both give rise to faster ECDLP algorithms, we asked whether combining the two methods could give an additional speed-up?

We tried combining the two methods for n-values of 27, 36 and 40 (using $m = 2$). We used values of k from 0 to one more than the optimal k-value in the hybrid method, and values of n' from $\lceil \frac{n}{m} \rceil$ to one less than the optimal value for the n' in the Delta method. Our data were generated in the same way as before. For $n = 27$, the best combination of k and Δ was simply to take $k = 0$ and n' optimal as determined in Section 4. For $n = 36$ and 40, the best times were obtained with $k = 1$ and n' one higher than optimal. However, these times were very similar to those for $k = 0$ and n' optimal.

In conclusion, combining the hybrid and Delta methods doesn't seem to offer any significant speedup compared to just using the Delta method.

6   Exploiting the Existence of Subfields

Another choice that we have when implementing Diem's algorithm is how to choose the factor basis. Recall that the factor basis $\mathcal{F}_V$ is the set of points on the elliptic curve whose x-coordinate lies in an n'-dimensional subspace V of the underlying finite field. So we can change the factor basis by changing the basis for V. The standard basis we used to generate our data so far is given by $\{1, z, \ldots, z^{n'-1}\}$, where z is a generator for $\mathbb{F}_{2^n}$. However if $m | n$, then one can choose a basis of the form $\{1, a, \ldots, a^{n'-1}\}$, where a is a generator for $\mathbb{F}_{2^{n/m}}$, so that V is an n'-dimensional subfield of $\mathbb{F}_{2^n}$ and $\mathcal{F}_V$ contains only points $(x, y) \in V \times V$. We do not expect that using this alternate basis will affect the probability that a random elliptic curve point can be written as a sum of m points in $\mathcal{F}_V$. Hence we will still need to do the same *number* of Gröbner basis calculations.

In this section we report on experiments for the case that $m = 2$ and n is even, so $n' = n/2$. Using a subfield-based basis for V, we performed the same experiments outlined in Section 4, but repeated our experiment until we had data for 50 solvable systems and 50 unsolvable systems. The reason that we used a greater number of systems was that the calculations required less time than they did for the Delta method. For even n values between 26 and 40, we set $m = 2$, $n' = n/2$ and calculated the average time required to do a Gröbner basis calculation. From theses calculations we estimated the expected time required to solve an ECDLP instance.

Gröbner basis calculations went much faster both for both systems with a solution and systems without a solution. For $n = 40$, the calculation took 0.02 seconds for systems with no solution and 0.08 seconds for systems with a solution. In comparison, similar calculations using the standard basis take 374.77 seconds and 388.09 seconds, respectively. Significant speedups were also observed for all lower n values.

We noted that using a subfield-based basis produced systems of polynomials with a much lower degree of regularity. For each n-value tested, every single solvable systems had a degree of regularity of 3. This is lower than the average degree of regularity observed when using the standard basis. Similar results hold for systems with no solution, where the average degree of regularity ranged between 2.1 and 2.3, depending on the value of n.

See Table 6.1 for a complete comparison of average degrees of regularity. Here the data for the standard basis are taken from our Delta method results.

Table 6.1. Average degrees of regularity for subfield-based basis and standard basis

			Average Degree of Regularity			
			Standard Basis		Subfield Basis	
n	m	n'	No Sol	Sol	No Sol	Sol
26	2	13	2.5	4.0	2.2	3.0
28	2	14	2.6	4.0	2.1	3.0
30	2	15	4.0	4.0	2.2	3.0
32	2	16	2.7	4.0	2.3	3.0
34	2	17	2.6	4.0	2.1	3.0
36	2	18	4.0	4.0	2.3	3.0
38	2	19	3.0	4.0	2.3	3.0
40	2	20	4.0	4.0	2.2	3.0

At present we do not have a good explanation for why the degree of regularity is so much lower. The decrease is not the result of any cancellation during the Weil descent, as the systems of polynomials produced have the same number of terms and same total degrees no matter which basis we chose. Nonetheless, the F4 algorithm is much better at finding low degree combinations of the polynomials when a subfield is used. It is normally able to determine if a system is solvable or not after only one iteration.

7   Bonus Track: Using the Magma "PairsLimit" Parameter

When using values of $m > 2$, the default Magma implementation of the F4 algorithm tends to start using massive amounts of memory if n is increased as high as even 19 or 20. Looking at the detailed output from the Magma implementation reveals a way to decrease the memory usage.

Let I be the ideal for which we are trying to find a Gröbner basis. The F4 algorithm goes through a sequence of steps in which it takes linear combinations of polynomials in the current basis for I, and adds some of them to the basis. Magma's default behaviour is to add all new polynomials having minimal degree among the new polynomials. In large systems, this can result in thousands of polynomials being added in a single step. However, it may turn out that only a fraction of these systems needed to be added to the basis in order to find a Gröbner basis. Thus by limiting the number of new polynomials that are added each steps, we can make sure that Magma's memory usage never increases by too much at once.

The Magma function "Gröbner Basis" takes an optional parameter "PairsLimit" that can be used to include at most k new pairs at each step. Experimental results show that setting an appropriate value for this parameter can significantly improve the running time. See Table 7.1 for running time data based on $n = 19$, $n' = 6$, $m = 3$ which uses a single system for each value of "PairsLimit".

Table 7.1. Effect of PairsLimit

n	m	n'	PairsLimit	Running Time (s)	Memory (MB)
19	3	6	500	305.5	2091
19	3	6	1000	238.8	2483
19	3	6	1750	197.1	2468
19	3	6	2500	234.4	3009

Unfortunately, there is no known formula for determining optimum values to use for "PairsLimit". Using too high of a value can result in high memory usage and can also cause the final F4 step to take a very long time. However, using too low of a value can result in needing a large number of steps before a Gröbner basis is found.

Some experimentally determined values that give low running times have been chosen as defaults when working with $m > 2$. The benefits described above are also present when using $m = 2$, although not to as large of an extent. Note that using the "PairsLimit" parameter is primarily helpful when using n' values that require working with systems that take lots of time and memory to solve. As such, using this parameter likely can't help improve times for the Delta method, since the systems that result from optimal Δ values can be solved very quickly and with very little memory.

8 Conclusion

Further research, both experimentally and theoretically is needed to determine the true complexity of the ECDLP index calculus method based on Semaev polynomials, Weil descent and Gröbner basis methods!

Acknowledgments. We used the Computer Algebra System Magma in our work. We would like to thank Koray Karabina and Brandon Weir for their helpful advice and comments. We are grateful to Christophe Petit for commenting on an earlier version of this paper. We would also like to thank NSERC of Canada for providing partial funding for this research.

References

1. Bettale, L., Faugère, J.-C., Perret, L.: Hybrid approach for solving multivariate systems over finite fields. J. Math. Crypt. 3, 177–197 (2009)
2. Collins, G.: The calculation of multivariate polynomial resultants. Journal of the Association of Computing Machinery 18, 515–522 (1971)
3. Diem, C.: On the discrete logarithm problem in elliptic curves. Compositio Mathematica 147(1), 75–104 (2011)
4. Faugère, J.-C.: A new efficient algorithm for computing Gröbner bases (F4). Journal of Pure and Applied Algebra 139, 61–88 (1999)
5. Faugère, J.-C.: A new efficient algorithm for computing Gröbner bases without reduction to zero (F5). In: Proceedings of the 2002 International Symposium on Symbolic and Algebraic Computation (ISSAC), pp. 75–83. ACM Press (2002)
6. Faugère, J.-C., Perret, L., Petit, C., Renault, G.: Improving the complexity of index calculus algorithms in elliptic curves over binary fields. In: Pointcheval, D., Johansson, T. (eds.) EUROCRYPT 2012. LNCS, vol. 7237, pp. 27–44. Springer, Heidelberg (2012)
7. Gaudry, P.: Index calculus for abelian varieties of small dimension and the elliptic curve discrete logarithm problem. J. Symbolic Computation 44(12), 1690–1702 (2009)
8. Bettale, L.: Hybrid approach for solving multivariate polynomial systems over finite fields, http://www-polsys.lip6.fr/~bettale/hybrid
9. Petit, C., Quisquater, J.-J.: On polynomial systems arising from a Weil descent. In: Wang, X., Sako, K. (eds.) ASIACRYPT 2012. LNCS, vol. 7658, pp. 451–466. Springer, Heidelberg (2012)
10. Semaev, I.: Summation polynomials and the discrete logarithm problem on elliptic curves, Cryptology ePrint Archive Report 2004/031 (2004), http://eprint.iacr.org/2004/031/

An Experiment of Number Field Sieve
for Discrete Logarithm Problem over $\mathrm{GF}(p^{12})$

Kenichiro Hayasaka[1], Kazumaro Aoki[2],
Tetsutaro Kobayashi[2], and Tsuyoshi Takagi[3]

[1] Graduate School of Mathematics, Kyushu University, 744, Motooka,
Nishi-ku, Fukuoka 819-0395, Japan
[2] NTT Secure Platform Laboratories, 3-9-11 Midori-cho,
Musashino-shi, Tokyo, 180-8585, Japan
[3] Institute of Mathematics for Industry, Kyushu University, 744, Motooka,
Nishi-ku, Fukuoka, 819-0395, Japan

Abstract. The security of pairing-based cryptography is based on the
hardness of the discrete logarithm problem (DLP) over finite field $\mathrm{GF}(p^n)$.
For example, the security of the optimal Ate pairing using BN curves,
which is one of the most efficient algorithms for computing paring, is
based on the hardness of DLP over $\mathrm{GF}(p^{12})$. Joux et al. proposed the
number field sieve over $\mathrm{GF}(p^n)$ as an extension of the number field sieve
that can efficiently solve the DLP over prime field $\mathrm{GF}(p)$. Two implementations of the number field sieve over $\mathrm{GF}(p^3)$ and $\mathrm{GF}(p^6)$ have been
proposed, but there is no report on that over $\mathrm{GF}(p^{12})$ of extension degree 12. In the sieving step of the number field sieve over $\mathrm{GF}(p)$ we
perform the sieving of two dimensions, but we have to deal with more
than two dimensions in the case of number field sieves over $\mathrm{GF}(p^{12})$. In
this paper we construct a lattice sieve of more than two dimensions, and
discuss its parameter sizes such as the dimension of sieving and the size
of sieving region from some experiments of the multi-dimensional sieving.
Using the parameters suitable for efficient implementation of the number
field sieve, we have solved the DLP over $\mathrm{GF}(p^{12})$ of 203 bits in about 43
hours using a PC of 16 CPU cores.

Keywords: pairing, discrete logarithm problem, number field sieve, extension field, lattice sieve.

1 Introduction

Pairing-based cryptography has been attracted due to the novel cryptographic
protocols such as ID-based encryption, functional encryption, etc. Many efficient
implementations of pairing have been reported, and one of the most efficient algorithms for computing pairing is the optimal Ate pairing [21] using BN curves [4].
The security of pairing-based cryptography using BN curves is based on the
hardness of the discrete logarithm problem (DLP) over finite field $\mathrm{GF}(p^{12})$.

The asymptotically fastest algorithm for solving the DLP over prime field
$\mathrm{GF}(p)$ is the number field sieve [19,7]. At CRYPTO 2006, Joux et al. extended

M. Fischlin and S. Katzenbeisser (Eds.): Buchmann Festschrift, LNCS 8260, pp. 108–120, 2013.

the number field sieve to the case of extension field $GF(p^n)$ of degree n and characteristic p [8]. The complexity of solving the DLP over finite field $GF(p^{12})$ of 3072 bits by the number field sieve is estimated approximately as 2^{128} [4]. There are two experimental reports on the implementation of the number field sieve over extension degree $GF(p^n)$ of $n = 3$ [8] and $n = 6$ [22,23]. However, to the best of our knowledge, there is no experimental report on the hardness of the DLP over finite field $GF(p^{12})$ by the number field sieve. In order to correctly estimate the security of the pairing-based cryptography we need some experimental evaluations of number field sieve over finite field $GF(p^{12})$.

The number field sieve over extension field $GF(p^n)$ has a substantially different sieving step from that over prime field $GF(p)$. There are two sieving algorithms, called the line sieve and lattice sieve [17]. The large-scale implementation of the number field sieve over prime field $GF(p)$ deploys the lattice sieve of two dimensions, but we have to construct the lattice sieve of more than two dimensions for the number field sieve over extension field $GF(p^{12})$. The currently known reports on the multi-dimensional sieving have discussed only the case of three dimensions [22,23].

In this paper, we propose the lattice sieving of more than 2 dimensions for the number field sieve over extension fields $GF(p)$ by naturally extending the lattice sieve of two dimensions. We implemented the proposed multi-dimensional lattice sieve over extension field $GF(p^{12})$ of 203 bits, and we show some experimental data for accelerating the number field sieve by choosing the suitable dimensions and sizes of the sieving region. Consequently we have solved the DLP over extension field $GF(p^{12})$ of 203 bits by the number field sieve using a PC of 16 CPU cores in about 43 hours.

2   Number Field Sieve over $GF(p^n)$ [8]

In this section we give an overview of the number field sieve over extension field $GF(p^n)$ proposed by Joux et al. at CRYPTO 2006 [8]. The number field sieve consists of three steps: polynomial selection, searching relations, and linear algebra. We explain each step of the number field sieve in the following.

2.1   Discrete Logarithm Problem over $GF(p^n)$

We denote by $GF(p^n)^*$ the multiplicative group of finite field of cardinality p^n, where p is a prime number and n is an extension degree. Let γ be a generator of $GF(p^n)^*$. The discrete logarithm problem (DLP) over finite field $GF(p^n)$ tries to find the non-negative smallest integer x that satisfies $\gamma^x = \delta$ for a given δ in $GF(p^n)^*$. This discrete logarithm x is written as $\log_\gamma \delta$ in this paper.

2.2   Polynomial Selection

In the polynomial selection of the number field sieve proposed by Joux et al., we generate two polynomials $f_1, f_2 \in \mathbb{Z}[X] \setminus \{0\}$ that satisfy the following conditions.

1. $f_1 \neq f_2$,
2. $\deg f_1 = n$,
3. f_1 is irreducible in $\mathrm{GF}(p)$,
4. $f_1 \mid f_2 \bmod p$.

From the conditions there exists $v \in \mathrm{GF}(p^n)$ such that $f_1(v) = f_2(v) = 0$ in $\mathrm{GF}(p^n)$. Let α_1 and $\alpha_2 \in \mathbb{C}$ be the root of $f_1(X) = 0$ and $f_2(X) = 0$, respectively. Let $\mathcal{O}_1$ and $\mathcal{O}_2$ be the ring of integers of the number field $\mathbb{Q}(\alpha_1)$ and $\mathbb{Q}(\alpha_2)$, respectively. There are homormophism maps

$$
\begin{aligned}
\phi_1 : \mathbb{Z}[\alpha_1] &\to \mathrm{GF}(p^n), \ \alpha_1 \mapsto v, \\
\phi_2 : \mathbb{Z}[\alpha_2] &\to \mathrm{GF}(p^n), \ \alpha_2 \mapsto v.
\end{aligned}
\tag{1}
$$

2.3 Searching Relations

In the step of searching relations, we try to find many relations of certain polynomials of degree $t \geq 1$. Let $B_1, B_2 \in \mathbb{R}_{>0}$ be the smoothness bound associated with polynomials f_1, f_2 in Section 2.2. We define the factor base $\mathcal{B}_1, \mathcal{B}_2$ as follows.

$$
\mathcal{B}_i = \{(q, g) \mid q : \text{prime}, q \leq B_i, \text{ and irreducible monic } g \in \mathbb{Z}[X] : g \mid f_i \bmod q\}
$$

In this paper we represent polynomial $h_a(X) = a_0 + a_1 X + \cdots + a_t X^t \in \mathbb{Z}[X]$ as a vector $a = (a_0, a_1, \ldots, a_t)^{\mathrm{T}} \in \mathbb{Z}^{t+1}$. For a given $H = (H_0, H_1, \ldots, H_t) \in \mathbb{R}_{>0}^{t+1}$, we define the $(t + 1)$-dimensional region $\mathcal{H}_a(H)$ as

$$
\mathcal{H}_a(H) = \{(a_0, a_1, \ldots, a_t)^{\mathrm{T}} \in \mathbb{Z}^{t+1} \mid |a_i| \leq H_i \ (0 \leq i \leq t), \ a_t \geq 0\}.
$$

Here H and $\mathcal{H}_a$ are called as the sieving interval and the sieving region, respectively. Next, the norm of $h_a(\alpha_i)$ is defined by $N(h_a(\alpha_i)) = |\mathrm{Res}(h_a, f_i)|$, where $\mathrm{Res}(h_a, f_i)$ is the resultant of $h_a(X)$ and $f_i(X)$ for $i = 1, 2$. In the step of searching relations, for given sieving interval H and smoothness bound B_1, B_2, we try to find $a \in \mathbb{Z}^{t+1}$ (called the hit tuple) that satisfies the following conditions.

1. $N(h_a(\alpha_1))$ is B_1-smooth,
2. $N(h_a(\alpha_2))$ is B_2-smooth,
3. $\gcd(a_0, a_1, ..., a_t) = 1$,

where B-smooth is the integer whose prime factors are at most B. Denote by S the set of all hit tuples gathered in searching relations. In order to solve the correct discrete logarithm, the size of S is chosen as

$$
\sharp S \geq \sharp \mathcal{B}_1 + \sharp \mathcal{B}_2 + 2n.
\tag{2}
$$

Next a hit tuple $a \in S$ has relationship $(h_a(\alpha_i))\mathcal{O}_i = \prod_{\mathfrak{q} \in \mathcal{B}_i} \mathfrak{q}^{\varepsilon_\mathfrak{q}}$ for $i = 1, 2$. We can compute $\varepsilon_\mathfrak{q}$ from the prime decomposition of norm $N(h_a(\alpha_i)) = \prod_{q:\text{prime}, q \leq B_i} q^{e_q}$ for $q \nmid [\mathcal{O}_i : \mathbb{Z}[\alpha_i]]$. Let r_i be the torsion-free rank of $\mathcal{O}_i$ for

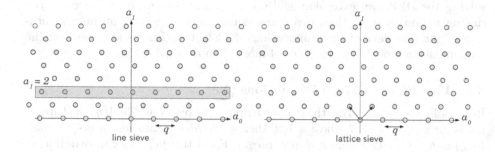

Fig. 1. Line sieve and lattice sieve in two dimensions

$i = 1, 2$. From $\phi_1(h_a(\alpha_1)) = \phi_2(h_a(\alpha_2))$ using homomorphism map (1), we obtain the following relation of the discrete logarithm

$$\sum_{q \in \mathcal{B}_1} \varepsilon_q \log \phi_1(q) + \sum_{j=1}^{r_1} \lambda_j(h_a(\alpha_1)) \log \Lambda_{1,j} \equiv$$

$$\sum_{q \in \mathcal{B}_2} \varepsilon_q \log \phi_2(q) + \sum_{j=1}^{r_2} \lambda_j(h_a(\alpha_2)) \log \Lambda_{2,j} \pmod{p^n - 1},$$

where $\log \phi_i(q)$ and $\log \Lambda_{i,j}$ are called the virtual logarithms [7,20] and $\lambda_j(h_a(\alpha_i))$ is the character map proposed by Schirokauer [19]. Consequently, we can compute $\log \phi_i(q), \log \Lambda_{i,j} \pmod{p^n - 1}$ by solving the linear algorithm obtained from the relations.

3   Searching Relations by Sieving in Multi Dimensions

In this section we discuss how to search the hit tuple in the sieving region of $t + 1$ dimensions from Section 2.3.

Sieving methods try to find elements $(a_0, a_1, \ldots, a_t)$ in the sieving region $\mathcal{H}_a$ whose norm is divisible by prime number q smaller than the smoothness bound. There are two different sieving methods, called the line sieve and lattice sieve [17,6]. The line sieve searches elements $(a_0, a_1, \ldots, a_t)$ by repeatedly adding a_0 to q for fixed $(a_1, \ldots, a_t)$. The lattice sieve generates a lattice of elements whose norm is divisible by q, and then finds elements $a \in \mathcal{H}_a$ whose norm is divisible by prime r in the lattice.

The number field sieve for solving the DLP over prime field GF(p) [19,7] or for factoring integers [13] utilizes the lattice sieve of two dimensions, namely $t = 1$. Fig. 1 shows a figure of the line sieve and lattice sieve in two dimensions. Currently the lattice sieve is often used for solving the DLP or factorization problem by the number field sieve of the size of more than 500 bits.

On the other hands, the lattice sieve of two dimensions can not efficiently accumulate sufficient number of smooth elements for the number field sieve for solving the DLP over extension field $GF(p^{12})$, and thus we have to extend the sieving region to more than two dimensions. Zajac presented an implementation of the line sieve of three dimensions [22,23], but there is no report on the implementation of the lattice sieve of more than dimension two.

3.1 Line Sieve in Multi Dimensions [23]

In the following we describe the line sieve presented by Zajac [23]. If $q \mid N(h_a(\alpha_i))$ holds for a prime $q < B_i$ and $i = 1, 2$, then $q \mid N(h(\alpha_i))$ satisfies for polynomials $h(X) = h_a(X) + kq$ where k is any integer. From this fact, we can search a hit tuple a divisible by q in the sieving region without performing the division of integers. Similarly, for $\mathfrak{q} = (q, g) \in B_i$ $(i = 1, 2)$, we have relationship

$$g \mid h_a \bmod q \implies q^{\deg g} \mid N(h_a(\alpha_i)). \tag{3}$$

Then we can find a candidate of hit tuple $a \in \mathcal{H}_a$ whose norm $N(h_a(\alpha_i))$ is B_i-smooth by repeatedly adding $L[a]$ to $\deg g \log q$ for the elements which satisfy the left-hand side of equation (3) for $\forall \mathfrak{q} \in B_i$ $(i = 1, 2)$. The candidate a is confirmed by checking that $N(h_a(\alpha_i))$ is B_i-smooth using the trial division (or other advanced methods), and then we obtain a hit tuple $a \in \mathcal{H}_a$.

Let I_d be an identity matrix of size $d \times d$. The set of all polynomials in $\mathbb{Z}[X]$ of degree less than $t + 1$ that satisfy the left-hand size of equation (3) is generated by the integer linear combination of the columns of the following matrix:

$$\left(\begin{array}{c|ccc} & g_0 & & 0 \\ qI_{\deg g} & g_1 & \ddots & \\ & \vdots & \ddots & g_0 \\ \hline & g_{\deg g} & & g_1 \\ 0 & & \ddots & \vdots \\ & 0 & & g_{\deg g} \end{array} \right), \tag{4}$$

where $g_0, g_1, \ldots, g_{\deg g}$ is the coefficient of the polynomial $g = \sum_{j=0}^{\deg g} g_j X^j$, respectively.

From $g_{\deg g} = 1$, we can convert the $(\deg g + 1)$-th column to $(t+1)$-th column of this matrix (4) by the integer linear combination of columns as follows:

$$M_{\mathfrak{q}} = \left(\begin{array}{c|c} qI_{\deg g} & T_{\mathfrak{q}} \\ \hline 0 & I_{t - \deg g + 1} \end{array} \right), \tag{5}$$

where $T_{\mathfrak{q}}$ is a $g \times g$ integer matrix. Conversely, for any $c = (c_0, c_1, \ldots, c_t)^{\mathrm{T}} \in \mathbb{Z}^{t+1}$ the relation $a = M_{\mathfrak{q}} c$ satisfies the left-hand side of equation (3). Therefore, for $\deg g \times (t - \deg g + 1)$ matrix $T_{\mathfrak{q}}$ and $c \in \mathbb{Z}^{t+1}$, we can represent $M_{\mathfrak{q}}$ as follows.

$$
\begin{pmatrix} a_0 \\ a_1 \\ \vdots \\ a_{\deg g - 1} \end{pmatrix} = q \begin{pmatrix} c_0 \\ c_1 \\ \vdots \\ c_{\deg g - 1} \end{pmatrix} + T_q \begin{pmatrix} a_{\deg g} \\ a_{\deg g + 1} \\ \vdots \\ a_t \end{pmatrix}. \tag{6}
$$

Here set $(u_0, \ldots, u_{\deg g - 1}) = T_q (a_{\deg g}, \ldots, a_t)^{\mathrm{T}}$ for the input $a_{\deg g}, \ldots, a_t$. Then we can search a that satisfies the left-hand size of equation (3) by repeatedly adding $u_0, \ldots, u_{\deg g - 1}$ to q in sieving region $\mathcal{H}_a$ for $(u_0, \ldots, u_{\deg g - 1}, a_{\deg g}, \ldots, a_t)$.

4   Proposed Lattice Sieve in Multi Dimensions

In this section we propose a lattice sieve in the sieving region of multi dimensions by extending the lattice sieve of two dimensions used for the number field sieve over prime field $\mathrm{GF}(p)$ [17,2].

The lattice sieve tries to find a candidate of hit tuples in the lattice whose elements are divisible by $q \in \mathcal{B}_i$ (called the special-Q). For $q = (q, g) \in \mathcal{B}_i$, let M_q be the matrix of equation (5), and let M_q^{LLL} be the matrix generated by LLL reduction algorithm [14] from M_q.

In this paper we call the search space of $t + 1$ dimensions for hit tuple $a \in \mathcal{H}_a$ as the a-space. On the other hand, the $(t + 1)$-dimensional lattice M_q^{LLL}, which is generated by $M_q^{\mathrm{LLL}} c$ for $c \in \mathbb{Z}^{t+1}$, is called as the c-space. Moreover, for the sieving interval $H_c \in \mathbb{R}_{>0}$, we define the sieving region over the c-space by

$$
\mathcal{H}_c(H_c) = \{(c_0, c_1, \ldots, c_t)^{\mathrm{T}} \in \mathbb{Z}^{t+1} \mid |c_i| \le H_c \ (0 \le i \le t), c_t \ge 0\}.
$$

The lattice sieve for the special-Q searches the candidates of the hit tuple in the sieving region $\mathcal{H}_c$ in the c-space.

Next we construct the matrix M_r from the element $r = (r, h) \in \mathcal{B}_i$ that is different from q in the factor base. By the same method for generating M_q from q, we can obtain equation (6) corresponding to M_r, and by reducing vector $r(c_0, \ldots, c_{\deg h - 1})^{\mathrm{T}}$ modulo r the next equation yields

$$
\begin{pmatrix} a_0 \\ a_1 \\ \vdots \\ a_{\deg h - 1} \end{pmatrix} \equiv T_r \begin{pmatrix} a_{\deg h} \\ a_{\deg h + 1} \\ \vdots \\ a_t \end{pmatrix} \pmod{r}. \tag{7}
$$

Here we decompose the $(t + 1) \times (t + 1)$ matrix M_q^{LLL} into the $\deg h \times (t + 1)$ matrix $M_{q,1}^{\mathrm{LLL}}$ and the $(t - \deg h + 1) \times (t + 1)$ matrix $M_{q,2}^{\mathrm{LLL}}$ as follows.

$$
M_q^{\mathrm{LLL}} = \begin{pmatrix} M_{q,1}^{\mathrm{LLL}} \\ M_{q,2}^{\mathrm{LLL}} \end{pmatrix}. \tag{8}
$$

Algorithm 1. Proposed Lattice Sieve in Multi Dimensions

Require: special-Q where $\mathcal{Q} \subset \mathcal{B}_j$, factor base $\mathcal{B}_1, \mathcal{B}_2$, sieving region $\mathcal{H}_c(H_c)$.
Ensure: S is the set of candidates of hit tuples.

1: $k \leftarrow (j \bmod 2) + 1$.
2: **for all** $\mathfrak{q} = (q, g) \in \mathcal{Q}$ **do**
3:　　Compute matrix $M_\mathfrak{q}^{\mathrm{LLL}}$ from $\mathfrak{q}$ using LLL.
4:　　$L[c] \leftarrow 0$ for $\forall c \in \mathcal{H}_c$.
5:　　$D[c] \leftarrow \log N(h_a(\alpha_j))$ where $a = M_\mathfrak{q}^{\mathrm{LLL}}c$ for $\forall c \in \mathcal{H}_c$.
6:　　**for all** $\mathfrak{r} = (r, h) \in \mathcal{B}_j$ s.t. $r < q$ **do**
7:　　　　Compute lattice $M_{\mathfrak{q},\mathfrak{r}}$ of $\mathfrak{r}$ over the c-space from $\mathfrak{q}$.
8:　　　　$L[c] \leftarrow L[c] + \deg h \log r$　s.t. $c \in \mathcal{H}_c, c \in M_{\mathfrak{q},\mathfrak{r}}$.
9:　　**end for**
10:　　**for all** $c \in \mathcal{H}_c$ **do**
11:　　　　**if** $L[c] + \deg g \log q - D[c]$ is small **then**
12:　　　　　$L[c] \leftarrow 0$
13:　　　　**else**
14:　　　　　$L[c] \leftarrow -\infty$
15:　　　　**end if**
16:　　**end for**
17:　　$D[c] \leftarrow \log N(h_a(\alpha_k))$ where $a = M_\mathfrak{q}^{\mathrm{LLL}}c$ for $\forall c \in \mathcal{H}_c$.
18:　　**for all** $\mathfrak{r} = (r, h) \in \mathcal{B}_k$ **do**
19:　　　　Compute lattice $M_{\mathfrak{q},\mathfrak{r}}$ of $\mathfrak{r}$ over the c-space from $\mathfrak{q}$.
20:　　　　$L[c] \leftarrow L[c] + \deg h \log r$　s.t. $c \in \mathcal{H}_c, c \in M_{\mathfrak{q},\mathfrak{r}}$.
21:　　**end for**
22:　　**for all** $c \in \mathcal{H}_c$ s.t. $L[c] - D[c]$ is small **do**
23:　　　　$S \leftarrow S \cup M_\mathfrak{q}^{\mathrm{LLL}}c$
24:　　**end for**
25: **end for**
26: **return** S

The set of all elements a divisible by $\mathfrak{q}$ is represented by $a = M_\mathfrak{q}^{\mathrm{LLL}}c$ for $c \in \mathbb{Z}^{t+1}$, namely

$$\begin{pmatrix} a_0 \\ a_1 \\ \vdots \\ a_{\deg h - 1} \end{pmatrix} = M_{\mathfrak{q},1}^{\mathrm{LLL}}\, c, \qquad \begin{pmatrix} a_{\deg h} \\ a_{\deg h + 1} \\ \vdots \\ a_t \end{pmatrix} = M_{\mathfrak{q},2}^{\mathrm{LLL}}\, c. \tag{9}$$

Therefore, from equations (7) and (9) we obtain

$$(M_{\mathfrak{q},1}^{\mathrm{LLL}} - T_\mathfrak{r}\, M_{\mathfrak{q},2}^{\mathrm{LLL}})\, c \equiv 0 \pmod{r}. \tag{10}$$

Next let $M_{\mathfrak{q},\mathfrak{r}}$ be the lattice generated by c from equation (10), namely $M_{\mathfrak{q},\mathfrak{r}}$ is the kernel of linear map $(M_{\mathfrak{q},1}^{\mathrm{LLL}} - T_\mathfrak{r}\, M_{\mathfrak{q},2}^{\mathrm{LLL}})$. Note that $a = M_\mathfrak{q}^{\mathrm{LLL}} M_{\mathfrak{q},\mathfrak{r}}\, e$ for any $e = (e_0, e_1, \ldots, e_t) \in \mathbb{Z}^{t+1}$ satisfies the left-hand size of equation (3) for both $\mathfrak{q}$ and $\mathfrak{r}$. We can compute $M_{\mathfrak{q},\mathfrak{r}}$ from the matrix $M_{\mathfrak{q},1}^{\mathrm{LLL}} - T_\mathfrak{r}\, M_{\mathfrak{q},2}^{\mathrm{LLL}}$ corresponding to equation (10).

From the above observations, we present the proposed lattice sieve for the special-Q as Algorithm 1.

5   How to Select Parameters t, H, B_1, B_2

In this section we explain how to select the parameters of the lattice sieve in Section 2.3 for given two polynomials f_1, f_2 in the polynomial selection in Section 2.2. In particular, we discuss the suitable size of dimension $t + 1$, sieving interval H, and the smoothness bound B_1, B_2 that satisfy the equation (2) for the number field sieve over extension degree $\mathrm{GF}(p^{12})$. If we select the parameters that accelerate both the searching relation step and the linear algebra step simultaneously, then the total running time of the number field sieve becomes faster.

5.1   Selection of t

Denote by V_H the size of sieving region $\mathcal{H}_a(H)$, namely $V_H = 2^t \prod_{j=0}^{t} H_j$. We extend the estimation of the average norm in the two-dimensional lattice sieve [15,1] to our multi-dimensional case. The average norm $N_{\mathrm{ave}}(h_a(\alpha_i))$ of polynomial f_i ($i = 1, 2$) in the lattice sieve of $t + 1$ dimensions is evaluated by the following equation.

$$N_{\mathrm{ave}}(h_a(\alpha_i)) = \frac{\sqrt{\int_0^{H_t} \int_{-H_{t-1}}^{H_{t-1}} \cdots \int_{-H_0}^{H_0} (\mathrm{Res}(h_a, f_i))^2 \, da_0 \ldots da_t}}{V_H}$$

Moreover, we approximate the probability $\rho(x, y)$ that the integers smaller than x are y-smooth as $(\log_y x)^{-\log_y x}$, and we assume that the total size of factor bases B_1, B_2 is $V_B = \pi(B_1) + \pi(B_2)$ where $\pi(B_i)$ is the number of primes smaller than or equal to B_i ($i = 1, 2$). Let R be the number of relations in the sieving region $\mathcal{H}_a(H)$, and then R is calculated by

$$R = \rho(N_{\mathrm{ave}}(h_a(\alpha_1)), B_1)\rho(N_{\mathrm{ave}}(h_a(\alpha_2)), B_2)V_H. \tag{11}$$

Here we have to find the parameters that satisfy (2), i.e., $R \geq V_B + 2n$. Fig. 2 shows the minimal V_B that satisfies $R > V_B$ (we neglect $2n$ due to $V_B \gg n$) for V_H in the lattice sieve of $t + 1$ dimensions in the extension field $\mathrm{GF}(p^{12})$ of 203, 514 and 3075 bits, respectively. In order to reduce the time of searching such bound V_B we set $H_0 = H_1 - \cdots - H_t$ and $B_1 = B_2$. From Fig. 2, we can select smaller sizes of sieving region V_H and factor base V_B that satisfy equation (2) using 8 dimension (6 or 4 dimensions) for extension field $\mathrm{GF}(p^{12})$ of 203 bits (514 or 3075 bits), respectively.

5.2   Selection of H and B_1, B_2

In the following we discuss the choice of the sieving interval H and the smoothness bound B_1, B_2 so that the running time of the number field sieve becomes faster.

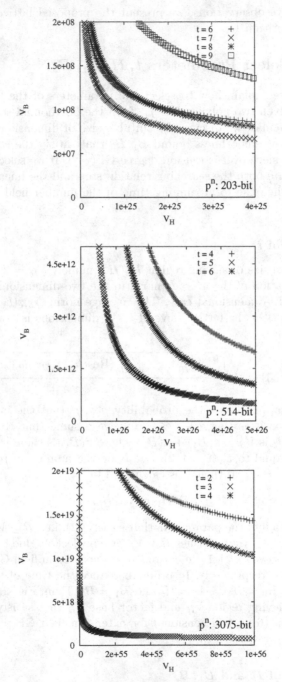

Fig. 2. V_H and V_B are the size of sieving region and the factor base of multi-dimensional lattice sieve for the number field sieve over extension field $GF(p^{12})$ of 203 bits (top), 514 bits (middle) and 3075 bits (bottom), respectively

Table 1. Comparison of known experiments of the number field sieve over extension field GF(p^n)

Finite Field	GF(p^3)	GF(p^6)	GF(p^{12})
Authors	Joux et al. [8]	Zajac [22]	Ours
Year	2006	2008	2012
CPU	Alpha (1.15GHz) × 8	Sempron (2.01GHz) × 8	Xeon (2.93GHz) × 4
Days	19 days	5 days	2 days
Bit Length	394	242	203
Sieving	2-dim. lattice sieve	3-dim. line sieve	**7-dim. lattice sieve**

For the fixed size of sieving interval V_H and factor base V_B, we first select the sieving interval H and then the smoothness bound B_1, B_2. The sieving interval H is chosen that the probability of $\rho(N_{\text{ave}}(h_a(\alpha_1)), B_1)\rho(N_{\text{ave}}(h_a(\alpha_2)), B_2)$ arisen from the hit tuple of equation (11) is maximum for fixed B_1, B_2 with $B_1 = B_2$. For the above H, we then select the smoothness bound B_1, B_2 so that the number of relations in equation (11) become maximum.

6   Our Experiment on Number Field Sieve over GF(p^{12})

In this section, we report our experiment on solving the discrete logarithm problem over extension field GF(p^{12}) of 203 bits using the number field sieve in Section 2. We chose the characteristic $p = 122663$ of 17 bits, namely the cardinality of the extension field GF(p^{12}) is

$$p^{12} = 116028047901493489912893641612452600729095851402664913077940811.$$

The computational environment in our experiment is as follows. We used one PC equipped with four CPUs (Intel Xeon X7350 2.93 GHz; Core2 micro architecture; 16 cores in total) and 64 GBytes of RAM. We utilize gmp-5.0.5 for the arithmetic of multi-precision integers, openmpi-1.6 for parallel implementation between processes, pari-2.5.1 [16] for the decomposition of ideals in the number field, and ntl-5.5.2 for the computation of lattice reduction using LLL. We use C++ with compiler gcc-4.7.1 on Linux OS (64 bits).

Table 1 presents the experimental data in our implementation and the previous ones of the number field sieve over extension field GF(p^n).

6.1   Polynomial Selection

In order to select two polynomials f_1, f_2 in Section 2.2, we use the polynomial selection similar to the previous experiments [8] and [22]. At first an irreducible polynomial $f_1 \in \mathbb{Z}[X]$ of degree 12 with small coefficients is chosen, and then we set $f_2 = f_1 + p$ or $f_2 = f_1 - p$.

In this paper, Murphy's α function [15,3] is used for selecting a more suitable pair of polynomials f_1, f_2. If the Murphy's α function of polynomial f_i ($i = 1, 2$)

is smaller, then the norm of hit pair $N(h_a(\alpha_i))$ $(i = 1, 2)$ is expected to become smoother, namely it is divisible by small prime divisors with higher probability. The coefficient of polynomial f_1 is searched in the range of ± 10, and then the sum of the Murphy's α of the following polynomials f_1, f_2 is smallest among the range of our search.

$$f_1(X) = X^{12} - 3X^4 + 9X^3 - 9X^2 - 9X + 2,$$
$$f_2(X) = X^{12} - 3X^4 + 9X^3 - 9X^2 - 9X - 122661.$$

6.2   Searching Relations

We used the method in Section 5 to select the dimension $t + 1$, sieving interval H and smoothness bound B_1, B_2 for our implementation of the lattice sieve.

In the estimation of Section 5.1, the suitable dimension of the lattice sieve for extension field $\text{GF}(p^{12})$ of 203 bits was estimated as eight. We perform some experiments of the lattice sieve of 6, 7 and 8 dimensions for random special-Q with fixed V_H and V_B. From this experiments, the lattice sieve of 7 dimensions yields the largest number of relations for one special-Q, and then we select $H = (443, 427, 304, 140, 70, 24, 9)$ and smoothness bounds $B_1 = 114547, B_2 = 148859$.

We run the lattice sieve using the above polynomials f_1, f_2 and parameters t, H, B_1, B_2. Our experiment has generated 32,241 hit pairs in about 42 hours using only 6 cores in our computational environment. This is about 1.3 times larger than the sufficient number of relations $\sharp B_1 + \sharp B_2 + 2n$.

6.3   Linear Algebra

From the hit pairs in the searching relations we construct a matrix of linear equations modulo $\ell = 6118607636866573789$ (63 bits) that is the maximum prime divisor of $p^{12} - 1$. The size of the matrix is 32241×24463, and it is shrunk to 16579×15073 by the filtering process such as eliminating duplicated relations. Then we solve it by Lanczos method [12,11].

We found the solutions of the linear equations in about 25 minutes using the 16 cores in our computational environment, and the virtual logarithms $\log \phi_i(\mathfrak{q})$ and $\log \Lambda_{i,j}$ were obtained.

Finally we present an example of solving the DLP in extension field $\text{GF}(p^{12}) \cong \text{GF}(p)[X]/f_1(X)$ of $p = 122663$. Let $\gamma = X^2 + X - 7$ be a generator of the multiplicative group $\text{GF}(p^{12})^*$. Let $\delta = X^2 - 5X + 7$ be a target element of computing the discrete logarithm $\log_\gamma \delta$ in $\text{GF}(p^{12})^*$. Note that both γ and δ are B_1-smooth. From the above virtual logarithms of the B_1-smooth elements, we can compute $\log \delta = 3540036734608022534$ and $\log \gamma = 3897708711757659596$, and thus the discrete logarithm $\log_\gamma \delta \mod \ell$ can be obtained by $\log \delta / \log \gamma \equiv 3161374319443177763 \mod \ell$.

7   Conclusion

In this paper we presented an implementation of the number field sieve for solving the DLP over extension field $\text{GF}(p^n)$ that underpins the security of pairing-based cryptography. Especially we proposed the implementation of the lattice sieve of more than two dimensions. In our experiment, we discussed the size of dimension, sieve region and smoothness bound suitable for the number field sieve over extension field $GF(p^{12})$. Finally we have solved the DLP over extension field $GF(p^{12})$ of 203 bits using a PC of 16 CPU core in about 43 hours.

In the future, we discuss how to select the sieve region for the DLP over extension field $GF(p^{12})$ of larger bits. We also extend the efficient lattice sieve of two dimensions proposed by Franke et al. [6,10] to the lattice sieve of more than two dimensions.

References

1. Aoki, K.: Sieving region, and relationship between numbers of required relations and factor bases on the number field sieve, Technical Report of IEICE, ISEC 104(53), 23–28 (2004) (in Japanese)
2. Aoki, K., Kida, Y., Ueda, H.: A trial of GNFS implementation (Part VI): lattice sieve, Technical Report of IEICE, ISEC 104(315), 9–14 (2004) (in Japanese)
3. Aoki, K., Ueda, H., Uchiyama, S.: Evaluation report on integer factoring problems. In: Investigation Reports on Cryptographic Techniques in FY 2003, no.0202-1 (2004) (in Japanese), http://www.cryptrec.go.jp/english/estimation.html
4. Barreto, P.S.L.M., Naehrig, M.: Pairing-friendly elliptic curves of prime order. In: Preneel, B., Tavares, S. (eds.) SAC 2005. LNCS, vol. 3897, pp. 319–331. Springer, Heidelberg (2006)
5. Cohen, H.: A course in computational algebraic number theory. In: Graduate Texts in Math., vol. 138, Springer (1993)
6. Franke, J., Kleinjung, T.: Continued fractions and lattice sieve. In: Workshop Record of SHARCS (2005),
   http://www.ruhr-uni-bochum.de/itsc/tanja/
   SHARCS/talks/FrankeKleinjung.pdf
7. Joux, A., Lercier, R.: Improvements to the general number field sieve for discrete logarithms in prime fields. A comparison with the Gaussian integer method. Math. Comp. 72, 953–967 (2003)
8. Joux, A., Lercier, R., Smart, N.P., Vercauteren, F.: The number field sieve in the medium prime case. In: Dwork, C. (ed.) CRYPTO 2006. LNCS, vol. 4117, pp. 326–344. Springer, Heidelberg (2006)
9. Kleinjung, T., et al.: Discrete logarithms in GF(p) - 160 digits, email to the NM-BRTHRY mailing list (2007),
   http://listserv.nodak.edu/cgi-bin/
   wa.exe?A2=ind0702&L=nmbrthry&T=0&P=194
10. Kleinjung, T., Aoki, K., Franke, J., Lenstra, A.K., Thomé, E., Bos, J.W., Gaudry, P., Kruppa, A., Montgomery, P.L., Osvik, D.A., te Riele, H., Timofeev, A., Zimmermann, P.: Factorization of a 768-bit RSA modulus. In: Rabin, T. (ed.) CRYPTO 2010. LNCS, vol. 6223, pp. 333–350. Springer, Heidelberg (2010)
11. LaMacchia, B.A., Odlyzko, A.M.: Solving large sparse linear systems over finite fields. In: Menezes, A., Vanstone, S.A. (eds.) CRYPTO 1990. LNCS, vol. 537, pp. 109–133. Springer, Heidelberg (1991)
12. Lanczos, C.: Solution of systems of linear equations by minimized iterations. J. Res. Nat. Bur. Stand. 49, 33–53 (1952)

13. Lenstra, A.K., Lenstra, H.W.: The Development of the Number Field Sieve. Lecture Notes in Math., vol. 1554. Springer (1993)
14. Lenstra, A.K., Lenstra, H.W., Lovász, L.: Factoring polynomials with rational coefficients. Math. Ann. 261, 515–534 (1982)
15. Murphy, B.: Polynomial selection for the number field sieve integer factorisation algorithm, PhD thesis, The Australian National University (1999)
16. PARI/GP, version 2.5.3, Bordeaux (2012), http://pari.math.u-bordeaux.fr/
17. Pollard, J.M.: The lattice sieve. In: [13], pp. 43–49
18. Pomerance, C., Smith, J.: Reduction of huge, sparse matrices over finite fields via created catastrophes. Experiment. Math. 1, 89–94 (1992)
19. Schirokauer, O.: Discrete logarithms and local units. Philos. Trans. Roy. Soc. London Ser. A 345, 409–424 (1993)
20. Schirokauer, O.: Virtual logarithms. J. Algorithms 57, 140–147 (2005)
21. Vercauteren, F.: Optimal pairings. IEEE Transactions on Information Theory 56, 455–461 (2010)
22. Zajac, P.: Discrete logarithm problem in degree six finite fields, PhD thesis, Slovak University of Technology (2008), http://www.kaivt.elf.stuba.sk/kaivt/Vyskum/XTRDL
23. Zajac, P.: On the use of the lattice sieve in the 3D NFS. Tatra Mt. Math. Publ. 45, 161–172 (2010)

Universal Security

From Bits and Mips to Pools, Lakes – and Beyond

for Johannes Buchmann's sixtieth birthday

Arjen K. Lenstra[1], Thorsten Kleinjung[1], and Emmanuel Thomé[2]

[1] EPFL IC LACAL, Station 14, 1015 Lausanne, Switzerland
[2] INRIA CNRS LORIA, Équipe CARAMEL - bâtiment A,
615 rue du jardin botanique, 54602 Villers-lès-Nancy Cedex, France

Abstract. The relation between cryptographic key lengths and security depends on the cryptosystem used. This leads to confusion and to insecure parameter choices. In this note a universal security measure is proposed that puts all cryptographic primitives on the same footing, thereby making it easier to get comparable security across the board.

Current Security Levels

The security of a cryptographic primitive is measured by the effort to break it. If that effort is 2^k, measured in some agreed upon unit, then it offers *k-bit security* and it is said to have *security level* k. Although this looks easy, it turns out to be impractical. For symmetric cryptosystems it is indeed easy: with a k-bit key they are supposed to offer k-bit security, measured in the most basic application of the cryptosystem, else they are no good. Cryptographic hash functions are harder: if the hash length is h they should offer and cannot offer more than $h/2$-bit security, measured in basic applications of the hash function. Given comparable speeds of the systems in similar environments – both in software or both in hardware – their security levels can easily be compared in a meaningful manner.

With public key primitives it gets confusing. For RSA the runtimes to factor an RSA modulus can be measured using experiments on small numbers. The results are then extrapolated to derive estimates for larger moduli: if factoring an n-bit modulus takes time T, then the time to factor an m-bit modulus (for m not much bigger than n) is estimated as $\frac{N(m)}{N(n)}T$, where $N(k) = \exp(1.923(\ln 2^k)^{1/3}(\ln\ln 2^k)^{2/3})$. In 1988 the first factorization of a 100-digit modulus required "100 MIPS years": equivalent to a century of computing time on a computer that performs a million "instructions" per second. In 1999, using a better factoring method, a 512-bit challenge modulus took about 40 years of computing time on a then current core (running at several hundred MHz, in four months in parallel on many cores), which was, somewhat questionably, estimated as 8400 MIPS years. The 2009 factorization of a 768-bit modulus took roughly a

M. Fischlin and S. Katzenbeisser (Eds.): Buchmann Festschrift, LNCS 8260, pp. 121–124, 2013.

year on 2000 cores running at 2GHz. With $\frac{N(1024)}{N(768)} \approx 1200$, a 1024-bit modulus can be factored within two million 2GHz core-years. Further extrapolation is trickier, but it is reasonable to estimate that 2048-bit RSA moduli are a billion times harder to factor than 1024-bit ones. Using generic figures for the speed of software implementations, it follows that 768-, 1024-, and 2048-bit RSA have approximate security levels 66, 76, and 106, respectively.

For discrete logarithm public key cryptosystems additional parameters further confuse the picture. Let G be a well-chosen prime order q group for which the group operation is 2^c times slower than a basic application of a symmetric cryptosystem (or cryptographic hash function). It may be assumed that $0 \le c \le 10$. Discrete logarithms in G offer $c + \log_2 \sqrt{q}$-bit security, assuming a large enough finite field F if $G \subset F^*$, the size of which is estimated in the same way as the security of RSA, with a small and uncertain twist $\tau > 0$ and with the potential of all kinds of trouble if an extension field is used. Thus, if q is a 160-bit prime and G an elliptic curve group, the security level is $c+80$. If, for a similar q-value, $G \subset F^*$ for a cardinality p finite field F, then the security offered by F^* needs to be taken into account as well: the security level is $\min(c + 80, 76 + \tau) = 76 + \min(c + 4, \tau)$ if p is a 1024-bit prime, and $\min(c + 80, 106 + \tau) = c + 80$ for 2048-bit p.

Lattice-based public key cryptosystems still escape proper analysis. Despite enthusiastically riding post-quantum waves and fully homomorphic hot air balloons, they have not found wide-spread application and are not further discussed.

Summarizing the above, for symmetric cryptosystems the key length and the security level are the same, for cryptographic hash functions the security level is half the hash length, for RSA the security level is an obscure but small and quickly decreasing fraction of the modulus length, and for discrete logarithms it is half the size of the largest prime dividing the group order, unless one works in a multiplicative group of a finite field in which case extra care needs to be taken. This is easy enough for those in the know, but incomprehensible to most: it is not uncommon for people to proudly declare that they are using 128-bit RSA moduli for their 128-bit security cryptosystems – or indeed for crypto-practitioners to use 256-bit AES, SHA-512, and 512-bit RSA thinking they get an overall security level of $\min(256, \frac{512}{2}, \frac{512}{2}) = 256$.

Intuitive Security Levels

The problem is that for all these cryptosystems key length and security level are measured in bits, but that the relationship between the two varies wildly – from trivial, to simple, to rather contrived. If this is not well understood, it is tempting to use just the trivial and simple relationships and to forget about the complicated ones, with potentially disastrous consequences. This has occurred in practice, on multiple occasions and involving major corporations.

To address this problem the relationships between key length and security level must be put on the same footing for all cryptosystems. Because the complicated relationships cannot be simplified, all must be made equally complicated. Below a new definition of security level is proposed that does away with "k-bit security" for "security level k". It has the additional advantage that it allows a more intuitive interpretation of what security actually means.

The new approach was inspired by a remark made by the third author during his presentation of the factorization of the 768-bit RSA challenge at Crypto 2010: *We estimate that the energy required for the factorization would have sufficed to bring two 20°C Olympic size swimming pools to a boil.* This amount of energy was estimated as half a million kWh. Thus, a cryptosystem is said to offer *pool security* if breaking it requires as much energy as it takes to boil a single Olympic size swimming pool (i.e., 2500 cubic meters of water). It follows that 65-bit symmetric cryptosystems, 130-bit cryptographic hash functions, and 745-bit RSA currently all offer pool security.

Larger quantities of water need to be considered to fully appreciate the security offered by practically relevant cryptosystems. On average, The Netherlands enjoys a healthy daily average of 82 million cubic meters of rain (i.e., 2^{15} pools); with $65 + 15 = 80$ and $\frac{N(1130)}{N(768)} \approx 2^{14}$ it is found that 80-bit symmetric cryptosystems, 160-bit cryptographic hash functions, and 1130-bit RSA offer *rain security* or *dagelijkse neerslagverdampingsenergiebehoeftezekerheid*. Equivalently, with each German citizen boiling a cubic meter of water, one may refer to rain security as *German security* or *JedeRtausendliterbierverdampfungssicherheit*[1].

Slightly more than 2^{25} pools suffice to fill the lake of Geneva (89 cubic kilometers of water), so 90-bit symmetric cryptosystems, 180-bit cryptographic hashes, and 1440-bit RSA all offer *lake security* or *sécurité lémanique*. Boiling all water on the planet (including all starfish) amounts to about 2^{24} lakes of Geneva and leads to *global security*: 114-bit symmetric cryptosystems, 228-bit cryptographic hashes, and 2380-bit RSA. This needs to be done 16 thousand times to break AES-128, SHA-256, or 3064-bit RSA.

With all water evaporated and no bodies of water left to fire the imagination, the above new security levels have run out of steam. *Solar security* can still be defined as offered by cryptosystems that can be broken in a year requiring that year's total solar energy: 140-bit symmetric cryptosystems, 280-bit cryptographic hash functions, and 3730-bit RSA. Reaching further in an intuitively appealing manner requires a different approach.

To Infinity – and Beyond

There is a big gap between the energy required to bring a gram of water to a boil (as usual starting at 20°C), namely $93 \cdot 10^{-6}$ kWh, and the mass-energy of that same gram of water, namely 25 million kWh. With $\frac{25 \cdot 10^{6}}{93 \cdot 10^{-6}} \approx 2^{38}$, cryptanalysis suddenly looks a lot easier – maybe a bit too easy. Breaking AES-128, SHA-256, or 3064-bit RSA using the mass-energy of the lake of Geneva (where $90 + 38 = 128$) still looks more or less reasonable, but that is mostly because it is

---

[1] The "2+"-bit difference between *bringing to a boil* and *Verdampfung* (which Alexander Kruppa kindly reminded us of) easily falls within the margin of Moore's eternal law which ensures that keys requiring evaporation a few years ago can now be boiled.

inconceivable. What about a 0.02 gram scrap of paper having the mass-energy (namely $0.02 \cdot 25 \cdot 10^6$ kWh = 500 000 kWh) needed to break 768-bit RSA? It implies *paper-thin security* for 1024-bit RSA: five A4 sheets of 80g/m^2 paper together weigh 25 grams, $25/0.02 \approx \frac{N(1024)}{N(768)}$, and therefore have enough mass-energy to break 1024-bit RSA. This could have been cooked up by Certicom's PR department, but does not feel intuitively right.

Finally, with the estimated mass-energy of the observable universe 2^{190} times what is required to boil two pools, $190 + 66 = 256$-bit symmetric cryptosystems, 512-bit cryptographic hash functions, and 14954-bit RSA offer *universal security*.

Summary

Determining the newly proposed security levels requires a trained professional. This considerably raises the bar for uneducated guesswork and thereby significantly diminishes the risk of insecure parameter choices. Worldwide adoption of these intuitive security levels will have a beneficial effect on Internet security.

The table lists the new[2] security levels[3], with "sea" referring to the Mediterranean Sea. For metrically-challenged prism-enabled readers it is noted that using the same numbers of gallons or cubic miles as the table's liters or cubic kilometers, respectively, requires adding two to the figures in the "symmetric" column, adding four to the figures in the "hash" column, and replacing n in the "RSA" column by the m for which $\frac{N(m)}{N(n)} \approx 4$. Note, however, that this requires longer showers, a larger lake, etc., which is only appropriate.

Table 1. Intuitive security levels

		bit-lengths		
security level	volume of water to bring to a boil	symmetric key	cryptographic hash	RSA modulus
teaspoon security	0.0025 liter	35	70	242
shower security	80 liter	50	100	453
pool security	2 500 000 liter	65	130	745
rain security	0.082 km^3	80	160	1130
lake security	89 km^3	90	180	1440
sea security	3 750 000 km^3	105	210	1990
global security	1 400 000 000 km^3	114	228	2380
solar security	-	140	280	3730

Acknowledgements. We gratefully acknowledge usage of Wikipedia and gladly blame its contributors for any errors in our data.

---

[2] Unfortunately, but not surprisingly, cloud security is not well-defined.

[3] Going further down the scale than in Table 1 would lead to *meat security*, the amount of computing humans can be expected to do, but we rather see computers as meat's next step on the evolutionary ladder
(`instruct.westvalley.edu/lafave/meat_in_space.html`)

Identities for Embedded Systems
Enabled by Physical Unclonable Functions

Dominik Merli[1], Georg Sigl[2], and Claudia Eckert[3]

[1] Fraunhofer Research Institution for Applied and Integrated Security (AISEC)
Munich, Germany
dominik.merli@aisec.fraunhofer.de
[2] Institute for Security in Information Technology
Technische Universität München, Munich, Germany
sigl@tum.de
[3] Department of Computer Science, Chair for IT Security
Technische Universität München, Munich, Germany
eckertc@in.tum.de

Abstract. Embedded systems, such as automotive control units, industrial automation systems, RFID tags or mobile devices are dominated by integrated circuits implementing their functionality. Since these systems operate in increasingly networked or untrusted environments, their protection against attacks and malicious manipulations becomes a critical security issue. Physical Unclonable Functions (PUFs) represent an interesting solution to enable security on embedded systems, since they allow identification and authentication of CMOS devices without non-volatile memory. In this paper, we explain benefits and applications of PUFs and give an overview of popular implementations. Further, we show that PUFs face hardware as well as modeling attacks. Therefore, specific analyses and hardening has to be performed, in order to establish PUFs as a reliable security primitive for embedded systems.

Keywords: Physical Unclonable Functions, Applications, Implementations, Attacks, Countermeasures.

1 Introduction

Today, embedded systems are present in a large number of products used in a variety of applications and areas. Beginning from lightweight RFID tags serving in logistic applications, to mobile devices enabling personalized payment and multimedia services, to industrial and transportation systems. One example is the in-vehicle network of a modern automobile, which connects tens of control units by automotive bus systems or wireless connections as, e.g., for tire pressure monitoring systems or maintenance devices. Further, industrial automation systems consist of hundreds of sensors and actuators, which are connected by to control systems, corporate IT systems and more and more also to the Internet.

During the last years, time after time, several attacks against embedded systems were discovered threatening people and cooperations. Attackers were able

M. Fischlin and S. Katzenbeisser (Eds.): Buchmann Festschrift, LNCS 8260, pp. 125–138, 2013.

to impersonate automotive bus participants, gain access to secure vehicle information, and control critical functions like braking or accelerating [15]. Also, the sophisticated Stuxnet malware demonstrated how to exploited several security weaknesses in order to take control of an industrial automation system [7].

Security solutions to protect the integrity of embedded systems require unique identification and authentication of system components. If these requirements can be met, unauthorized communications can be detected and eliminated before the manipulations can take effect. One way to achieve secure identities for electronic components, is the usage of Non-Volatile Memory (NVM), which is programmed by the component manufacturer with a unique identification number, i.e. a secret key. However, on-chip NVM is expensive in production costs, because special processes are necessary. One-Time Programmables (OTPs), e.g., fuses burned by a laser during manufacturing, can also be used to embed a device-unique bit string, but these structures can easily be identified and read-out by optical inspection.

Physical Unclonable Functions (PUFs) are physical structures, which deliver a device-specific response based on physical properties when stimulated by a challenge. They are cheaper in production than costly secure non-volatile memory and put a higher barrier to attackers trying to extract the secret bits by optical/physical analyses than OTPs do. During the last decade, several implementations of PUFs have been introduced [9,26,10,4] and analyzed, which can be implemented in standard CMOS technology. Components, such as, ring oscillators [26], SRAM cells [10], delay paths [9] and bistable rings [4] can be used to derive a chip-unique bit pattern, because the electrical properties of these structures are prone to uncontrollable random process variations during manufacturing. Because these unique component properties are similar to human features in biometric systems, PUFs are also referred to as 'fingerprints of microchips' or 'biometrics of material'.

Further, PUFs can offer tamper protection, if more expensive structures are exploited, e.g., special chip coatings [27] or integrated optical sensors [6], evaluating light effects with a smartcard.

Besides replacing an NVM by a combination of a PUF and a key generation module, it is also possible to achieve challenge-response authentication of hardware modules integrating a PUF. Even authentication of devices containing absolutely no secret information can be accomplished by PUFs.

This paper is organized as follows. In Section 2, we show scenarios in which PUFs can be exploited to either gain in cost effectiveness or security. We describe popular CMOS implementations of PUFs in Section 3, followed by an explanation which approaches have been introduced to achieve tamper resistance by PUFs, in Section 4. Further, we demonstrate possible attack scenarios in Section 5 and stress the necessity to specifically analyze and harden PUFs against these threats. We conclude in Section 6.

2   Applications of PUFs

PUFs are mainly targeted for secret key generation or challenge-response authentication in Applications Specific Integrated Circuits (ASICs) and Field Programmable Gate Arrays (FPGAs).

2.1   Key Generation

One of the most important use cases for PUFs is the extraction or the embedding of secret keys. In the first case, PUF bits are directly used to generate a device-unique secret key. For asymmetric cryptography, this can be exploited in a way that the private key will never leave the device and will thus be only known to the device. In the embedding case, an externally generated key is bound to PUF bits enabling the storage of a master key based on chip-unique physical properties.

Since PUFs perform physical measurements, which always contain a certain amount of measurement noise, and since their properties slightly change over temperature, voltage and other environmental influences, error-tolerant algorithms are necessary to derive a stable and reliable electronic fingerprint. Some PUFs already implement a first step of compensation, e.g., the Ring Oscillator PUF [26] utilizes relative frequency comparison to mitigate frequency deviations caused by temperature or voltage variations. However, usually, additional mechanism based on error-correcting codes are necessary. During an enrollment phase, these algorithms generate helper data containing redundancy in order to be able to restore the enrolled key at any later point in time.

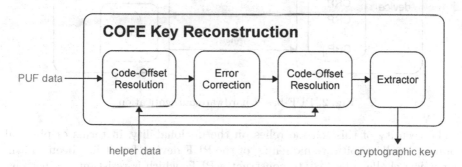

Fig. 1. Key reconstruction process

Dodis et al. [5] introduced Code-Offset Fuzzy Extractors (COFEs) in 2004, which have since been regarded as the standard architecture to reliably extract secret bits from PUF responses. Its helper data consists of an XOR code-offset between a random codeword of an error-correcting code and the secret PUF bits. As shown in Figure 1, a conventional COFE PUF key extraction system consists of error correction module able to correct noisy bits and a subsequent extractor module which ensures a uniformly distributed key output.

A popular method for key embedding into noisy data is Index-Based Syndrome coding (IBS) [28], which was proposed by Yu and Devadas in 2010. There, the helper data stores the indices of the most reliable PUF bits corresponding to the key bits to embed. Thereby, the error probability is reduced, but usually an additional (classic) error correction step is necessary to achieve high reliability. Hiller et al. were able to increase the efficiency of IBS by storing not only indices of corresponding bits, but also of complementary bits, improving its performance when no stable corresponding bits are available. The new method is called Complementary IBS (C-IBS) [11].

2.2   Hardware Authentication

The challenge-response behavior of PUFs is unique for each instance. Therefore, it can be exploited to authenticate a specific PUF device [9]. First, one collects a list of random Challenge-Response-Pairs (CRPs) from each device and saves the list to a secure database, as shown in Figure 2. Later, if the authenticity of a device has to be verified, one can apply randomly chosen challenges from the database to the PUF and verify the correctness of its responses. In contrast to key generation scenarios no on-chip error-correction is necessary, because the response verification process might also include error tolerance mechanisms.

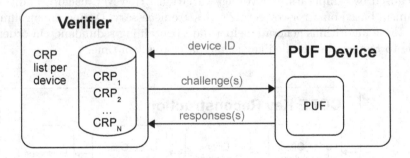

Fig. 2. PUF-based hardware authentication

The security of this scheme relies on the unclonability, in terms of physical cloning as well as software modeling, of the PUF device, but it has been shown, that it is a challenging task to construct a PUF, which is resistant against machine learning attacks [24].

This problem is circumvented in the approach of public PUFs [2], also called SIMPL systems [23]. There, the model of the PUF is publicly known and an efficient simulation algorithm is provided. To perform the authentication, a verifier chooses a random challenge and applies it to the PUF and the public simulation model. Afterwards, he not only checks if the result is the same, but also if the PUF responded within a specified time, in which only the correct hardware device is able to generate this response. In this case, attackers do not face the challenge of modeling a PUF, but of reproducing it physically.

3   CMOS Implementations

Since PUFs have gained interest in academia and industry, a main focus was set on physical structures implementable by standard CMOS circuits or even by logic cells within FPGAs. The most popular architectures are explained in the following sections.

3.1   Ring Oscillator PUF

The Ring Oscillator PUF (RO PUF), as shown in Figure 3, was introduced by Suh and Devadas in 2007 [26]. There, a set of ring oscillators with unique frequencies is used to generate a device-specific fingerprint. In the original proposal, the frequencies of two oscillators are measured and compared. Depending on their relation, i.e. which oscillator is faster, the PUF's output bit is set to 0 or 1. The relative comparison allows to partly compensate environmental influences like temperature variations. This architecture is mainly targeted for secret key generator and does not provide a large number of CRPs for hardware authentication.

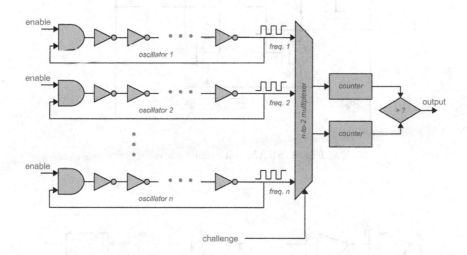

Fig. 3. Ring Oscillator PUF architecture

3.2   SRAM PUF

In 2007, Guajardo et al. presented a PUF idea [10] based on the start-up behavior of SRAM cells. When these memory cells, as shown in Figure 4, are powered up, they settle in a state depending on their physical characteristics, e.g., the threshold voltage of the implemented transistors. Assuming a balanced cell, minimal manufacturing variations define the start-up state, which can be exploited as a PUF. Capturing the initial values of several cells, it is possible to extract a device unique bit string, which can be used for key generation.

3.3   Arbiter PUF

The Arbiter PUF was the first PUF architecture based on CMOS gates. It was developed by Gassend et al. and published in 2002 [9]. It measures the propagation delay difference between two paths of delay elements and depending on which path is faster, a 0 or 1 is output. As shown in Figure 5, The input bits, namely the challenge, combine different delay elements to form several different combinations of paths. It was intensionally targeted for challenge-response hardware authentication. However, machine learning attacks have shown that Arbiter PUFs can be modeled in software with an rather small amount of CRPs [24].

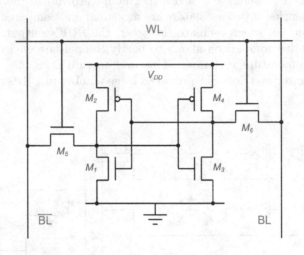

Fig. 4. SRAM cell architecture

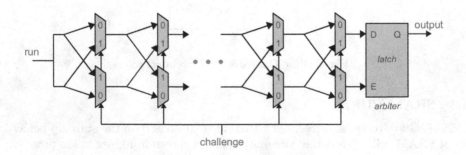

Fig. 5. Arbiter PUF architecture

3.4   Bistable Ring PUF

The architecture of a Bistable Ring PUF (BR PUF) [4] was introduced by Chen et al. in 2011. The basic principle is a bistable ring circuit consisting of an even number of inverters. This ring can take two possible stable states: [01...01] and [10...10]. If the inverters are replaced by a cell consisting of a demultiplexer, two NOR gates and a multiplexer, the ring circuit is turned into a PUF structure as shown in Figure 6. Each challenge bit controls the demultiplexer and the multiplexer of a cell and thereby defines the signal path to take. A reset signal connected to each NOR gate enables to reset the ring into an unstable all-zeros state. After releasing the reset signal, the ring oscillates until it falls into a stable state, defining its output bit as 0 or 1. This challenge-response behavior can be exploited for hardware authentication, but resistance against machine learning attacks still has to be demonstrated.

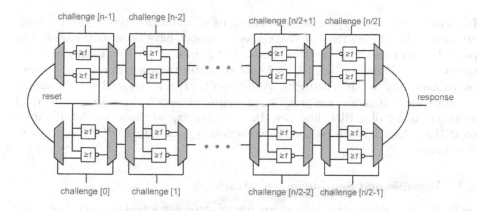

Fig. 6. Bistable Ring PUF architecture

4   Tamper Resistant Implementations

It is often argued that PUFs naturally achieve tamper resistance. While this might be true for the PUF itself, it does not cover the system it is embedded in. However, PUFs have the potential to provide tamper protection mechanism for microchips. Approaches aiming for that target are explained in the following sections.

4.1   Coating PUF

Tuyls et al. propose to built read-proof hardware modules by using a Coating PUF [27]. This tamper-sensitive PUF is established by covering an integrated circuit by a protective coating, which consists of a matrix material containing randomly distributed dielectric particles. The described prototype is build with an aluminophosphate matrix material and a mixture of particles made of TiO2 and TiN. These materials make the coating intransparent, hard and protective against chemical tampering. In order to measure the unique dielectric properties of the added particle layer, the top metal layer of the chip implements comb-shaped capacitive sensors. The measured capacitance values can be used to obtain a unique and robust secret key.

4.2   Optical Smartcard PUF

A recent idea, published by Esbach et al. [6] generates a PUF by the combination of security chip and plastic body of a smartcard. There, light sources and sensors, located on the top-side of the microchip, are used to measure minimal material variations of the smartcard body. This approach achieves a similar tamper resistance as the Coating PUF based on optical effects.

5   Attacks on PUFs

Implementations of cryptographic algorithms in embedded systems face a variety of attacks. During the last years, these methods have been continuously improved. They range from invasive attacks [1,14], to semi-invasive techniques [25], to side-channel analysis [12]. PUFs do not constitute an exception when it comes to vulnerability to these attacks. Additionally, PUFs targeted for challenge-response authentication are prone to modeling attacks trying to built an equal software model of a PUF instance. In the following sections, we detail threat to PUFs, which have to be considered seriously, when using PUFs as security primitives.

5.1   Invasive and Semi-invasive Attacks

Invasive and semi-invasive attacks are one of strongest attacks on cryptographic devices. In order to be able to perform these attacks, the chip package has to be

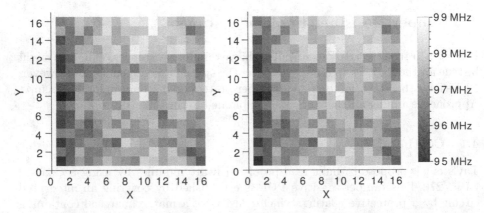

Fig. 7. Comparison of ring oscillator frequencies before (left) and after (right) decapsulation of the FPGA

removed mechanically or by means of etching. Afterwards, ultra-fine needles can be placed at the die surface to observe distinct signals, faults can be injected at specific locations by laser bolts, and optical structure analyses can be performed. As for the latter, PUFs are naturally more resistant than OTPs, because there secrets are hidden in sub-micron variations.

Sometimes, PUFs are per se considered to be tamper resistant, i.e., decapsulation of a chip would alter PUF characteristics and thereby implicitly destroy the original secret. However, this assumption is, if ever realistic, only true for PUF architectures described in Section 4. In order to verify our doubts about tamper resistance of PUFs solely implemented in CMOS technology, we measured the frequencies of 256 ring oscillators implemented on a Xilinx Spartan-3 FPGA in its original state and after removing the chip package and the backside leadframe copper plate. Figure 7 shows the matrices of ring oscillator frequencies before and after decapsulation. Thereby, we were able to show, that this strong manipulation of the chip's environment does not significantly influence frequencies of FPGA ring oscillators [19]. Therefore, we conclude, that PUFs based on CMOS circuits, as proposed to date, cannot be regarded as tamper resistant, because they do not enough involve their environment into their characteristic behavior. As a consequence, invasive and semi-invasive attacks are a threat for PUFs as for any other cryptographic device, except for the case of public PUFs, where there is no secret present in hardware at all.

5.2   Side-Channel Attacks on Key Generation

Since several years, side-channel analysis is a central topic for the security of cryptographic implementations in embedded systems. Attacks on device-internal secrets can be achieved by observation of the power consumption [13], the runtime behavior [12], and the electro-magnetic emission of a device [8,22]. Side-channel attacks also pose a threat to PUFs, even for tamper resistant PUFs, since these attacks do not require any device manipulation. One target can be post-processing algorithms like fuzzy extractors, which can be attacked by well-known methods like Simple Power Analysis (SPA) or Differential Power Analysis (DPA), but in addition, side-channels can also be exploited to directly characterize PUF properties or to analyze their intermediate signals.

We were able to show that the Toeplitz hashing [16] function, which is popular for efficient COFE implementations [3,17], can be attacked by SPA [20]. Depending on its input data, operations may be performed or not. Since the corrected PUF response bits constitute the input data for the Toeplitz hashing in fuzzy extractors, this attack enables the extraction of all PUF response bits and enableds the reconstruction of the key extracted from it. Further, we discovered that differential attacks on error correction modules or extractors are possible based on helper data manipulation [21].

To protect PUF-based key generation systems from first-order side-channel attacks, we proposed a codeword masking scheme [21], which enables the randomization of the key generation process while preserving the essential error correction capabilities. This masking scheme can also be continued throughout

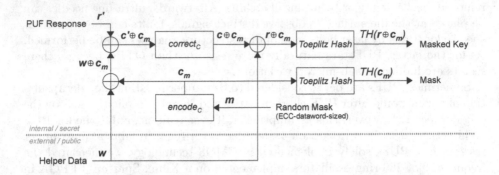

Fig. 8. Masked COFE implementation

Fig. 9. Setup for localized EM measurements

the Toeplitz hashing module, because of the linearity property of this extractor. Figure 8 shows how a masked COFE implementation can look like.

5.3   Localized Electromagnetic Analysis of RO PUFs

In PUF-enabled systems, not only key generation modules represent a target of attacks, but also the PUF measurement circuits themselves. We performed an attack on an RO PUF [19], exploiting its spatially limited electro-magnetic emission. Figure 9 shows our measurement setup, consisting of an EM probe with a 150m coil, an Xilinx Spartan-3 FPGA decapsulated from the backside and an automated X-Y-table. This enabled us to record EM traces over whole surface of the FPGA in 100m steps. A map indicating the localized EM emission corresponding to the RO PUF is shown in Figure 10.

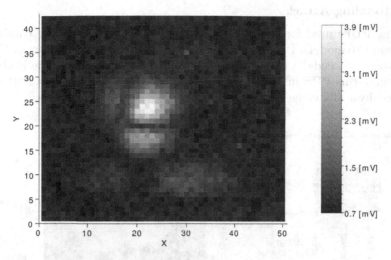

Fig. 10. Map indicating where ring oscillator frequencies can be observed

In an efficient RO PUF implementation, pairs of oscillators are compared step by step and always the second oscillator of the preceding comparison is the first oscillator in the subsequent comparison. Therefore, all frequencies of the RO PUF are observable, one after the other, within the indicated area of Figure 10. After this analysis, an attacker is in possession of a complete PUF model. We also proposed an enhanced RO PUF architecture [19] resisting these attacks by avoiding multiple usage of ring oscillators.

In a second, more advanced EM analysis of RO PUFs [18], we discovered that it is possible to separate RO PUF measurement components like counters and multiplexers by localized EM measurements. Figure 11 shows a floorplan of an RO PUF implementation with three multiplexers and three counters. Next to the floorplan, one can see the localized EM emission of the FPGA implementation. We were able to separate the emissions caused by the different multiplexers and counters, which allowed us to separately observe all measured RO PUF frequencies. With this information, an attacker can built a complete RO PUF model, which breaks its security.

In order to protect RO PUFs from the found vulnerabilities, we proposed two countermeasures. The first one aims at randomizing the RO PUF measurement process, i.e., the mapping of measurement components (multiplexers and counters) and the measured ROs is broken. Although this is an effective protection approach, it also requires a large area and resource overhead. The second countermeasure considers interleaved placement and routing of RO PUF implementations. Interleaving the basic digital building blocks of counters, multiplexers and ROs is a practical way of mixing the EM emission of all components in order to render the separation of these components impossible in practice.

5.4   Modeling Attacks

Attacking PUFs used for challenge-response authentication, mainly consists of the attempt to model a PUF's behavior with only a relatively small set of CRPs. If an attack succeeds in cloning a PUF, even if only in software, he is able to impersonate this PUF instance. Only for the case of public PUFs, an attackers needs to physically clone a PUF.

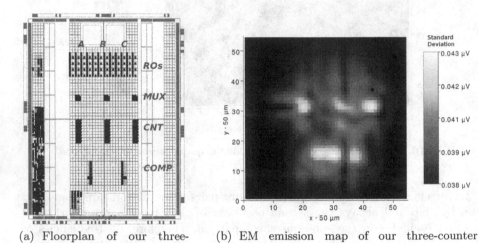

(a) Floorplan of our three-counter RO PUF design

(b) EM emission map of our three-counter RO PUF implementation

Fig. 11. Localized EM measurements allow for separation of RO PUF measurement components

In 2010, Rührmair et al. showed that several common PUF architectures including the Arbiter PUF and some of its more complex variants are susceptible to machine learning attacks [24]. Until now, no PUF construction was proposed which was shown to be resistant against machine learning attacks.

6   Conclusion

In conclusion, PUFs enable secure identities for microchips without requiring on-chip NVM. We explained how CMOS structures like ring oscillators, delay paths and memory cells can be exploited to generate cryptographic keys with the help of fuzzy extractors and key embedding algorithms. Additionally, if future PUF research leads to more complex and modeling-resistant architectures, challenge-response authentication based on PUFs can be achieved. We have also shown that tamper resistant PUF approaches have the potential to protect integrated circuits from physical attacks. However, PUFs are also exposed to a wide range of attacks, which have to be considered in detail when designing PUF-based systems. As shown by our presented experimental results, targeted analyses can reveal individual weaknesses of PUFs and constitute the basis for effective countermeasures.

References

1. Anderson, R.J.: Security Engineering: A Guide to Building Dependable Distributed Systems, 1st edn. John Wiley & Sons, Inc., New York (2001)
2. Beckmann, N., Potkonjak, M.: Hardware-based public-key cryptography with public physically unclonable functions. In: Katzenbeisser, S., Sadeghi, A.-R. (eds.) IH 2009. LNCS, vol. 5806, pp. 206–220. Springer, Heidelberg (2009)
3. Bösch, C., Guajardo, J., Sadeghi, A.-R., Shokrollahi, J., Tuyls, P.: Efficient helper data key extractor on FPGAs. In: Oswald, E., Rohatgi, P. (eds.) CHES 2008. LNCS, vol. 5154, pp. 181–197. Springer, Heidelberg (2008)
4. Chen, Q., Csaba, G., Lugli, P., Schlichtmann, U., Rührmair, U.: The bistable ring PUF: A new architecture for strong physical unclonable functions. In: IEEE Int. Symposium on Hardware-Oriented Security and Trust (June 2011)
5. Dodis, Y., Reyzin, L., Smith, A.: Fuzzy extractors: How to generate strong keys from biometrics and other noisy data. In: Cachin, C., Camenisch, J.L. (eds.) EUROCRYPT 2004. LNCS, vol. 3027, pp. 523–540. Springer, Heidelberg (2004)
6. Esbach, T., Fumy, W., Kulikovska, O., Merli, D., Schuster, D., Stumpf, F.: A new security architecture for smartcards utilizing PUFs. In: Proceedings of the 14th Information Security Solutions Europe Conference (ISSE 2012). Vieweg+Teubner Verlag (2012)
7. Falliere, N., Murchu, L.O., Chien, E.: W32.stuxnet dossier. Technical report, Symantex Security Response (February 2011)
8. Gandolfi, K., Mourtel, C., Olivier, F.: Electromagnetic analysis: Concrete results. In: Koç, Ç.K., Naccache, D., Paar, C. (eds.) CHES 2001. LNCS, vol. 2162, pp. 251–261. Springer, Heidelberg (2001)
9. Gassend, B., Clarke, D., van Dijk, M., Devadas, S.: Silicon physical random functions. In: CCS 2002: Proceedings of the 9th ACM Conference on Computer and Communications Security, pp. 148–160. ACM, New York (2002)
10. Guajardo, J., Kumar, S.S., Schrijen, G.-J., Tuyls, P.: FPGA intrinsic PUFs and their use for IP protection. In: Paillier, P., Verbauwhede, I. (eds.) CHES 2007. LNCS, vol. 4727, pp. 63–80. Springer, Heidelberg (2007)
11. Hiller, M., Merli, D., Stumpf, F., Sigl, G.: Complementary IBS: Application specific error correction for PUFs. In: 2012 IEEE International Symposium on Hardware-Oriented Security and Trust (HOST), pp. 1–6 (June 2012)
12. Kocher, P.C.: Timing attacks on implementations of diffie-hellman, RSA, DSS, and other systems. In: Koblitz, N. (ed.) CRYPTO 1996. LNCS, vol. 1109, pp. 104–113. Springer, Heidelberg (1996)
13. Kocher, P.C., Jaffe, J., Jun, B.: Differential power analysis. In: Wiener, M. (ed.) CRYPTO 1999. LNCS, vol. 1666, pp. 388–397. Springer, Heidelberg (1999)
14. Kömmerling, O., Kuhn, M.G.: Design principles for tamper-resistant smartcard processors. In: WOST 1999: Proceedings of the USENIX Workshop on Smartcard Technology on USENIX Workshop on Smartcard Technology, p. 2. USENIX Association, Berkeley (1999)
15. Koscher, K., Czeskis, A., Roesner, F., Patel, S., Kohno, T., Checkoway, S., McCoy, D., Kantor, B., Anderson, D., Shacham, H., Savage, S.: Experimental security analysis of a modern automobile. In: 2010 IEEE Symposium on Security and Privacy (SP), pp. 447–462 (May 2010)
16. Krawczyk, H.: LFSR-based hashing and authentication. In: Desmedt, Y.G. (ed.) CRYPTO 1994. LNCS, vol. 839, pp. 129–139. Springer, Heidelberg (1994)

17. Maes, R., Tuyls, P., Verbauwhede, I.: Low-overhead implementation of a soft deci-sion helper data algorithm for SRAM PUFs. In: Clavier, C., Gaj, K. (eds.) CHES 2009. LNCS, vol. 5747, pp. 332–347. Springer, Heidelberg (2009)
18. Merli, D., Heyszl, J., Heinz, B., Schuster, D., Stumpf, F., Sigl, G.: Localized elec-tromagnetic analysis of RO PUFs. In: Proceedings of the IEEE Int. Symposium of Hardware-Oriented Security and Trust. IEEE (June 2013)
19. Merli, D., Schuster, D., Stumpf, F., Sigl, G.: Semi-invasive EM attack on FPGA RO PUFs and countermeasures. In: 6th Workshop on Embedded Systems Security (WESS 2011), Taipei, Taiwan. ACM (October 2011)
20. Merli, D., Schuster, D., Stumpf, F., Sigl, G.: Side-channel analysis of PUFs and fuzzy extractors. In: McCune, J.M., Balacheff, B., Perrig, A., Sadeghi, A.-R., Sasse, A., Beres, Y. (eds.) Trust 2011. LNCS, vol. 6740, pp. 33–47. Springer, Heidelberg (2011)
21. Merli, D., Stumpf, F., Sigl, G.: Protecting PUF error correction by codeword mask-ing. Cryptology ePrint Archive, Report 2013/334 (2013), http://eprint.iacr.org/2013/334
22. Quisquater, J.-J., Samyde, D.: ElectroMagnetic Analysis (EMA): Measures and Counter-Measures for Smard Cards. In: Attali, S., Jensen, T. (eds.) E-smart 2001. LNCS, vol. 2140, pp. 200–210. Springer, Heidelberg (2001)
23. Rührmair, U.: Simpl systems: On a public key variant of physical unclonable func-tions. Technical report, Cryptology ePrint Archive, International Association for Cryptologic Research (2009)
24. Rührmair, U., Sehnke, F., Sölter, J., Dror, G., Devadas, S., Schmidhuber, J.: Mod-eling attacks on physical unclonable functions. In: Proceedings of the 17th ACM Conference on Computer and Communications Security, CCS 2010, pp. 237–249. ACM, New York (2010)
25. Skorobogatov, S.P.: Semi-invasive attacks – A new approach to hardware security analysis. Technical Report UCAM-CL-TR-630, University of Cambridge, Com-puter Laboratory (April 2005)
26. Suh, G.E., Devadas, S.: Physical unclonable functions for device authentication and secret key generation. In: 44th ACM/IEEE Design Automation Conference, DAC 2007, pp. 9–14 (2007)
27. Tuyls, P., Schrijen, G.-J., Škorić, B., van Geloven, J., Verhaegh, N., Wolters, R.: Read-proof hardware from protective coatings. In: Goubin, L., Matsui, M. (eds.) CHES 2006. LNCS, vol. 4249, pp. 369–383. Springer, Heidelberg (2006)
28. Yu, M.-D.M., Devadas, S.: Secure and robust error correction for physical unclon-able functions. IEEE Des. Test 27(1), 48–65 (2010)

When Should an Implementation Attack Be Viewed as Successful?

Werner Schindler[1,2]

[1] Bundesamt für Sicherheit in der Informationstechnik (BSI)
Godesberger Allee 185–189
53175 Bonn, Germany
Werner.Schindler@bsi.bund.de
[2] CASED (Center for Advanced Security Research Darmstadt)
Mornewegstraße 32
64289 Darmstadt, Germany
Werner.Schindler@cased.de

Dedicated to Johannes Buchmann on the occasion of his 60th birthday

Abstract. In this paper we address the problem when a side-channel attack or a fault attack should be counted successful in case the attack does not reveal all bits of the secret key but provides only partial information. Many interesting questions arise in this context, which demand advanced mathematical methods. The topic is illustrated by three well-known examples. The credo of this paper is that there is broad room and the need for fruitful collaboration between researchers dealing with implementation attacks and researchers from mathematical (algorithm-oriented) cryptography.

Keywords: Side-channel analysis, fault analysis, algebraic side-channel analysis, exponent blinding, lattice-based cryptography.

1 Introduction

In 1996 and in 1999 Paul Kocher introduced timing analysis [16] and power analysis [17]. Both were pioneering papers in the field of side-channel analysis, which excited the community since it had become clear that the security analysis of cryptographic devices and systems should not be divided into several pieces, which are treated separately and independently by mathematicians, computer scientists and engineers. Nearly at the same time Boneh, DeMillo and Lipton published a recognized paper in fault analysis [4].

Very soon after Kocher's publications side-channel analysis became an important research topic in both academia and industry. Some years later fault analysis moved into their focus, too. In the first years side-channel analysis, and there in particular power analysis, was almost exclusively a research domain for engineers. This seemed to be natural because electrical engineers have expertise in hardware and were able to perform experiments while this was usually not the case for mathematicians or computer scientists. Similar was true for fault attacks.

M. Fischlin and S. Katzenbeisser (Eds.): Buchmann Festschrift, LNCS 8260, pp. 139–150, 2013.

A drawback of this development was the effect that the side-channel leakage often was not exploited efficiently but large parts of the information were wasted. In the course of the years more and more mathematicians and computer scientists have worked on this field and have become involved in interdisciplinary research projects. The applied mathematical methods have become more advanced and sophisticated, which did not only lead to more efficient attacks and effective countermeasures but also to deeper insight into the nature of leakage and theoretical considerations, c.f. [1, 2, 5–8, 12, 14, 15, 18, 24–28, 30, 31] and many more. For many years side-channel analysis and fault analysis have been important components of security evaluations of high-security devices.

Of course, security implementations shall be immune against side-channel attacks and fault attacks. In scientific papers the secret key is usually discovered completely, possibly after a (feasible) brute force search. No doubts remain whether the attack has been successful or not. In real-world security evaluations the situation may be quite different: By a side-channel attack or a by fault attack the evaluator might gain partial information on the targeted key, e.g. the Hamming weight of some key bytes. The fundamental question is whether this information is more or less useless or can be exploited efficiently to recover the whole secret key. From a purely academic point of view, of course, an implementation should not leak any information at all. The evaluator, however, has to decide whether the target of evaluation may yet be viewed as secure or whether it should be counted as broken. Usually, he has to make his decision within a short period.

In the Sections 3 to 5 three well-known examples are discussed where partial information could successfully be exploited by non-obvious mathematical methods. All-in-all we will see that implementation attacks raise many interesting questions, which narrows the 'gap' to mathematical cryptography. In particular, closer cooperation between research groups from both fields is certainly fruitful. It might be noted that the author has already addressed this topic in a presentation held at CryptArchi 2013. The feedback further encouraged him to write this paper.

Johannes Buchmann's scientific work mainly belongs to mathematical cryptography. We'd like to point out that the jubilar is co-author of [19], which is the subject of our third example.

2   Implementation Attacks with Partial Information

Ideally, side-channel attacks and fault attacks on secure devices should not reveal any information on the secret key at all. Unfortunately, this goal cannot always be achieved.

Consider, for example, the case where the attacker learns the Hamming weights of all key bytes of the final AES round. This knowledge obviously reduces the workload for exhaustive search. The advantage can be quantified, and the optimal guessing strategy is nearly straight-forward. In contrast, consider an implementation attack, which reveals the Hamming weight of all key bytes of an RSA prime. Is this information really helpful? Exhaustive search remains infeasible anyway, and it is not clear whether this information can be used to speed up the factorization of the RSA module. Or would the Hamming weight of some bytes of an RSA exponent be helpful? We leave it to the reader to extend this list of natural (open) questions, which raise challenging and important research topics.

To illustrate the topic in the following we briefly sketch three well-known examples where the gained information could successfully be exploited.

3   A Fault Attack on DSA

The first example summarizes [20]. It describes a fault attack on chip, which computes DSA signatures [10]. Recall that a DSA signature is a pair (r, s) with

$$r \equiv \left(g^k(\bmod p)\right)(\bmod q) \tag{1}$$
$$s \equiv k^{-1}(h(m) + xr)(\bmod q). \tag{2}$$

As usually, x denotes the secret (long-term) key, p (1024 bit) and q (160 bit) are primes such that q devides $p - 1$. Further, $1 < g < p$ has multiplicative order q modulo p, and $h :=$ SHA-1 is the hash function. The public key is given by the quadruple $(p, q, g, y := g^x(\bmod p))$. The ephemeral key k is a 160-bit number. As is well-known N signatures $(r_1, s_1), \ldots, (r_N, s_N)$ give an underdetermined system of N linear equations over $\mathrm{GF}(q)$ in $(N + 1)$ unknowns $(x, k_1, \ldots, k_N)$. The corresponding ephemeral keys $k_1, \ldots, k_N$ must be kept secret, and it shall not be able to guess them with non-negligible probabability or even to compute them. This demands a strong random number generator (RNG). The quality of the used RNG will yet not be considered in the following.

In [20] the generation of an ephemeral key k_i demands 20 calls of the function ReadRandomByte(), which returns one random byte.

```
For (i=0; i<20; i++)  k[i] := ReadRandomByte();
```

Note that $k = (k[0], \ldots, k[19])$. The attacker tries to terminate the loop ahead of time by a power glitch. In case of success (at least) the least significant byte $k[19]$ of the ephemeral key k keeps its pre-defined integer value a, in [20] $a = 0$. By an SPA the attacker can easily check whether the loop has indeed been terminated early ($\leftarrow$ shorter execution time).

We now assume that the attacker was able to generate N signatures (r_1, s_1), $\ldots, (r_N, s_N)$ for which the least significant byte of the ephemeral key k_i equals $a = 0$. From an information-theoretical point of view this clearly allows to solve the system of linear equations if N is sufficiently large. However, this property might not trouble an evaluator since all practical cryptographic algorithms are not secure in terms of information theory. In fact, still $(152N + 160)$ unkonwn bits remain. The decisive question yet is whether the gained information can be *practically* exploited. As explained below the answer on this question is yes.

Let (r, s) be a DSA signature, for which the least significant $l = 8$ bits of k are known, i.e. $k = 2^l \cdot b + a$ with known a and unknown b. Since $k < q$ clearly $b < \frac{q}{2^l}$. Elementary transformations yield

$$s \equiv k^{-1}(h(m) + xr)(\mathrm{mod}\ q)$$
$$s \cdot k \equiv h(m) + xr(\mathrm{mod}\ q)$$
$$xr \equiv s \cdot k - h(m)(\mathrm{mod}\ q)$$
$$x \cdot rs^{-1}(2^l)^{-1} \equiv -s^{-1}(2^l)^{-1}h(m) + b + (2^l)^{-1}a(\mathrm{mod}\ q)$$
$$xt \equiv v' + b(\mathrm{mod}\ q) \tag{3}$$

with known $t := rs^{-1}(2^l)^{-1}(\mathrm{mod}\ q)$ and $v' := (a - s^{-1}h(m))(2^l)^{-1}(\mathrm{mod}\ q)$. (As usual $z(\mathrm{mod}\ q)$ is the smallest nonnegative number with the same remainder $(\mathrm{mod}\ q)$ as z.) Equation (3) implies $0 \leq (xt - v') + cq = b < \frac{q}{2^l}$ for a suitable integer c, and $-\frac{q}{2^{l+1}} \leq xt - v' - \frac{q}{2^{l+1}} + cq < \frac{q}{2^{l+1}}$. For $v := v' + \frac{q}{2^{l+1}}$ we finally obtain

$$| xt - v |_q := \min_{w \in Z}\{| xt - v + wq |\} \leq \frac{q}{2^{l+1}} \tag{4}$$

In other words: $v(\mathrm{mod}\ q)$ may be viewed as an 'approximator' of $xt(\mathrm{mod}\ q)$. Consequently,

$$| xt_j - v_j |_q \leq \frac{q}{2^{l+1}} \qquad \text{for } 1 \leq j \leq N \tag{5}$$

for all signatures $(r_1, s_1), \ldots, (r_N, s_N)$ with interrupted loop. Equation (5) is equivalent to

$$| (xt_j + c_j q) - v_j | \leq \frac{q}{2^{l+1}} \qquad \text{for } 1 \leq j \leq N \text{ and suitable } c_j \in Z. \tag{6}$$

The key observation is that the left-hand sides (6) is small compared to q. The remaing task is to recover the secret key x from these information. It has already been well-known for many years how to tackle this problem: One transforms the search for the long-term key x into a closest vector problem of an $(N + 1)$-dimensional lattice. (The terms $xt_j + c_j q$ are elements of this lattice.) The closest vector problem can be solved with well-known methods. Indeed, 27 signatures with known last bytes of the ephemeneral keys suffice to recover the long-term key x [20].

Altogether, [20] provides an example where partial information on the ephemeral keys could successfully be exploited in a non-trivial way. However, the transfer to a lattice problem was already well-known when [20] was written, c.f. e.g. [21]. The attack may be viewed as a nice interplay between mathematical cryptography and implementation attacks.

However, an interesting question remains: Would the authors have detected their attack without that prior knowledge? Or alternatively: Assume that a security evaluation in 1999 would have detected that it was possible to interrupt the generation of the ephemeral keys by a glitch attack. What would have been the security assessment of this target of evaluation?

4 RSA: Exponent Blinding as a Countermeasure against Side-Channel Analysis

Our second example summarizes the main aspects of [29]. It considers an RSA implementation (CRT, square & always multiply), which is assumed to be secure against SPA and single trace template attacks (i.e., template attacks with sample size 1 in the attack phase). Additionally, exponent blinding [16], Sect. 10, shall prevent even stronger power attacks (DPA, CPA, template attacks with > 1 power traces in the attack phase etc.). More precisely, the exponentiation modulo p does not apply the exponent $d_p := d(\mathrm{mod}\, p)$ itself but $d_p + r(p - 1)$ with random blinding factor r where $0 \le r < 2^R$. In particular, for bases $y_1, \ldots, y_N$ random (secret) R-bit blinding factors $r_1, \ldots, r_N$ are used, which yield different exponents $v_1 := d_p + r_1(p-1), \ldots, v_N := d_p + r_N(p-1)$ for the exponentiation modulo p. If the attacker knows d_p or any v_j he can easily factorize the modulus $n = pq$, c.f. [29], Remark 1.

It is absolutely clear that exponent blinding increases the resistance of an RSA implementation against side-channel attacks. A more difficult question is, however, to which degree. Problems of this type are interesting by themselves but also relevant in the context of security evaluations. It should be noted that in security evaluations with regard to the Common Criteria (CC), for instance, the hardware is often evaluated first while software countermeasures (e.g. exponent blinding) are considered later in a so-called composite evaluation. In those cases it would be desirable to be able to quantify (e.g.) the impact of exponent blinding in order to 'extend' the security assessment of the hardware evaluation. A reasonable argumentation might be the following: Since the blinding factors are unknown it is not possible to combine the side-channel information from the particular power traces, which finally implies the security against (all types of) power attacks.

This argumentation yet is false. Reference [11], e.g. describes a successful power attack if some parts of the secret exponents are known with certainty. This is a strong assumption indeed, which is typical for fault attacks but is usually not fulfilled for power attacks. A more realistic assumption for power attacks and for side-channel attacks in general is that all exponent bits are only known with some probability. Reference [29] tackles this problem.

The vector $(v_{j;k+R-1}, \ldots, v_{j;0})_2$ denotes the binary representation of the blinded exponent v_j where leading zero digits are allowed (k-bit prime p with R-bit blinding factors). On basis of an SPA or a single trace template attack the attacker (resp., the evaluator) guesses the particular exponent bits. Each bit guess is false with probability ϵ_b ('error rate'). That is, instead of v_j the attacker guesses

$$\widetilde{v}_j = (\widetilde{v}_{j;k+R-1}, \ldots, \widetilde{v}_{j;0})_2 = v_j \oplus e_j \qquad \text{for } j = 1, 2, \ldots, N. \qquad (7)$$

The term '$\oplus$' denotes the bitwise XOR addition, and the integer e_j expresses the guessing error for exponent v_j. The attacker may commit two types of guessing errors: He may decide for $\widetilde{v}_{j;i} = 1$ although $v_{j;i} = 0$ is correct, or he may decide for $\tilde{v}_{j;i} = 0$ although $v_{j;i} = 1$ is true. For simplicity we assume that both error probabilities are equal ($= \epsilon_b$) and that all guesses are independent. In the Subsections 4.1 and 4.2 we describe two different attack variants (c.f. [29] for details). We mention that these attacks can easily be transferred to RSA without CRT and to ECC (point multiplication with constant scalar) [29].

4.1   The Basic Attack

For $1 \le j < m \le N$ we consider the hamming weight of

$$\tilde{v}_j \oplus \tilde{v}_m = (v_j \oplus v_m) \oplus (e_j \oplus e_m). \tag{8}$$

Then

$$\mathrm{ham}(\tilde{v}_j \oplus \tilde{v}_m) = \begin{cases} \mathrm{ham}(e_j \oplus e_m) & \text{if } r_j = r_m \\ \mathrm{ham}(v_j \oplus v_m \oplus e_j \oplus e_m) & \text{if } r_j \ne r_m \end{cases} \tag{9}$$

For $r_j \ne r_m$ we may assume that the term $\mathrm{ham}(v_j \oplus v_m \oplus e_j \oplus e_m)$ is a realization of a normally distributed random variable with mean $(k + R)/2$ and variance $(k + R)/4$ (as for $\mathrm{ham}(z)$ for a $(k + R)$-bit random number z). In contrast, if $r_j = r_m$ the mean equals $2\epsilon_b(1 - \epsilon_b) < 2\epsilon_b$. (Numerical example: For $(k, R, \epsilon_b) = (1024, 16, 0.20)$ both means are 520 and 332.8, respectively.) This means that from (9) a distinguisher between the cases $r_j \ne r_m$ and $r_j = r_m$ can be derived, namely

$$\text{Decide for } (r_j = r_m) \text{ iff } \mathrm{ham}(\tilde{v}_j \oplus \tilde{v}_m) < \gamma \quad \text{for suitable boundary } \gamma. \tag{10}$$

The attacker applies decison rule (10) to divide the exponent guesses $\tilde{v}_1, \ldots, \tilde{v}_N$ into classes so that within each class all elements have identical (yet unknown) blinding factors (provided, of course, that all decisions (10) have been correct; otherwise some classes may be 'spoiled' by exponents with different blinding factors). More precisely, the attacker decides whether $\tilde{v}_j$ belongs to any of the classes, which already exist. (Their union equals $\{\tilde{v}_1, \ldots, \tilde{v}_{j-1}\}$.) If this is not the case a new class is opened, which contains $\tilde{v}_j$. The first phase of the basic attack ends as soon as one class ('winning class') contains t elements ('t-birthday'). Of course, all t elements of the winning class (should) belong to the same blinding factor. In the second phase of the basic attack all exponent bits are treated independently. One decides for that value, which has been guessed more often for the t power traces of the winning class (majority decision rule; usually t is odd). Remaining bit errors (at most a very small number) are corrected later by exhaustive search. Of course, the larger ϵ_b the larger t has to be selected, which increases the sample size N and in particular the number of decisions (10). Simulation experiments in [29] verify the exemplary sets of suitable parameters $(k, R, \epsilon_b, t) = (1024, 10, 0.1, 7)$ and $(k, R, \epsilon_b, t) = (1024, 10, 0.2, 17)$, for instance. The (admittedly artificially small) parameter $R = 10$ allows to tackle an error rate of $\epsilon_b = 0.28$. For increasing R the tolerable error rate decreases since Phase 1 requires more correct decisions. Heuristic arguments indicate that for $R = 16$ still $\epsilon_b = 0.25$ can be tolerated.

For medium-sized or even large blinding parameter R a sample size N in the order of 2^R (very coarse estimate) is needed before a t-birthday occurs. (If many false decisions occur this number further increases.) The second bottleneck is the number of decisions (10). Note that for $\widetilde{v}_j$ in average more decisions (10) than half of the number of already existing classes are required. Both the number of power traces and the number of decisions make the attack impractical for large R [29], Subsect. 2.3.

An effective countermeasure against the basic attack is to limit the maximum number of applications of the RSA key (e.g., for digital signatures) clearly below $2^{R/2}$, which even prevents 2-birthdays. This countermeasure prevents the basic attack rigorously. But does it indeed solve all problems?

4.2   The Enhanced Attack

The enhanced attack gets by with significantly less power traces and less computations than the basic attack. On the negative side the tolerable error rate ϵ_b decreases. Again, N denotes the number of power traces. Further, $u > 1$ denotes a small integer (let's say, $u \leq 4$) and $M_u := \{(j_1, \ldots, j_u) \mid 1 \leq j_1 \leq \cdots \leq j_u \leq N\}$. For each u-tuple $(j_1, \ldots, j_u) \in M_u$ we define the 'u-sum' $S_u(j_1, \ldots, j_u) := r_{j_1} + \cdots + r_{j_u}$. For $(j_1, \ldots, j_u), (i_1, \ldots, i_u) \in M_u$ we write $(j_1, \ldots, j_u) \sim (i_1, \ldots, i_u)$ iff $S_u(j_1, \ldots, j_u) = S_u(i_1, \ldots, i_u)$. In analogy to the basic attack we need to distinguish between the two cases $(j_1, \ldots, j_u) \sim (i_1, \ldots, i_u)$ and $(j_1, \ldots, j_u) \nsim (i_1, \ldots, i_u)$, Therefore, we apply the following decision rule:

$$\text{Decide for } (j_1, \ldots, j_u) \sim (i_1, \ldots, i_u) \text{ iff} \tag{11}$$
$$\text{ham}(\text{NAF}(\widetilde{v}_{j_1} + \cdots + \widetilde{v}_{j_u} - (\widetilde{v}_{i_1} + \cdots + \widetilde{v}_{i_u}))) < b_0$$
$$\text{for some suitable boundary } b_0.$$

The term $\text{NAF}(z)$ denotes the NAF representation of the integer z. The justification of (11) is analogous to the basic attack. More precisely, for $\nsim$ the term $\text{ham}(\text{NAF}(\widetilde{v}_{j_1} + \cdots + \widetilde{v}_{j_u} - (\widetilde{v}_{i_1} + \cdots + \widetilde{v}_{i_u})))$ may be viewed as a realization of a normally distributed random variable with mean $(k + R)/3$ and (rather small) variance $0.334 * (k + R)$ [29], Lemma 1(ii). In case of '$\sim$' we have

$$\text{ham}(\text{NAF}(\widetilde{v}_{j_1} + \cdots + \widetilde{v}_{j_u} - \widetilde{v}_{i_1} - \cdots - \widetilde{v}_{i_u})) \leq \tag{12}$$
$$\text{ham}(\text{NAF}(e_{j_1})) + \cdots + \text{ham}(\text{NAF}(e_{j_u})) + \cdots + \text{ham}(\text{NAF}(e_{i_u})) \leq$$
$$\text{total number of guessing errors for } v_{j_1}, \ldots, v_{j_u}, v_{i_1}, \ldots, v_{i_u}.$$

Consequently, the corresponding normal distribution has small mean (depending on ϵ_b, of course). Each decision for $(j_1, \ldots, j_u) \sim (i_1, \ldots, i_u)$ gives a linear equation $S_u(j_1, \ldots, j_u) = S_u(i_1, \ldots, i_u)$ in the blinding factors. The goal of Step 1 of the enhanced attack is to obtain a system of homogeneous linear equations over Z with co-rank 2, i.e. rank(kernel) = 2.

Example 1. The quintuples (k, R, u, N, ϵ_b) below constitute sets of parameters, which yield a system of linear equations with co-rank 2 with high probability (if not, additional power traces are needed): $(1024, 16, 2, 128, 0.13)$, $(1024, 16, 3, 30, 0.08)$, $(1024, 16, 4, 20, 0.06)$, $(1024, 32, 2, 4700, 0.10)$ etc.

If the kernel has rank 2 it is spanned by the vectors $(1, \ldots, 1)^t$ and $(r_1, \ldots, r_N)^t$. In Step 2 the attacker solves the system of linear equations, and finally in Step 3 an error-correcting algorithm reveals $\phi(p) = p - 1$. The by far most time-consuming part is Step 1.

The enhanced attack scales much better than the basic attack but due to the number of decisions sufficiently large R prevents the enhanced attack, too. For example, $(R, u) = (64, 4)$ requires about 2^{77} decisions while $(64, 2)$ and $(64, 3)$ are even more costly. This is an extremely large workload for a side-channel attack. Since the tolerable error rate ϵ_b is small [29] and since one may not be sure in advance whether the power attack yields sufficiently good guesses $\tilde{v}_1, \ldots, \tilde{v}_N$ one might view an attack against $R = 64$ as not practical. However, when using $R > 64$ (e.g., $R = 96$) one is on the safe side, anyway. Of course, strong hardware (enforcing large error rates ϵ_b) prevents both the basic and the enhanced attack, regardless of the parameter R.

5   Algebraic Side-Channel Analysis

The key of a block cipher can be expressed as the solution of a system of non-linear equations over $GF(2)$. From the AES encryption function and the key schedule together 1696 linear equations and 4600 quadratic equations in 3296 variables over $GF(2)$ can be derived [3], and the searched AES key is a subvector of its solution. Due to the simple algebraic structure of the AES cipher at the beginning of this millenium a lot of researchers tried to solve such systems of nonlinear equations, or more precisely to estimate the required workload. Often linearization techniques (XL, XSL) were used [3, 9] etc. Reference [22] describes an alternate interesting approach, which considers nonlinear equations over $GF(2^8)$. These approaches seemed to promising at that time but turned out to be ineffective later.

In fact, no efficient algorithm is known so far. Of course, the situation should become more favourable if additional equations are known, and in fact this is the central idea of algebraic side-channel analysis. These additional equations stem from a side-channel attack, e.g. from a cache attack or from a power attack [19, 32] etc.

Reference [19] applies power analysis against an unprotected (non-masked) AES implementation. In the first step single trace template attacks were applied, which allow to guess the Hamming weights of 84 bytes within each AES round (16 bytes before and 16 bytes after the S-Box layer, 52 bytes during the MixColumn operation). This approach assumes that the Hamming weight model describes the leakage of the targeted device sufficiently well. In [19] only the AES cipher but not the key schedule algorithm was targeted.

Applying the template attacks to two consecutive rounds i and $i + 1$, for instance, gives $2 \cdot 84 = 168$ Hamming weights, which results in 168 equations. In the first part of the paper the optimal case was treated where all guessed

Hamming weights are correct. Only these 168 equations were considered, aiming at the round key k_i. Applying the key schedule algorithm finally yields the searched AES key. (If the guess for k_i is not unique the knowledge of a (plaintext / ciphertext) pair allows to identify the correct key candidate.) To solve the equations J. Buchmann et al. in [19] - the lead author may excuse the unconventional citation due to the special circumstances of a festschrift - applied a combination of Gröbner basis techniques with SAT solvers, meaningfully denoted by IASCA (improved algebraic side-channel analysis). Experiments verified a success rate of $> 90\%$ (cf. [19], Subsect. III.B, Table IV). The experiments were conducted on a Sun X4440 server with four CPUs (Quad-Core AMD OpteronTM Processor 8356), each CPU running at 2.3 MHz, and 128 GB main memory were available. The upper time limit per trial was 100 sec.

Remark 1. Masking a block cipher is rather costly as it requires re-masking in each round. On the other hand usually power attacks target only the first or the last round(s). Hence the following approach might be reasonable: To save computation time and / or space the designer masks only the Rounds 1-4 and 7-10 (or maybe only Rounds 1-3 and 8-10) while Round 5 and Round 6 (resp., Round 4 to 7) remain unprotected. However, as we have just learnt applying the attack from [19] to the unmasked Rounds 5 and 6 reveals the key with high probability. In other words: One should definitely forget this idea! We point out that this scenario reminds of [13] where a partially masked DES implementation was targeted. Instead of the solution of nonlinear equations there differential cryptanalysis was supported. Of course, [13] could also have served as an adequate example for this section.

So far we have assumed that all Hamming weight guesses have been correct, which is an optimistic (and usually non-realistic) assumption in the context of single trace template attacks. In its second part [19] also allows wrong guesses, which of course increases the necessary number of Hamming weight guesses. The template attack was performed on an ATMega256-1 microcontroller, c.f. [19], Sect. IV for details.

In the first scenario (all guesses correct) IASCA was compared with two other methods, which had been proposed before, and also in the second scenario (false guesses allowed) IASCA was compared with a prior technique. The authors point out that IASCA is most efficient in both scenarios [19].

6   Conclusion

In Sections 3 to 5 three examples were discussed where non-obvious mathematical methods were successfully used to exploit information gained by an implementation attack. While the problems in first two scenarios might seem to be 'occasional' algebraic side-channel analysis has become a field of research for its own.

Besides their practical relevance, which was worked out in the introduction and in Section 2, problems arising from implementation attacks with partial information are interesting and challenging. They offer a broad area for research work that connects implementation attacks with mathematical (algorithm-oriented) cryptography.

References

1. Acıiçmez, O., Schindler, W.: A Vulnerability in RSA Implementations due to Instruction Cache Analysis and Its Demonstration on OpenSSL. In: Malkin, T. (ed.) CT-RSA 2008. LNCS, vol. 4964, pp. 256–273. Springer, Heidelberg (2008)
2. Archambeau, C., Peeters, E., Standaert, F.-X., Quisquater, J.-J.: Template Attacks in Principal Subspaces. In: Goubin, L., Matsui, M. (eds.) CHES 2006. LNCS, vol. 4249, pp. 1–14. Springer, Heidelberg (2006)
3. Biryukov, A., De Cannière, C.: Block Ciphers and Systems of Quadratic Equations. In: Johansson, T. (ed.) FSE 2003. LNCS, vol. 2887, pp. 274–289. Springer, Heidelberg (2003)
4. Boneh, D., DeMillo, R.A., Lipton, R.J.: On the Importance of Checking Cryptographic Protocols for Faults (Extended Abstract). In: Fumy, W. (ed.) EUROCRYPT 1997. LNCS, vol. 1233, pp. 37–51. Springer, Heidelberg (1997)
5. Brier, E., Clavier, C., Olivier, F.: Correlation Power Analysis with a Leakage Model. In: Joye, M., Quisquater, J.-J. (eds.) CHES 2004. LNCS, vol. 3156, pp. 16–29. Springer, Heidelberg (2004)
6. Brumley, D., Boneh, D.: Remote Timing Attacks are Practical. In: 12th Usenix Security Symposium. Usenix Association (2003)
7. Chari, S., Rao, J.R., Rohatgi, P.: Template Attacks. In: Kaliski Jr., B.S., Koç, Ç.K., Paar, C. (eds.) CHES 2002. LNCS, vol. 2523, pp. 13–28. Springer, Heidelberg (2003)
8. Coron, J.-S., Goubin, L.: On Boolean and Arithmetic Masking against Differential Power Analysis. In: Paar, C., Koç, Ç.K. (eds.) CHES 2000. LNCS, vol. 1965, pp. 231–237. Springer, Heidelberg (2000)
9. Courtois, N., Klimov, A., Patarin, J., Shamir, A.: Efficient Algorithms for Solving Overdefined Systems of Multivariate Polynomial Equations. In: Preneel, B. (ed.) EUROCRYPT 2000. LNCS, vol. 1807, pp. 392–407. Springer, Heidelberg (2000)
10. FIPS PUB 186-4: Digital Signature Standard (DSS). NIST (July 2013), http://nvlpubs.nist.gov/nistpubs/FIPS/NIST.FIPS.186-4.pdf
11. Fouque, P., Kunz-Jacques, S., Martinet, G., Muller, F., Valette, F.: Power Attack on Small RSA Public Exponent. In: Goubin, L., Matsui, M. (eds.) CHES 2006. LNCS, vol. 4249, pp. 339–353. Springer, Heidelberg (2006)
12. Gierlichs, B., Batina, L., Tuyls, P., Preneel, B.: Mutual Information Analysis - A Generic Side-Channel Distinguisher. In: Oswald, E., Rohatgi, P. (eds.) CHES 2008. LNCS, vol. 5154, pp. 426–442. Springer, Heidelberg (2008)
13. Handschuh, H., Preneel, B.: Blind Differential Cryptanalysis for Enhanced Power Attacks. In: Biham, E., Youssef, A.M. (eds.) SAC 2006. LNCS, vol. 4356, pp. 163–173. Springer, Heidelberg (2007)

14. Heuser, A., Kasper, M., Schindler, W., Stöttinger, M.: How a Symmetry Metric Assists Side-Channel Evaluation - A Novel Model Verification Method for Power Analysis. In: Kitsos, P. (ed.) 14th Euromicro Conference on Digital System Design, DSD 2011, pp. 674–681. IEEE Press (2011)
15. Kasper, M., Schindler, W., Stöttinger, M.: A Stochastic Method for Security Evaluation of Cryptographic FPGA Implementations. In: 2010 International Conference on Field-Programmable Technology, FPT 2010, pp. 146–153. IEEE Press, CFP10528_CDR (2010)
16. Kocher, P.C.: Timing Attacks on Implementations of Diffie-Hellman, RSA, DSS and Other Systems. In: Koblitz, N. (ed.) CRYPTO 1996. LNCS, vol. 1109, pp. 104–113. Springer, Heidelberg (1996)
17. Kocher, P., Jaffe, J., Jun, B.: Differential Power Analysis. In: Wiener, M. (ed.) CRYPTO 1999. LNCS, vol. 1666, pp. 388–397. Springer, Heidelberg (1999)
18. Lemke-Rust, K., Paar, C.: Analyzing Side Channel Leakage of Masked Implementations with Stochastic Methods. In: Biskup, J., López, J. (eds.) ESORICS 2007. LNCS, vol. 4734, pp. 454–468. Springer, Heidelberg (2007)
19. Mohamed, M.S.E., Bulygin, S., Zohner, M., Heuser, A., Walter, M., Buchmann, J.: Improved Algebraic Side-Channel Attack on AES. In: HOST 2012, pp. 156–161. IEEE Press (2012)
20. Naccache, D., Nguyên, P.Q., Tunstall, M., Whelan, C.: Experimenting with Faults, Lattices and the DSA. In: Vaudenay, S. (ed.) PKC 2005. LNCS, vol. 3386, pp. 16–28. Springer, Heidelberg (2005)
21. Mahassni, E.E., Nguyên, P.Q., Shparlinski, I.E.: The Insecurity of Nyberg-Rueppel and Other DSA-like Signature Schemes with Partially Known Nonces. In: Silverman, J.H. (ed.) CaLC 2001. LNCS, vol. 2146, pp. 97–109. Springer, Heidelberg (2001)
22. Murphy, S., Robshaw, M.: Essential Algebraic Structure within the AES. In: Yung, M. (ed.) CRYPTO 2002. LNCS, vol. 2442, pp. 1–16. Springer, Heidelberg (2002)
23. Peeters, E., Standaert, F.-X., Donckers, N., Quisquater, J.-J.: Improved Higher-Order Side-Channel Attacks with FPGA Experiments. In: Rao, J.R., Sunar, B. (eds.) CHES 2005. LNCS, vol. 3659, pp. 309–323. Springer, Heidelberg (2005)
24. Schindler, W.: A Timing Attack against RSA with the Chinese Remainder Theorem. In: Paar, C., Koç, Ç.K. (eds.) CHES 2000. LNCS, vol. 1965, pp. 109–125. Springer, Heidelberg (2000)
25. Schindler, W., Koeune, F., Quisquater, J.-J.: Improving Divide and Conquer Attacks Against Cryptosystems by Better Error Detection / Correction Strategies. In: Honary, B. (ed.) Cryptography and Coding 2001. LNCS, vol. 2260, pp. 245–267. Springer, Heidelberg (2001)
26. Schindler, W.: A Combined Timing and Power Attack. In: Naccache, D., Paillier, P. (eds.) PKC 2002. LNCS, vol. 2274, pp. 263–279. Springer, Heidelberg (2002)
27. Schindler, W., Lemke, K., Paar, C.: A Stochastic Model for Differential Side Channel Cryptanalysis. In: Rao, J.R., Sunar, B. (eds.) CHES 2005. LNCS, vol. 3659, pp. 30–46. Springer, Heidelberg (2005)
28. Schindler, W.: Advanced Stochastic Methods in Side Channel Analysis on Block Ciphers in the Presence of Masking. Math. Crypt. 2, 291–310 (2008)
29. Schindler, W., Itoh, K.: Exponent Blinding Does not Automatically Lift (Partial) SPA Resistance to Higher-Level Security. In: Lopez, J., Tsudik, G. (eds.) ACNS 2011. LNCS, vol. 6715, pp. 73–90. Springer, Heidelberg (2011)

30. Schramm, K., Paar, C.: Higher Order Masking of the AES. In: Pointcheval, D. (ed.) CT-RSA 2006. LNCS, vol. 3860, pp. 208–225. Springer, Heidelberg (2006)
31. Standaert, F.-X., Malkin, T.G., Yung, M.: A Unified Framework for the Analysis of Side-Channel Key Recovery Attacks. In: Joux, A. (ed.) EUROCRYPT 2009. LNCS, vol. 5479, pp. 443–461. Springer, Heidelberg (2009)
32. Zhao, X., Zhang, F., Guo, S., Wang, T., Shi, Z., Liu, H., Ji, K.: MDASCA: An Enhanced Algebraic Side-Channel Attack for Error Tolerance and New Leakage Model Exploitation. In: Schindler, W., Huss, S.A. (eds.) COSADE 2012. LNCS, vol. 7275, pp. 231–248. Springer, Heidelberg (2012)

AMASIVE: An Adaptable and Modular Autonomous Side-Channel Vulnerability Evaluation Framework

Sorin A. Huss[1], Marc Stöttinger[2], and Michael Zohner[3]

[1] Integrated Circuits and Systems Lab, Computer Science Dept.,
Technische Universität Darmstadt, Germany
huss@vlsi.informatik.tu-darmstadt.de
[2] Physical Analysis and Cryptographic Engineering, SPMS,
Nanyang Technological University, Singapore
mstottinger@ntu.edu.sg
[3] EC-SPRIDE,
Engineering Cryptographic Protocols Group, Germany
michael.zohner@ec-spride.de

Abstract. Over the last decades computer aided engineering (CAE) tools have been developed and improved in order to ensure a short time-to-market in the chip design business. Up to now, these design tools do not yet support a design strategy for the development of side-channel resistant hardware implementations. In this chapter we introduce a novel engineering framework named AMASIVE (*Adaptable Modular Autonomous SIde-Channel Vulnerability Evaluator*), which supports the designer in implementing side-channel hardened devices. An attacker model is introduced for the analysis and the evaluation of a given cryptographic design in regard to application-specific vulnerabilities and exploitations. We demonstrate its application to a hardware implementation of the block cipher PRESENT.

1 Introduction

Nowadays, embedded devices find application in an increasing part of everyday life. Due to the on-going progress of the fabrication technology integrated devices get more and more powerful in terms of computational throughput as well as complexity. However, the increasing number of embedded devices also gives rise to potential unintended exploitations. In order to secure the devices against misuse and to create a secure communication environment, various protocols and cryptographic schemes have been developed.

In the last two decades another attack type emerged, targeting directly at implementations that process sensitive data. This kind of assaults is called *side-channel analysis attacks* and exploit physical observables that depend upon the processed data. Side-channel attacks are especially dangerous due to their general applicability to cryptographic schemes and to their non-invasive nature. Thus, a side-channel attack can be mounted on various implementations and a compromised secret is very difficult to be detected.

M. Fischlin and S. Katzenbeisser (Eds.): Buchmann Festschrift, LNCS 8260, pp. 151–165, 2013.
© Springer-Verlag Berlin Heidelberg 2013

During a side-channel evaluation, the hardware designer provides all necessary information to the security engineer. For the evaluation phase, the security analyst has to select an appropriate attack method that depends on the threat scenario of the circuit's application scenario. Additionally, he or she has to construct a physical model of the device that is as precise as possible and describes the expected exploitable information leakage. Both the model and the prototype are then taken to the evaluation phase. This phase aims at identifying exploitable side-channel leakage and, if possible, also its origin. If a successful attack is identified, this information can then be used to restart the design cycle in order to improve, i. e., harden, the hardware module. Such a design flow requires at least a designer, who is skilled in both in side-channel analysis and hardware design. However, professionals who have expertise in both areas are hard to find.

In this chapter we investigate the question how to add an automated and supportive side-channel evaluation to the conventional CAE toolset. We thereby propose a framework that supports both roles in the design process, the security engineer as well as the hardware engineer, in performing side-channel analysis and evaluation. The proposed design framework is able to automatically analyze a given HDL design description of a cryptographic module in terms of information and data flow. This chapter is structured as follows: In Section 2 we provide some background information to side-channel analysis and to related work followed by Section 3, which introduces the core idea of AMASIVE. The main focus of this chapter is to focus on the analysis section of the AMASIVE framework. After detailing of the model building and the automative analysis of the data we provide as a comprehensive application example a FPGA-based hardware implementation of the PRESENT block cipher.

2   Background and Related Work

In the academic cryptographic community, implementation attacks were first introduced in the nineties. Such attacks use weaknesses in the implementation of a cryptographic algorithm or application that handles security-sensitive information. One class of implementation attacks are the so-called side-channel attacks, which we further detail in this section.

2.1   Side-Channel Attacks

Side-channel attacks (SCA) are a class of non-invasive implementation attacks, which exploit physical observables of an implementation of a cryptographic algorithm in order to extract secret information, for instance the secret key of an encryption scheme. They are especially threatening since it is hard to notice a successful side-channel attack. These kinds of attacks are mounted during the operation of a device, i. e., runtime, while the cryptographic module is running in normal operation mode without any violation to its operational specification. The most commonly exploited observables are run-time of execution, dynamic power consumption, and electromagnetic emission. We refer to [4] for a more detailed introduction to the basic side-channel analysis methods.

Power analysis attacks exploit the data-dependent switching activity of a cryptographic implementation in order to extract information of intermediate values of internal operations or of the current internal state of the algorithm. The hypotheses base on the expected values of the observed physical quantity caused by the adversary's focussed intermediate value changes or state transitions. The construction of theses hypotheses does not require detailed information about the implementation in general and targets only parts of the complete secret key[1] following the *divide and conquer* principle.

2.2 State of the Art

The reasons for a rather late consideration of side-channel vulnerabilities are manifold such as the short time to market, unpredictability of side-channel properties during the design process, or an yet unidentified side-channel information leakage of the algorithm. Side-channel vulnerabilities are often detected too late to be taken into account when deciding on the architecture of the cryptographic module. This situation is rather unfortunate, because the designer has access to an enormous amount of knowledge about the device properties. So, she or he is in the position to perform a much more accurate side-channel analysis than an outside attacker ever could. However, identifying, combining, and utilizing this comprehensive knowledge is a difficult task and requires experience in side-channel analysis next to hardware design skills. Recently, there have been a few contributions that aim at a timely support of securing software and hardware implementations of cryptographic algorithms against side-channel attacks.

In [7] Standaert et al. proposed an unified framework to cope with side-channel issues of algorithm implementations. These authors introduce various formalisms and methods to identify side-channel leakage in order to compare the vulnerability of several design variants. Countermeasures for side-channel attacks to software implementations is tackled by two approaches, which provide automated code analysis and countermeasure integration, cf. [1,5,6]. The work in [1] applies a side-channel attack to the assembled code and utilizes an estimation of the mutual information [7] to evaluate and to highlight vulnerabilities of a code section. The second approach, proposed by Moss et al. [5,6], features an automatic insertion of masks by evaluating the secrecy requirements of an intermediate value in the program code.

In this contribution we introduce a novel approach, which supports the designer in a comprehensive way when developing side-channel resistant hardware devices. The AMASIVE framework is aimed to automatically identify side-channel vulnerabilities at different stages in the design flow of embedded systems, to suggest countermeasures, and to quantify the effectivity of the proposed countermeasures. Thereby, the designer is notified early about potential vulnerabilities and can thus assess the costs of implementing various countermeasures in the design. To the best of our knowledge, the AMASIVE framework is the first approach to automatically identifying vulnerabilities early in the hardware

---

[1] So-called subkeys.

design process without requiring a detailed knowledge about the implemented cryptographic algorithm. The strengths of this design tool set are its modularized structure and the adaptability to various implementations by exploiting several attacking procedures, leakage models, and side-channel distinguishers. Similar to [7], we consider an attacker model for our analysis in order to secure an implementation especially with respect to its threat scenario.

3   The AMASIVE Concept

The core idea of the AMASIVE framework relies on a multistage approach of security analysis and dedicated user-controlled, automatic countermeasure insertions, cf. Figure 1. The security analysis process can further be subdivided into *information collection*, *graph representation*, and *vulnerability analysis* phase, respectively. In summary, the information collection phase combines all relevant design information that are then used to construct an intermediate graph representation, which in turn forms the basis for the vulnerability analysis. Its output is then forwarded to the designer and the countermeasure generation module.

Depending on the state of the design process, more detailed information about the design can be included in the graph representation and considered in the subsequent security analysis. Thus, while potential side-channel vulnerabilities can be vague at the beginning of the design process, they can become more concrete at later stages.The designer is informed about potential vulnerabilities and can thus take them into account at an early stage in the design process.

In the following we first detail the security analysis and then present how AMASIVE autonomously formulate hypothesis for detecting side-channel flaws. Since the advocated concept is general and can be applied in the context of both different architectures and design tools, we first discuss it from a high-level point of view as depicted in Fig. 1 and then discuss a detailed instantiation of a cryptographic module using VHDL code as the information source.

3.1   Automated Side-Channel Analysis

AMASIVE identifies side-channel vulnerabilities from three steps as depicted in Figure 2. First, various information is collected from an arbitrary set of sources, such as the given implementation of a crypto-system or directly from its designer (cf. Figure 2a). Secondly, this information is combined into a graph, which represents both the design and an attacker model, which specifies the attacker's capabilities (cf. Figure 2b). Third, both the graph and the attacker model are used as the foundation to identify and to then highlight side-channel vulnerabilities directly in the design (cf. Figure 2c). The results from this analysis phase can directly be used either for an actual side-channel attack or harden support of the designer in considering and selecting countermeasure in order to harden the design.

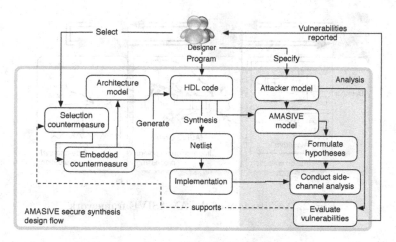

Fig. 1. Overview on the AMASIVE work-flow

The graph and the attacker model are the basic parts of the security analysis and represent the relation between the device and the attacker in the real world. In the following, we start with a general overview on the information collection, the graph and the attacker model, and the concept of the security analysis. A more detailed presentation on the realization of these phases is given in Section 3.5.

3.2   Information Collection

In the information collection phase the basic data that are required for the vulnerability analysis are collected from various sources. These sources depend on the type of the implementation and on the design tools being used. For example, for a VHDL design the required information can be obtained directly from the VHDL-based source code, the related simulation results, and finally the synthesized net list. Additionally, we require the designer to specify the envisaged capabilities of the attacker, such as his available side-channel attack methods, the maximum number of traces that can be measured, and the computational power at hand.

We structure the type of information that we require for the subsequent analysis into two categories: structural and functional information. Structural information describes the information flow of the algorithm as well as the interconnection of the related modules in a hardware design. This type of information is aimed to construct the graph representation of the algorithm during the graph representation phase. In contrast, functional information describes the operations that are performed when processing the data. This type of information is used to evaluate the complexity and to generate an adequate side-channel hypothesis during the vulnerability analysis. Note that both types of information can be provided at a different granularity, thus allowing the vulnerability analysis to

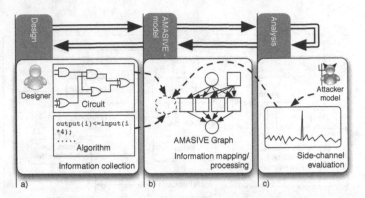

Fig. 2. Architecture of the AMASIVE framework

vary in precision. For instance, if the VHDL code is already written but not directly synthesizable in full detail, at least the interconnection of the components can already be extracted and potential information leaks can be highlighted. When the design is synthesizable and a target platform is available, the design extraction can be implemented physically and a complete side-channel analysis, using the pre-determined information leaks, may be performed.

3.3 Models

In order to make the security analysis independent of a specific implementation, we rely on the collected information to construct two abstract models. The first model is a graph representation of the design that models the control and data-flow of the implementation at hand. The second model specifies the capabilities of the attacker and can be adjusted without any relation to the envisaged implementation. Both models interact during the subsequent security analysis in order to determine potential side-channel vulnerabilities.

Graph Representation Model. The AMASIVE graph as outlined in Figure 2b) models the device in terms of its architectural and algorithmic features and forms the basis for both the security analysis and the communication with the designer. We define the graph G as a relation between a set of nodes V, which represent execution modules, and a set of edges E, which connect nodes. Nodes can have input edges that specify the origin of the processed data and output edges that specify the resulting values. Since the AMASIVE graph is used as the foundation for a later side-channel evaluation, we define it such that it may contain all relevant information for the analysis. That is, we model how the values are processed, at which steps in the algorithm a secret is being accessed, and by which instructions during the execution secret values are stored. Accordingly, we distinguish between three kinds of nodes: *operations*, *entropy sources*, and *registers*. Operations describe modules, which modify the processed data, such as an SBox module or an XOR gate. Entropy sources model secret values, inserted

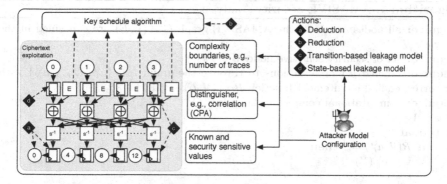

Fig. 3. Scheme of the attacker model configuration

during the execution of the algorithm, i.e., keys or random numbers. Lastly, registers are elements in an architecture that store data and thus pose a potential point of data leakage. Additionally, both the start and end nodes of the graph are marked in order to identify both the input to and the output of an algorithm implemented in hardware.

Attacker Model. The attacker model is used to model attack scenarios in which the attacker aims at recovering security sensitive information under the assumption of various properties. This model is then taken into consideration to evaluate the security strength of the design. Figure 3 depicts an example for the attacker model as well as its assignable properties. We assign, similar to the notation in [8], a local view $\mathbb{V} \subseteq (V \times \mathbb{R})$ to the attacker model that sets all nodes of the previously introduced graph $G = (V, E)$ in relation to an entropy level, which in turn specifies the uncertainty of the attacker regarding the value of the node. Furthermore, we create a set $\mathbb{S} \subseteq \mathcal{P}(V)$ consisting of subsets of security sensitive nodes, which may lead to a key recovery when they are known to the attacker. With the known and security sensitive nodes marked, the attacker model is aware of both the starting and the target points in the graph. However, the attacker model still requires a set of actions $\mathbb{A}$ for traversing the graph. These actions represent the performable assaults of the attacker model. An action $a \in \mathbb{A}$ is defined as a function $a : \mathbb{V} \mapsto \mathbb{V}$. A requirement function $R_G : (\mathbb{V}, \mathbb{A}) \mapsto \{0, 1\}$ checks if the requirements for performing any action $a \in \mathbb{A}$ in the attacker view $\mathbb{V}$ are satisfied. R_G is defined as

$$R_G(\mathbb{V}, a) = \begin{cases} 1 \text{ , if requirements hold for } \mathbb{V} \\ 0 \text{ , else.} \end{cases} \tag{1}$$

The strength of the attacker can be increased or decreased by adding or removing known nodes and actions, or by changing the complexity boundaries. Note that the attacker model is independent of the constructed graph and can thus be varied without requiring any change in the graph.

Algorithm 1. AMASIVE Security Analysis

Require: all nodes $n \in V$ of the AMASIVE graph $G = (V, E)$ are contained in the view $\mathbb{V}$

Require: set of security sensitive nodes $\mathbb{S} \subseteq \mathcal{P}(V)$

Require: ordered list of N actions to traverse the graph $\mathbb{A} = \{a_1^G, a_2^G, ..., a_N^G\}$ with corresponding requirement function $R_G : (\mathbb{V}, \mathbb{A}) \mapsto \{0, 1\}$

Require: computational complexity boundary C

1:   $i = 1$
2:   **repeat**
3:     **if** $R(\mathbb{V}, a_i^G) = 1$ **then**
4:       $\mathbb{V} = a_i^G(\mathbb{V})$
5:       $i = 1$
6:     **else**
7:       $i = i + 1$
8:     **end if**
9:   **until** $i > N$
10: Attacker wins if $\exists \{s_1, s_2, ..., s_p\} \in \mathbb{S}$ with $(s_j, \epsilon_j) \in \mathbb{V}$ such that $(\sum_{j=1}^{p} \epsilon_j) < C$

3.4   Security Analysis

We express the security analysis as a game where the goal of the attacker is to recover a set of security sensitive nodes from the graph $G = (V, E)$. The attacker is given a set of initially known nodes $\mathfrak{I}$ by adding these nodes with an entropy of zero to his view $\mathbb{V}$:

$$\mathbb{V} = \bigcup_{\forall n \in \mathfrak{I}} \{(n, 0)\}. \tag{2}$$

Every node n, contained in the set $(V \backslash \mathfrak{I})$, is added to $\mathbb{V}^+$ together with its respective number of bits n_b, as entropy:

$$\mathbb{V}^+ = \mathbb{V} \cup \bigcup_{\forall n \in (V \backslash \mathfrak{I})} \{(n, n_b)\}. \tag{3}$$

Subsequently, the attacker tries to apply actions to recover information about more nodes. Algorithm 1 details the security analysis game. The attacker wins the game if the complexity for guessing a group of security sensitive nodes is lower than his pre-defined computational complexity bound. If the attacker wins the game, then the designer is notified by highlighting the recovered nodes and by visualizing the corresponding sequence of actions that led to the node's recovery.

3.5   Diagnosis Component

In the following we detail an implementation of AMASIVE, which analyzes FPGA designs represented in VHDL code. First, we describe the sources from which we collect the required information for the analysis. Then we outline the construction of the graph. Lastly, we detail the attacker model and the subsequent steps performed during the security analysis.

Information Collection. First, we need information about the hardware design, which is the structure of the algorithm mapped to the hardware device and a functional description of the operations. Second, we need information about the attacker capabilities to create the attacker model. All the needed information is obtainable from the following sources: VHDL source code, its simulation results, and the designer. The VHDL source code is parsed in order to extract the structure information of the algorithm. Additionally defined constrains and added flags in the VHDL source code increases the readability of the code to generate the graph. For instance, each VHDL instance is modeled as a node element in the graph. In this scheme the graph can contain the four different node elements introduced above.

The AMASIVE framework is able to map the complete functionality of the investigated algorithm to the graph model. Therefore, simulation procedures and the tools from the Xilinx, Inc. design suite ISE aimed to related SRAM-based FPGA implementations are exercised to generate a functional lookup table. Hence, the complete functionality of the VHDL code can then be mimicked by the graph due to simulation procedures and automated learning. We only query the designer if information can not be obtained autonomously from the given code and related simulation results. In the current version of AMASIVE, we ask the designer to specify the attacker model by only stating the public and security sensitive values and by defining the security constraints.

Constructing the Models. In order to construct the models, we first need to instantiate all elements of the envisaged XML description of the graph as nodes. The instantiation ensures that each implemented module is also represented in the graph. When each entity is instantiated, loops in the VHDL code are unrolled. From the instantiation of the entities we obtain a node for every execution of a module on the FPGA.

In the next step the nodes are connected using the channel elements. The connected nodes result in a graph that represents the steps in the execution of the algorithm. Subsequently, permutations are dissolved, since on an FPGA they are represented as a permutation of wires. The dissolving is done by using the permutation lookup table, gained from the simulation of the VHDL design, and connecting the outputs of the previous and the inputs of the next elements of the permutation. The last step in the process of constructing the graph is the assignment of non-trivial properties to the nodes. These attributes address features that have to be computed from a combination of directly readable features. Non-trivial properties are, for instance, the invertibility of functions, the mapping properties of functions (e.g., bijectivity), and the vulnerability of a function with respect to side-channel attacks.

Graph Analysis. After the construction of the graph, the security analysis starts. As the first member from the set of distinguishers to be used in for the security analysis we provide the Correlation Power Analysis (CPA) distinguisher.

Algorithm 2. Hypothesis function identification for HW model

Require: view $\mathbb{V}$ for the graph $G = (V, E)$
Require: computational complexity boundary C
 1: Identify operation node op, where node i and entropy node e are input nodes, o is
    the output node, $\{(i, \epsilon_i), (e, \epsilon_e), (o, \epsilon_e)\} \subseteq \mathbb{V}$ holds, and $\epsilon_e > 0$
 2: **if** $\epsilon_i = 0$ **and** $\epsilon_o > 0$ **then**
 3:     hypothesis function $\mathfrak{h} = HW(op\,(i, e))$
 4: **else if** $\epsilon_i > 0$ **and** $\epsilon_o = 0$ **and** op is invertible **then**
 5:     hypothesis function $\mathfrak{h} = HW(op^{-1}(o, e))$
 6: **else**
 7:     **return** error: no suitable hypothesis found
 8: **end if**
 9: **if** $\epsilon_e < C$ **then**
10:     **return** hypothesis function $\mathfrak{h}$
11: **else**
12:     **return** error: no suitable hypothesis found
13: **end if**

An essential step of the CPA is to make appropriate leakage predictions by formulating hypotheses about exploitable intermediate values or register transitions in the algorithm, cf. [3]. The identification of a suitable hypothesis function can currently be performed for the Hamming weight (HW) model and for distance-based leakage models such as the Hamming distance (HD) model. Note that additional side-channel attacks, such as the template attack, can easily be integrated into the analysis phase of AMASIVE.

The attacker model has then to be defined using the specifications of the information collection phase in order to start the security analysis of the graph. The view of the attacker is initialized by assigning to all public nodes the entropy value zero and to all non-public nodes their corresponding bit-size as entropy values. These security sensitive nodes are assembled into groups and are then added to the set of security sensitive nodes. Subsequently, the actions for the attacker model are assigned. Currently, we have defined four actions: *deduction*, *reduction*, *HW-leakage* exploitation (using the Hamming weight model), and *HD-leakage* exploitation (using the Hamming distance model). The framework exercises these actions in order to automatically analyze the resulting graph.

The simplest action, the deduction, evaluates an operation or propagates a value and sets the entropy value of the corresponding node in the attacker view to zero. The deduction requires either the input or the output of a node to be known and the node to be invertible. The reduction infers information about a node and reduces its entropy. A reduction currently requires an associated node to have a reduced complexity or an operation to have certain features such as being non-surjective.

Algorithm 3. Hypothesis function identification for HD model

Require: view $\mathbb{V}$ for the graph $G = (V, E)$
Require: computational complexity boundary C
 1: Identify register transition from register node $(p, 0) \in \mathbb{V}$ to register node $(r, \epsilon_r) \in \mathbb{V}$
   with entropy source $(e, \epsilon_e) \in \mathbb{V}$ along the path and $\epsilon_r, \epsilon_e > 0$
 2: **if** a suitable register transition is identified **then**
 3:   **for all** edges i from/to r **do**
 4:     traverse graph along i until node $(n, 0) \in \mathbb{V}$ is reached
 5:     add path from i to n to hypothesis function $\mathfrak{h}_t$
 6:   **end for**
 7:   hypothesis function $\mathfrak{h} = HD(p, \mathfrak{h}_t)$
 8:   compute total entropy E of $\mathfrak{h}$ by summing up all entropy source nodes in $\mathfrak{h}$
 9:   **if** $E < C$ **then**
10:     **return** $\mathfrak{h}$
11:   **end if**
12: **end if**
13: **return** error: no suitable hypothesis found

A successful power attack sets the entropy of the corresponding node in the attacker view to zero. The requirement of a power attack based on the Hamming weight model seems to be a suitable hypothesis function according to the literature. The process of finding such a function is denoted in Algorithm 2.

A power attack based on the Hamming distance model sets the entropy values of all nodes, included in the hypothesis function, to zero too. However, finding the required hypothesis function for the Hamming distance model is somewhat more complex due to the requirement of a register transition. Algorithm 3 depicts the procedure of finding a hypothesis function for the Hamming distance model.

The security analysis of the graph is then performed as stated in Algorithm 1 using the all available actions. If the attacker wins the game of security analysis, then the actions and nodes, which led to the recovery of the secret, are presented to the designer. Furthermore, in order to determine the practical feasibility of the attack, a CPA is performed for each identified hypothesis function. Every such function is written into a C source file to be subsequently executed by a MATLAB script that implements the CPA. The feasibility of each hypothesis function is then verified by checking whether the required number of measurements is within the complexity boundary. If the CPA succeeds within the pre-defined complexity boundary, then the node is deemed recoverable and its entropy is updated according to the results of the CPA. Otherwise, the node is deemed not recoverable, and all consecutive actions have to be re-evaluated.

When all side-channel attacks have been performed and their feasibility has been evaluated, the designer is notified whether the mounted attack was successful and how high the complexity for recovering each node was. The designer can then decide for which nodes he or she should consider to implement countermeasures. Afterwards, he or she can then rerun the security analysis in order to quantify the achieved security improvement.

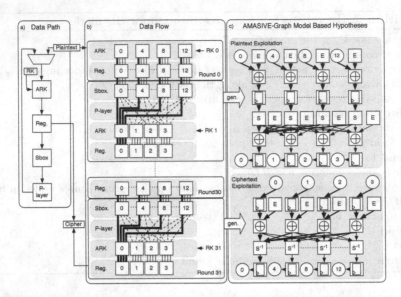

Fig. 4. PRESENT block cipher represented as an AMASIVE graph

4   Evaluation Results for the Block Cipher PRESENT

In this section we demonstrate the analysis of the block cipher PRESENT [2] by exercising the advocated AMASIVE framework. We focus on the analysis of just the first and the last round, because only at these rounds a side-channel analysis can be applied. Figure 4 a) depicts the block diagram of a PRESENT implementation.

Information Collection. At first, the workflow starts with the analysis of PRESENT block cipher by parsing its VHDL representation and by producing an XML description, which contains the instantiated modules AddRoundKey, S-box, P-layer, and their communication channels. We obtain the following elements in the XML description: two operations (AddRoundKey (ARK) and S-box), a permutation (P-layer), a register (Reg), an entropy source (Entropy), and the channels, which connect these elements. Next, we simulate the VHDL code of the design in order to obtain a functional description of the operations. Thus, we obtain a 16×16 lookup table for the S-box and a 256×16 lookup table for the ARK operation. The permutation P-layer is stored in an 64×2 lookup table. Finally, we define our attacker model. We assume the input and output of the algorithm to be known, since this is a common scenario in side-channel analysis. Subsequently, we mark all entropy elements as security sensitive and assume that the knowledge of all entropy elements of each round leads to the recovery of the secret key. We allow our attacker to be able to run a brute-force attack up to a complexity, which equals to a 2^{32} problem, and to perform up to 10.000 measurements. Then, we grant the attacker the ability to execute a deduction, a reduction, and a CPA using the Hamming distance model.

Security Analysis of the Unprotected Implementation. After generating the graph, we perform the security analysis utilizing the pre-defined attacker model. Figure 4 b) depicts the first and last round of the resulting graph and may be used for visualization during the security analysis. Since we have multiple nodes of one type in each round, we denote the i-th node n of round d as n_i^d. In the following, we denote the addroundkey nodes as x, the S-box nodes as s, the register nodes as r, the entropy nodes as e, the Input nodes as i, and the Output nodes as o, respectively. Also, we use their corresponding capital letters, together with a round index, to denote the set of all associated nodes of a round.

We begin the analysis by initializing the view $\mathbb{V}$ of the attacker using the input nodes *Input* and output nodes *Output*:

$$\mathbb{V} = \bigcup\nolimits_{\forall n \in (Input \cup Output)} \{(n, 0)\}.$$

The remaining nodes are added to $\mathbb{V}$ taking their corresponding bit-size as the entropy values. Then we add the Entropy nodes to the security sensitive nodes by grouping all nodes $e^d \in V$ of one round into:

$$\mathbb{S} = \bigcup\nolimits_{0 \leq d < 32} \{\{e_0^d, e_1^d, ..., e_{15}^d\}\}.$$

Finally, we determine and prioritize the available actions. The priority of the actions is as follows: deduction ded_G, reduction red_G, and CPA using the Hamming distance leakage model HD_G.

After the initialization we start the security analysis of the AMASIVE graph. The first deduction ded_G on $\mathbb{V}$ yields

$$\mathbb{V}^+ = ded_G(\mathbb{V}) = \{(n, \epsilon) \in \mathbb{V} | n \notin R^{31}\} \cup (R^{31} \times 0).$$

In other words, we can simply recover the value of the last round's register nodes $R^{31} = \{r_0^{31}, r_1^{31}, ..., r_{15}^{31}\}$ by propagating the value of the output. We set $\mathbb{V} = \mathbb{V}^+$ and since ded_G successfully recovered some nodes, we can again perform this action on the updated $\mathbb{V}$ resulting in:

$$\mathbb{V}^+ = ded_G(\mathbb{V}) = \{(n, \epsilon) \in \mathbb{V} | n \notin X^{31}\} \cup (X^{31} \times 0).$$

Thus, the second ded_G operation yields the value of the ARK nodes of the last round $X^{31} = \{x_0^{31}, x_1^{31}, ..., x_{15}^{31}\}$ by simply propagating the value from the recovered Reg nodes R^{31}.

We set $\mathbb{V} = \mathbb{V}^+$ again and observe that we can not apply ded_G to $\mathbb{V}$, since no value can be propagated and no operation may be evaluated. Thus, we verify the requirements for the reduction and also observe that it can not be applied to $\mathbb{V}$. Thus, we verify the requirement for the CPA using the Hamming distance model HD_G. Indeed, by applying Algorithm 3 to $\mathbb{V}$, we obtain hypotheses for all nodes in E^0 and E^{31}, which are visualized in Figure 4 c). In the following we use the short hand notation $n_{j,b}$ to refer to the b-th most significant bit of n_j. The hypothesis function $\mathfrak{h}_{e_0^0}$ for element $e_0^0 \in E^0$ is given by:

$$\mathfrak{h}_{e_0^0} = HD(r_0^0, r_0^1) = HD(S(i_0 \oplus e_0^0), (x_{0,0}^1 || x_{1,0}^1 || x_{2,0}^1 || x_{3,0}^1)), \tag{4}$$

where $x_{j,b}^1 = S(i_j \oplus e_j^0)_b \oplus e_{j,b}^1$ for $0 \le j < 16$ holds. Analyzing the hypothesis function of e_0^0 shows that we also have to make hypotheses for e_1^0, e_2^0, e_3^0, and e_0^1 too. This raises the complexity of a CPA to $(2^4)^5 = 2^{20}$, which still lies with in the predefined complexity boundary of 2^{32}. Thus, we obtain a successful Hamming distance hypothesis and are now able to set the entropy values for the nodes e_0^0, e_1^0, e_2^0, e_3^0, and e_0^1 to zero.

The hypothesis function $\mathfrak{h}_{e_0^{31}}$ for $e_0^{31} \in E^{31}$ is given by:

$$\mathfrak{h}_{e_0^{31}} = HD(r_0^{30}, r_0^{31}) = HD(S^{-1}(s_{0,0}^{31}||s_{4,0}^{31}||s_{8,0}^{31}||s_{12,0}^{31}), r_0^{31})), \tag{5}$$

where $s_{j,b}^{31} = o_{j,b} \oplus e_{j,b}^{31}$ holds. In contrast to the hypothesis $\mathfrak{h}_{e_0^0}$, $\mathfrak{h}_{e_0^{31}}$ only requires a complexity of 4 bit, making an attack more likely to succeed with a smaller number of measurements.

Since we found a hypothesis for HD_G, we can update $\mathbb{V}$:

$$\mathbb{E} = E^0 \cup E^1 \cup E^{31}$$
$$\mathbb{V}^+ = HD_G(\mathbb{V}) = \{(n, \epsilon) \in \mathbb{V} | n \notin \mathbb{E}\} \cup (\mathbb{E} \times 0). \tag{6}$$

We apply ded_G again on the updated $\mathbb{V}$, which may result in the recovery of X^0 and X^{31}. However, we stop the analysis at this point, since we have collected sufficient information. The current set of $\mathbb{V}$ contains the round keys of the first, second, and last round, respectively. Due to the key-schedule of PRESENT, using one round key, we can already recover the actual key by guessing the 2^{16} remaining bits.

Power Analysis Attack on the Unprotected Implementation. In order to verify the feasibility of the attack, we mounted a complete CPA using the hypotheses $\mathfrak{h}_{e_i^{31}}$ on a block cipher implementation on top of a SASEBO-II FPGA, which executes the described PRESENT algorithm. We performed 10.000 power measurements and used the Pearson correlation coefficient as the comparison method between the hypotheses and power traces. The resulting correlation of the attack on the last round of PRESENT using $\mathfrak{h}_{e_0^{31}}$ is depicted in Figure 5 and confirms the correctness of the generated hypothesis (hypothesis of the correct

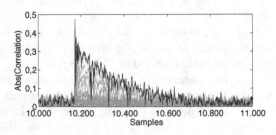

Fig. 5. CPA of an unprotected PRESENT implementation with hypothesis $\mathfrak{h}_{e_0^{31}}$

subkey is drawn in black). After only 3400 captured traces the complete round key may be recovered and thus the secret key of the block cipher was revealed at a computational effort of just 2^{16}.

5 Conclusion

In this paper we presented an adaptable framework for side-channel security analysis called AMASIVE. This framework represents the cryptographic algorithm as a dedicated graph and exploits a sophisticated and adaptable attacker model in order to determine side-channel vulnerabilities of an implementation denoted in VHDL code. In contrast to existing side-channel security tool sets, we focus on adaptability and extensibility in order to support the analysis of a wide range of crypto-systems. In addition, AMASIVE autonomously analyzes a given HDL description of the design under analysis and also generates attacking vectors in form of side-channel hypotheses We evaluated the feasibility of the presented framework by analyzing a PRESENT hardware implementation in order to demonstrate that this framework can handle complex circuit structures instead of just some operation components of cryptographic algorithm s such as an SBox.

Acknowledgements. The work presented in this chapter was supported in part by the German Federal Ministry of Education and Research (BMBF) in the project RESIST under grant number 01IS10027A.

References

1. Bayrak, A.G., Regazzoni, F., Brisk, P., Standaert, F.X., Ienne, P.: A first step towards automatic application of power analysis countermeasures. In: Stok, L., Dutt, N.D., Hassoun, S. (eds.) DAC, pp. 230–235. ACM (2011)
2. Bogdanov, A., Knudsen, L.R., Leander, G., Paar, C., Poschmann, A., Robshaw, M.J.B., Seurin, Y., Vikkelsoe, C.: Present: An ultra-lightweight block cipher. In: Paillier, P., Verbauwhede, I. (eds.) CHES 2007. LNCS, vol. 4727, pp. 450–466. Springer, Heidelberg (2007)
3. Elaabid, M.A., Guilley, S.: Practical improvements of profiled side-channel attacks on a hardware crypto-accelerator. In: Bernstein, D.J., Lange, T. (eds.) AFRICACRYPT 2010. LNCS, vol. 6055, pp. 243–260. Springer, Heidelberg (2010)
4. Mangard, S., Popp, T., Oswald, E.: Power Analysis Attacks - Revealing the Secrets of Smart Cards. Springer (2007)
5. Moss, A., Oswald, E., Page, D., Tunstall, M.: Automatic insertion of dpa countermeasures. IACR Cryptology ePrint Archive 2011, 412 (2011)
6. Moss, A., Oswald, E., Page, D., Tunstall, M.: Compiler assisted masking. In: Prouff, E., Schaumont, P. (eds.) CHES 2012. LNCS, vol. 7428, pp. 58–75. Springer, Heidelberg (2012)
7. Standaert, F.X., Malkin, T., Yung, M.: A unified framework for the analysis of side-channel key recovery attacks. In: Joux, A. (ed.) EUROCRYPT 2009. LNCS, vol. 5479, pp. 443–461. Springer, Heidelberg (2009)
8. Zohner, M., Stöttinger, M., Huss, S.A., Stein, O.: An adaptable, modular, and autonomous side-channel vulnerability evaluator. In: HOST, pp. 43–48. IEEE (2012)

A Performance Boost for Hash-Based Signatures

Thomas Eisenbarth[1], Ingo von Maurich[2], Christof Paar[2], and Xin Ye[1]

[1] Worcester Polytechnic Institute, Worcester, MA, USA
[2] Horst Görtz Institute for IT-Security, Ruhr-University Bochum, Germany
{teisenbarth,xye}@wpi.edu, {ingo.vonmaurich,christof.paar}@rub.de

Abstract. Digital signatures have become a key component of many embedded system solutions and are facing strong security and efficiency requirements. In this work, algorithmic improvements for the authentication path computation decrease the average signature computation time by close to 50 % when compared to state-of-the-art algorithms. The proposed scheme is implemented on an Intel Core i7 CPU and an AVR ATxmega microcontroller with optimized versions for the respective target platform. The theoretical algorithmic improvements are verified and cryptographic hardware accelerators are used to achieve competitive performance.

Keywords: hash-based cryptography, signatures, software, microcontroller, post-quantum cryptography, embedded security.

1 Motivation

Today, digital signatures have become important for many embedded systems. Examples include firmware updates (a usage which is applicable to virtually every networked embedded system), identification (e.g., with smart cards), electronic payment (e.g., with smart phones via the NFC interface), or protection against product counterfeiting. Unfortunately, conventional digital signature schemes are very computational intensive and are a challenge on embedded microprocessors. Exploring signature schemes that are efficient on embedded platforms is thus of great interest and can offer advantage over the prevailing choices of (EC-)DSA and RSA.

Prior work by Rohde et al. [19] as well as by Hülsing et al. [11] suggest that the Merkle Signature Scheme (MSS) in combination with Winternitz One-Time Signatures (W-OTS) is a possible choice for a time-limited signature scheme and can be efficiently implemented in embedded systems. We analyze and extend the proposal by Rohde et al. and propose several modifications that lead to significant performance improvements.

Contribution. We implement the proposed signature scheme on two widespread platforms (Intel Core i7 CPU and a low-cost AVR 8-bit microcontroller) and make use of available cryptographic hardware accelerators to gain maximum efficiency. Furthermore, we propose an improved algorithm for the authentication

M. Fischlin and S. Katzenbeisser (Eds.): Buchmann Festschrift, LNCS 8260, pp. 166–182, 2013.

path computation of a Merkle tree. At the same time we decrease the average computation time by close to 50% compared to the most efficient authentication path computation algorithm at the price of a slightly increased memory consumption. Explicit formulas are developed to quantify the number of times each leaf of the Merkle tree is computed during the authentication path computation. The drawback of current authentication path computation algorithms is the unbalanced number of computations per leaf. Our improved algorithm mitigates this issue by reducing the number of computations for often used leaves and allows for more efficient computation of the authentication path.

2   Hash-Based Signatures

In the following we describe the foundations of the Merkle signature scheme. It was introduced in [18] and a detailed description of MSS can be found in [7]. Details about the implementation inspiring our work are given in [19]. We use Winternitz one-time signatures [8] for message signing. The one-time keys are generated using a PRNG to minimize storage requirements as proposed in [19].

The following components use an at least second preimage resistant, undetectable n-bit one-way function f and a cryptographic m-bit hash function g:

$$f : \{0,1\}^n \to \{0,1\}^n , \quad g : \{0,1\}^* \to \{0,1\}^m$$

2.1   The Merkle Signature Scheme

Given a One-Time Signature Scheme (OTSS) a tree height H is chosen to allow for the creation of 2^H signatures that are verifiable with the same verification key. Let the nodes of the Merkle tree be denoted as $\nu_h[s]$ with $h \in \{0, \dots, H\}$ being the height of the node and $s \in \{0, \dots, 2^{H-h} - 1\}$ being the node index on height h.

Key Generation. The 2^H leaves of the Merkle tree are defined to be digests $g(Y_i)$ of one-time verification keys Y_i. Starting from the leaves, the MSS verification key which is the root node of the Merkle tree $\nu_H[0]$ is generated following

$$\nu_{h+1}[i] = g(\nu_h[2i] \,||\, \nu_h[2i+1]), \quad 0 \le h < H, \ 0 \le i < 2^{H-h-1},$$

meaning that a parent node is generated by hashing the concatenation of its two child nodes.

Signature Generation. A Merkle signature $\sigma_s(d)$ of a digest $d = g(M)$ of a message M consists of a signature index s, a one-time signature σ_{OTS}, a one-time verification key Y_s, and an authentication path $(\text{AUTH}_0, \dots, \text{AUTH}_{H-1})$ that allows the verification of the one-time signature with respect to the public MSS verification key, hence

$$\sigma_s(d) = (s, \sigma_{\text{OTS}}, Y_s, (\text{AUTH}_0, \dots, \text{AUTH}_{H-1})).$$

The signature index $s \in \{0, \ldots, 2^H - 1\}$ is incremented with every issued signature. The OTSS is applied using signature key X_s to generate the signature $\sigma_{\text{OTS}} = \text{Sign}_{\text{OTS}}(d, X_s)$ of the message digest d. The authentication path for the sth leaf are all sibling nodes AUTH_h, $h \in \{0, \ldots, H - 1\}$ on the path from leaf $\nu_0[s]$ to the root node $\nu_H[0]$. It enables the verifier to recompute the root node of the Merkle tree and authenticates the current one-time signature.

We would like to stress that the signature generation reflects the structure of an online/offline signature scheme. The authentication path only depends on the OTSS verification key Y_s which is known prior to the message and hence can be precomputed.

Signature Verification. Given a message digest $d = g(M)$ and a signature $\sigma_s(d)$ the verifier checks the one-time signature σ_{OTS} with the underlying one-time signature verification algorithm $\text{Verify}_{\text{OTS}}(d, \sigma_s(d))$. In addition, the root node is reconstructed using the provided authentication path

$$\phi_{h+1} = \begin{cases} g\left(\phi_h \,||\, \text{AUTH}_h\right), & \text{if } \lfloor s/2^h \rfloor \equiv 0 \bmod 2 \\ g\left(\text{AUTH}_h \,||\, \phi_h\right), & \text{if } \lfloor s/2^h \rfloor \equiv 1 \bmod 2 \end{cases}, \quad \phi_0 = \nu_0[s], \quad h = 0, \ldots, H - 1.$$

If the one-time signature σ_{OTS} is successfully verified and ϕ_H is equal to $\nu_H[0]$ the MSS signature is accepted.

2.2 Winternitz One-Time Signatures

Winternitz OTS [8] are a convenient choice for the one-time signature scheme, as they reduce the overall signature length. The Winternitz parameter $w \geq 2$ determines how many bits are signed simultaneously and t determines of how many random n-bit strings x_i the Winternitz signature keys consist.

$$t = t_1 + t_2, \quad t_1 = \left\lceil \frac{n}{w} \right\rceil, \quad t_2 = \left\lceil \frac{\lfloor \log_2 t_1 \rfloor + 1 + w}{w} \right\rceil$$

Key Generation. A W-OTS signature key $X = (x_0, \ldots, x_{t-1})$ is generated by selecting t random bit strings $x_i \in \{0,1\}^n$, $0 \leq i < t$. The W-OTS verification key $Y = g(y_0 \,||\, \ldots \,||\, y_{t-1})$ is computed from the signature key by applying f $2^w - 1$ times to each x_i giving $y_i = f^{2^w - 1}(x_i)$, $0 \leq i < t$ and computing the hash of the concatenated y_i's. Note, the superscript denotes multiple executions of f, e.g., $f^2(x_i) = f(f(x_i))$ and $f^0(x_i) = x_i$.

Signature Generation. A signature for a message M is created by signing its digest $d = g(M)$ under key X. Digest d is divided into t_1 blocks $b_0, \ldots, b_{t_1-1}$ of length w and a checksum $c = \sum_{i=0}^{t_1-1} (2^w - b_i)$ is computed. Checksum c is divided into t_2 blocks $b_{t_1}, \ldots, b_{t-1}$ of length w (zero-padding to the left is applied if c or d are no multiples of w). The W-OTS signature $\sigma_{\text{W-OTS}} = (\sigma_0, \ldots, \sigma_{t-1})$ is computed with $\sigma_i = f^{b_i}(x_i)$, $0 \leq i < t$.

Signature Verification. Given a message digest $d = g(M)$, a signature $\sigma_{\text{W-OTS}}$ and a verification key Y_s the verifier generates blocks $b_0, \ldots, b_{t-1}$ from d as in signature generation and reconstructs

$$Y_s' = g\left(f^{2^w - 1 - b_0}(\sigma_0) \,\|\, \ldots \,\|\, f^{2^w - 1 - b_{t-1}}(\sigma_{t-1})\right).$$

If Y_s' equals Y_s the signature is valid, otherwise it has to be rejected. When using W-OTS signatures in MSS, transmitting Y_s and comparing Y_s to Y_s' can be omitted. Y_s' can simply be used together with the nodes of the authentication path to recompute the root of the Merkle tree. If the recomputed root equals the MSS public key, then Y_s' is a valid OTS verification key.

2.3 Private Key Generation

Storing 2^H one-time signature or verification keys can be an infeasible task, especially on constrained implementation platforms. Generating keys on-the-fly by using a PRNG significantly reduces the required storage space (cf. [19]).

Each W-OTS signature key $X_i = (x_0, \ldots, x_{t-1})$, $0 \leq i < 2^H$ is generated by the PRNG from a seed $\text{SEED}_{\text{W-OTS}_i}$. These seeds in turn are also generated by the PRNG from a initial randomly selected seed $\text{SEED}_0 \in_R \{0,1\}^n$ which serves as the MSS signature key. On input of k_i the PRNG outputs a random string r_{i+1} and an updated seed k_{i+1}.

$$\text{PRNG} : \{0,1\}^n \to \{0,1\}^n \times \{0,1\}^n, k_i \to (k_{i+1}, r_{i+1}) \tag{1}$$

Starting from the initial SEED_0 the seeds for the signature keys $\text{SEED}_{\text{W-OTS}_i}$ are created by

$$(\text{SEED}_{i+1}, \text{SEED}_{\text{W-OTS}_i}) \leftarrow \text{PRNG}(\text{SEED}_i), \quad 0 \leq i < 2^H.$$

The t n-bit strings of the i-th W-OTS signature key $X_i = (x_0, \ldots, x_{t-1})$, $0 \leq i < 2^H$ are then generated by

$$(\text{SEED}_{\text{W-OTS}_i}, x_j) \leftarrow \text{PRNG}(\text{SEED}_{\text{W-OTS}_i}), \quad 0 \leq j < t.$$

2.4 Authentication Path Computation

Creating an authentication path for a specific leaf s can be accomplished by storing all tree nodes in memory and looking up the required nodes when needed. However, because of the exponential growth of nodes in tree height H this approach becomes infeasible for reasonable practical applications. Hence, algorithms for efficient on-the-fly authentication path computation during signature generation are required.

The currently best known algorithm for on-the-fly computation of authentication nodes is the BDS algorithm [7] (Algo. 3, cf. Appendix). It makes use of several treehash algorithm instances TREEHASH_h for heights $0 \leq h \leq H - K - 1$. The treehash algorithm was introduced in [18] and modified in [21]. It allows to efficiently create (parts of) Merkle trees. In the BDS algorithm each instance is initialized with a leaf index s to which it computes the corresponding node value. Each instance is updated until the required authentication node is computed. During a treehash update the next leaf is created and parent nodes are computed if possible.

The generation of the authentication path is split up into two parts that go alongside with the key and signature generation of MSS. During key generation all treehash instances TREEHASH$_h$ are initialized with ν_h [3] and the first authentication path stored is AUTH$_h = \nu_h$ [1], $0 \le h \le H - 1$.

The BDS algorithm generates left authentication nodes either by computing the leaf value or by one hash-function evaluation of the concatenation of two previously computed nodes that are held in memory. Right authentication nodes in contrast are computed from the leaf up, which is computationally more expensive. Since right nodes close to the top are expensive to compute a positive integer $K \ge 2, (H - K$ even) decides how many of these nodes are stored in RETAIN$_h$, $H - K \le h \le H - 2$ during key generation.

Authentication nodes change every 2^h steps for height h. During signature generation the treehash instances are updated and if a authentication node from a treehash instance is used, the instance is re-initialized to compute the next authentication node for that height.

2.5 Security of MSS

The security properties of the signature scheme described above is discussed in [7]. Specifically, the work shows that the Lamport-Diffie one-time signatures [14] are existentially unforgeable under an adaptive chosen message attack (i.e., CMA-secure), if the chosen one-way function is preimage resistant. The employed Merkle signature scheme is also CMA-secure if the underlying OTS is CMA-secure and if the underlying hash function is collision resistant. For increased efficiency (and shorter signatures) we chose Winternitz OTS rather than the classic Lamport-Diffie OTS. The security of the Winternitz one-time signatures is discussed in [5,8,10]. The findings in [5] and [10] show that Winternitz OTS are CMA-secure if used with pseudo-random functions or collision-resistant, undetectable one-way functions, respectively. The level of bit security lost by using a small Winternitz-parameter is in both cases rather small. In our case, the biggest Winternitz parameter is $w = 4$, hence we still provide a security level of approx. 95 bits for a 128-bit PRF or 116 bits for W-OTS+ [10]. Related discussions for a similar MSS scheme can also be found in [6].

3   Optimized Authentication Path Computation

Since the Merkle-tree is not stored, the parts of the Merkle tree needed for the authentication path must be generated. One optimized algorithm for this purpose is the BDS algorithm [7]. Its design goal was to minimize costly leaf computations. In the following we describe further optimizations that, at the expense of a slightly higher memory consumption, reduce the number of computations for the leaves that are computed most often in the BDS algorithm. We thereby reduce the overall number of leaf calculations and hence the overall computation time by close to 50%.

3.1 Authentication Path Computation

The authentication path consists of nodes of the Merkle tree. For the computation of upcoming authentication nodes we use several stacks of nodes for different heights of the tree. Treehash instances TREEHASH_h are used for heights $0 \le h \le H - K - 1$. Each instance is initialized with a leaf index s and is updated in Algo. 3 until the required authentication node is computed. During a treehash update the next leaf is created and parent nodes are computed by hashing previously created nodes if possible. Authentication nodes change every 2^h steps for height h and if an authentication node is used from a treehash instance, this instance is re-initialized to compute the following authentication node for that height.

Preliminaries. The total number of leaf computations that occur during execution of Algo. 3 can be calculated by counting all invocations of LEAFCALC, a function that on input s outputs leaf $\nu_0[s]$. As mentioned in [7] it is possible to omit LEAFCALC in Step 3 of Algo. 3 since the sth W-OTS key pair is used to sign the current message, hence the verification key can be computed from the signature and one additional hash computation yields leaf $\nu_0[s]$. If a different OTSS is used the verification key is part of the OTS and can be hashed to create $\nu_0[s]$. This saves 2^{H-1} LEAFCALC invocations. Careful analysis of Algo. 3 leads to the total number of leaf computations in the BDS algorithm

$$N_{H,K_{\text{total}}} = \sum_{h=0}^{H-K-1} \left(2^{H-1} - 2^{h+1}\right) = (H - K)2^{H-1} - 2^{H-K+1} + 2.$$

In order to count the necessary computations for a specific leaf s during execution of Algo. 3 we have to consider all occurrences of s as parameter of LEAFCALC, except for when s is a left leaf (Step 3 of Algo. 3), as explained above. To determine if leaf s is computed in treehash instance TREEHASH_h we make the following observation: TREEHASH_0 computes leaves $(5),(7),(9),\ldots$, TREEHASH_1 computes leaves $(10,11),(14,15),\ldots$, TREEHASH_2 computes leaves $(20,21,22,23),(28,29,30,31),\ldots$ and so forth. Hence, the total number of computations for leaf s is given by

$$N_{H,K}(s) = \sum_{h=0}^{H-K-1} \left\lfloor \frac{s \bmod 2^{h+1}}{2^h} \right\rfloor \cdot \left\lceil \frac{\left\lfloor \frac{s}{5 \cdot 2^h} \right\rfloor}{2^H} \right\rceil$$

Drawbacks. A drawback of the BDS algorithm (Algo. 3) is that it does not balance the computation of leaf nodes. There are leaves that are calculated various times, while others are barely touched. On average each leaf of the Merkle tree is computed $\overline{N_{H,K}} = N_{H,K_{\text{total}}}/2^H \approx \frac{1}{2}(H - K)$ times. However, the computations per leaf deviate from the average as shown in Fig. 1 for a Merkle tree $(H = 10, K = 2)$ with 1024 leaves.

3.2   Balanced Authentication Path Computation

Since the rightmost nodes of each treehash instance are calculated most frequently, we propose to cache and reuse them for balancing the leaf computations. We use an array RIGHTNODES to store those nodes. Note, the root of each treehash instance and the complete treehash instance TREEHASH$_0$ are not stored since lower treehash instances do not require those nodes. This leads to a significantly reduced computation time, at the cost of an increased memory consumption.

From TREEHASH$_1$ we store node ν_0 [7], from TREEHASH$_2$ we store nodes ν_1 [7] and ν_0 [15] and so on. More generally, we store h nodes $\nu_j \left[2^{2+h-j} - 1\right]$, $j = 0, \dots, h - 1$ for each instance TREEHASH$_h$, $1 \leq h \leq H - K - 1$. The required storage space is

$$S_{\text{RightNodes}}(H, K) = \sum_{h=1}^{H-K-1} h = \binom{H - K}{2} = \triangle_{H-K-1}.$$

Table 1 lists the storage requirements for common $H - K$ values. The initialization of the RIGHTNODES array is done during the computation of the public key of the Merkle tree. The updated initial setup is formalized in Algo. 2.

Table 1. Storage space required by the RIGHTNODES array where the rightmost nodes of each treehash instance TREEHASH$_h$, $h = 1, \dots, H - K - 1$ are stored for reusage by lower treehash instances

$H - K$	$\triangle_{H-K-1}$	128-bit digest	160-bit digest	256-bit digest
6	15	240 byte	300 byte	480 byte
8	28	448 byte	560 byte	896 byte
10	45	720 byte	900 byte	1440 byte
12	66	1056 byte	1320 byte	2112 byte
14	91	1456 byte	1820 byte	2912 byte
16	120	1920 byte	2400 byte	3840 byte
18	153	2448 byte	3060 byte	4896 byte

In Step 5 of Algo. 3 the treehash instances receive updates if they are initialized and not finished. In every update one leaf is computed and higher nodes are generated if possible by hashing concatenated nodes from the stack.

During the last update before the treehash instance is finished, the rightmost leaf of this treehash instance is computed and all other rightmost nodes of this treehash instance are consecutively generated. If the leaf index $s \equiv 2^h - 1 \mod 2^h$ in instance TREEHASH$_h$, we store the following nodes in the RIGHTNODES array starting from offset $h(h-1)/2$. An adapted version of the treehash update algorithm is given in Algo. 1.

In every second re-initialization of treehash instances TREEHASH$_h$, $h = 0, \ldots,$ $H - K - 2$ the authentication node can be copied from the RIGHTNODES array because it has been computed before by treehash instance TREEHASH$_{h+1}$. If $s + 1 \equiv 0 \mod 2^{h+2}$ the authentication node can be copied from the RIGHTN-ODES array and if $s + 1 \equiv 2^{h+1} \mod 2^{h+2}$ the authentication node has to be computed. If we can reuse nodes, we not only copy the authentication node (root of TREEHASH$_h$) but also its rightmost child nodes from RIGHTNODES, so they can be reused for instances TREEHASH$_j$, $j < h$. This improvement can be easily integrated into the BDS algorithm by modifying Step 4c) accordingly.

Comparison. In order to quantify our improvements, we give the total amount of leaf computations and show how to determine the leaf computations for a specific leaf s. As before, each instance TREEHASH$_h$ computes 2^h leaves until they are finished. The re-initializations however are halved for treehash instances TREEHASH$_h$, $h = 0, \ldots, H - K - 2$, to $2^{H-h-2} - 1$ re-initializations because in half of all cases previously computed nodes can be copied from the RIGHTN-ODES array and the LEAFCALC computations are skipped. Hence, the number of calls to LEAFCALC from each TREEHASH$_h$ instance is $2^{H-2} - 2^h$. The tree-hash instance TREEHASH$_{H-K-1}$ cannot copy nodes from higher instances since it is the topmost treehash instance. It calls LEAFCALC as before, resulting in $2^{H-1} - 2^{H-K}$ computations. The total number of leaf computations is

$$N'_{H,K_{\text{total}}} = \sum_{h=0}^{H-K-2} \left(2^{H-2} - 2^h\right) + 2^{H-1} - 2^{H-K}$$
$$= (H - K + 1) 2^{H-2} - 3 \cdot 2^{H-K-1} + 1.$$

When compared to $N_{H,K_{\text{total}}}$ of the BDS algorithm this is nearly a 50% reduction.

To retrieve the number of leaf computations in the improved version for a specific leaf s we have to check whether s is a left or a right leaf. If s is even, it is a left leaf and can be computed from the current one-time signature or verification key as mentioned in Section 3.1 for Step 3 of Algo. 3. If s is odd, it is a right leaf thus LEAFCALC is not executed directly. To determine if s is computed in treehash instance TREEHASH$_h$, $h = 0, \ldots, H - K - 2$, we have to consider that in half of all cases it is copied and not computed. For this purpose we construct function $\delta'_{H,K}(s)$ that returns the number of times leaf s is computed in treehash instances TREEHASH$_h$, $h = 0, \ldots, H - K - 2$.

$$\delta'_{H,K}(s) = \sum_{h=0}^{H-K-2} \left\lfloor \frac{s \mod 2^{h+1}}{2^h} \right\rfloor \cdot \left\lceil \frac{\left\lfloor \frac{s}{5 \cdot 2^h} \right\rfloor}{2^H} \right\rceil \cdot \left(1 - \left\lfloor \frac{s \mod 2^{h+2}}{2^{h+1}} \right\rfloor\right)$$

The topmost treehash instance TREEHASH_{H-K-1} cannot copy nodes from the RIGHTNODES array because the required nodes have not been computed so far. Thus, we have to count the number of computations for this instance as in the unoptimized version. The total number of times leaf s is generated during the computation of all authentication nodes can now be summed up to

$$N'_{H,K}(s) = \left\lfloor \frac{s \bmod 2^{H-K}}{2^{H-K-1}} \right\rfloor \cdot \left\lceil \frac{\left\lfloor \frac{s}{5 \cdot 2^{H-K-1}} \right\rfloor}{2^H} \right\rceil + \delta'_{H,K}(s).$$

On average each leaf is now computed $\overline{N'_{H,K}} = N'_{H,K_{\text{total}}}/2^H \approx \frac{1}{4}(H - K + 1)$ times. The reduced number of computations for each leaf is shown in Fig. 2. Visual comparison between Fig. 1 and Fig. 2 already gives an intuition of the reduction and balancing of leaf computations. For further comparison see Fig. 3. Table 2 compares the total number of leaf computations, how often a leaf has to be computed in the worst-case, and the average number of leaf computations for common heights $H = \{10, 16, 20\}$ and $K = \{2, 4\}$. The total number of leaf computations as well as the average computations per leaf are decreased by about $38 - 48\%$ for the chosen parameters of H and K. Both the worst-case computation time as well as the average signature computation time are decreased. This means that, for example, battery-powered devices greatly profit from the reduced overall computation time which directly relates to the overall power consumption.

Since all but the topmost treehash instance only need to be computed every second time, the number of updates per signature (Algo.3, Step 5) can be reduced from $\lceil (H - K)/2 \rceil$ to $\lceil (H - K + 1)/4 \rceil$. As a result, the average update time is much better balanced than in Algo. 3 and the worst case computation time is also improved. The BDS algorithm needs to store $3H + \lfloor H/2 \rfloor - 3K + 2^K - 2$ tree nodes and $2(H - K) + 1$ PRNG seeds as signature key. Due to storing the rightmost nodes our improved algorithm increases the number of tree nodes that have to be stored by $\binom{H-K}{2}$. Even if the additional memory is used to increase K for the original BDS algorithm, the speedup is still significant. E.g., comparing our $(H, K) = (16, 4)$ to BDS$(16, 6)$ gives comparable storage requirements, but

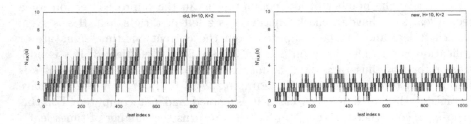

Fig. 1. Number of times each leaf is computed by the original BDS algorithm for a Merkle tree of height $H = 10$ and $K = 2$ **Fig. 2.** Number of times each leaf is computed by our variation for a Merkle tree of height $H = 10$ and $K = 2$

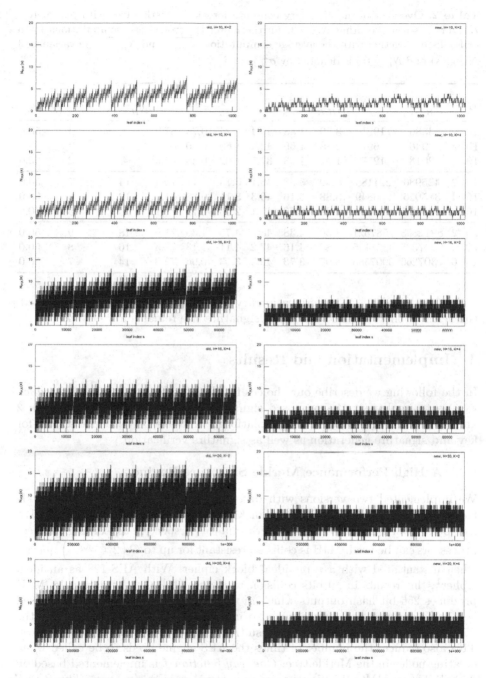

Fig. 3. Comparison of $N_{H,K}(s)$ and $N'_{H,K}(s)$ for $H = \{10, 16, 20\}$ and $K = \{2, 4\}$ for all leaves s of the respective tree

Table 2. Overview of the necessary computations for a Merkle tree with parameters H and K when executing Algo. 3. Furthermore, the worst-case computations for a leaf is listed together with the average computations $\overline{N_{H,K}}$ and $\overline{N'_{H,K}}$. The variance of $N_{H,K}(s)$ and $N'_{H,K}(s)$ is denoted by $\sigma^2_{H,K}$ and $\sigma'^2_{H,K}$.

H	K	$N_{H,K_{\text{tot}}}$	$N'_{H,K_{\text{tot}}}$	$\overline{N_{H,K}}$	$\overline{N'_{H,K}}$	%	$\sigma^2_{H,K}$	$\sigma'^2_{H,K}$	%	max. $N_{H,K}(s)$	max. $N'_{H,K}(s)$	%
10	2	3586	1921	3.50	1.88	46.4	2.24	0.73	67.3	8	4	50.0
10	4	2946	1697	2.88	1.66	42.4	1.60	0.50	68.5	6	3	50.0
10	6	2018	1257	1.97	1.23	37.7	1.02	0.33	67.9	4	2	50.0
16	2	425986	221185	6.50	3.38	48.1	3.75	1.11	70.4	14	7	50.0
16	4	385026	206849	5.88	3.16	46.3	3.11	0.88	71.6	12	6	50.0
16	6	325634	178689	4.97	2.73	45.1	2.53	0.71	72.1	10	5	50.0
20	2	8912898	4587521	8.50	4.38	48.5	4.75	1.36	71.4	18	9	50.0
20	4	8257538	4358145	7.88	4.16	47.2	4.11	1.13	72.5	16	8	50.0
20	6	7307266	3907585	6.97	3.73	46.5	3.53	0.96	72.9	14	7	50.0

still a speedup of 36%. The verification key and signature sizes remain unaffected: the verification key size is m and the signature size remains at $t \cdot n + H \cdot m$.

4   Implementation and Results

In the following we describe our choices for the cryptographic primitives which we use to implement the proposed signature scheme described in Sections 2 and 3. We then detail on the target platforms and give performance figures for key and signature generation as well as signature verification.

4.1   A High Performance Merkle Signature Engine

We implemented two versions with different *hash functions* g for the Merkle tree. Both versions use AES-128 in an MJH construction [15]. Using AES-128 as block cipher is favorable from a performance perspective as existing AES co-processors can be used. MJH is collision resistant for up to $\mathcal{O}\left(2^{\frac{2n}{3}-\log n}\right)$ queries when instantiated with a n-bit ideal block cipher. With AES-128 as an ideal cipher[1], this results in 80 bits collision resistance [15]. On the downside, MJH produces 256-bit hash outputs which in the MSS setting leads to an increased key and signature size. Hence, we also implement a version that shortens the 256-bit output of MJH to 160-bit, resulting in smaller key and signature sizes. This also reduces the number of times the AES-engine needs to be called when creating nodes in the Merkle tree. *One-way function* f is implemented based on AES-128 in an MMO [16,17] construction: $f(x_i) := \text{AES}_{\text{IV}}(x_i) \oplus x_i$. The *PRNG*

---

[1] Please note that with recent cryptanalytic breakthroughs against the AES family of ciphers [4], it is doubtful that AES-128 is indistinguishable from an ideal block cipher.

defined in (1) is implemented based on the leakage-2-limiting PRNG proposed in [20]. In particular, $\mathrm{PRNG}(k_i) := (\mathrm{AES}_{k_i}(0^{128}), \mathrm{AES}_{k_i}(0^{127}||1))$, where AES_{k_i} denotes the AES-128 with a 128-bit key k_i. A discussion of the leakage properties of this implementation can be found in [9].

4.2 Implementation Platforms

We implement the signature scheme on two different platforms. On the one side we choose a lightweight and low-cost 8-bit Atmel ATxmega microcontroller and on the other side a powerful Intel Core i7 notebook CPU.

Intel Core i7-2620M 64-bit CPU. Intel's off-the-shelf Core i7-2620M 64-bit Sandy Bridge notebook CPU [12] features two cores running at 2.70 GHz (with Turbo Boost technology up to 3.40 GHz). For accurate measurement, we disabled Turbo Boost and hyper-threading during our benchmarks. The CPU incorporates the recent extensions to the x86 instruction set that improve the performance when en-/decrypting data using AES. The extension is called AES-NI and consists of six additional instructions [13]. All standardized key lengths (128 bit, 192 bit, 256 bit) are supported for a block size of 128 bit.

Atmel AVR ATxmega128A1 8-bit Microcontroller. We are using the Atmel evaluation board AVR XPLAIN [3] that features an ATxmega128A1 microcontroller [1,2], as it is a typical example for a low-power embedded platform.. The ATxmega offers hardware accelerators for DES and AES and is clocked at 32 MHz. The hardware acceleration is limited to AES with 128-bit key and block size.

4.3 Performance Results

In the following we give performance figures of the signature scheme for selected Merkle tree heights H and parameters K and w on both platforms.

CPU Performance. On the Intel CPU we measure the time it takes to create the root node of the Merkle tree, i.e., the verification key generation. We iterate over all leaves and sign random messages to measure the average computation time that is needed to create a valid MSS signature. Additionally, we measure the time it takes to verify an MSS signature. Signature computation includes creating the signing key, performing a one-time signature with the created signing key, and generating the next authentication path (the last step can be removed, as it can be precomputed at any time between two signing operations). The measurement is done for tree height $H = 16$ with $K = 2$ and $w = 2$. Note, due to the binary tree structure the root node computation can be parallelized if more than one CPU core is available, which would bring down the required computation time by roughly the factor of cores used. We compare our results against the originally proposed signature scheme [19] in Table 3.

Table 3. Performance figures of a Merkle tree with parameters $H = 16, K = 2, w = 2$ on an Intel i7 CPU and $H = 10, K = 2, w = 2$ on an ATxmega microcontroller. f is implemented using a hardware-accelerated AES-128 (AES-NI instructions, ATxmega crypto accelerator) in MMO construction. g is implemented using AES-128 in an MJH-256 construction and with the output truncated to 160 bit. The Intel CPU was clocked at 2.7 GHz and the ATxmega at 32 MHz.

Hash g		MJH-256 w/ AES-128			MJH-160 w/ AES-128		
Target		[19]	our	impr.	[19]	our	impr.
Core i7	KEYGEN	6546.9 ms	6037.5 ms	8%	4218.7 ms	3886.3 ms	8%
Core i7	SIGN	743.9 us	401.3 us	46%	487.1 us	256.2 us	47%
Core i7	VERIFY	76.1 us	78.1 us	-3%	50.8 us	49.3 us	3%
AVR	SIGN	110.0 ms	64.9 ms	41%	70.7 ms	41.7 ms	41%
AVR	VERIFY	18.4 ms	18.4 ms	0%	11.0 ms	11.0 ms	0%

Compared to the previous results of [19] our improved algorithm in combination with the exchanged PRNG yields on average a performance gain of 46-47 % for signature generation. The new PRNG improves the computation time on average by 8%, the algorithmic changes to the authentication path computation algorithm yield 38-39% points.

When generating verification keys an 8% improvement can be observed. This is due to the exchanged PRNG which uses a hardware-accelerated AES-engine since our algorithmic improvements do not affect key generation. Signature verification is more or less stable, regardless of cipher/algorithm combinations and is about a factor of 5 faster than signature generation.

Microcontroller Performance. On the microcontroller we measure the average computation time that is needed to create a valid MSS signature (including next authentication path computation) and the time it takes to verify an MSS signature. We omit the verification key generation since for reasonable tree heights it is an infeasible task for the microcontroller. Verification keys have to be computed once on a computer platform when initializing the microcontroller. The code was compiled using `avr-gcc` version 3.3.0. We found optimization stage `-O2` to provide the best tradeoff between runtime and code size.

The results on the microcontroller are in accordance with the results observed on the Intel CPU. The average signature generation time improves by 41 % when using our proposed changes. Signature verification remains stable and is four times faster than signature generation. The memory consumption is listed in Table 4. Compared to the setting of [19] we need more flash and SRAM memory due to the additional storage for the RIGHTNODES array.

Table 5 compares key and signature sizes for different MSS implementations. Note that the increased signature sizes for [11] enable on-card key generation.

Table 4. Required memory on the ATxmega128A1 microcontroller. In total 128 kByte flash memory and 8 kByte SRAM are available on this device. Memory consumption is reported in bytes and includes the verification and signature keys.

		MJH-256 w/ AES-128 [19]		our		MJH-160 w/ AES-128 [19]		our	
H	K	Flash	SRAM	Flash	SRAM	Flash	SRAM	Flash	SRAM
10	2	10,608	1,486	12,070	2,382	10,204	1,066	11,352	1,626
10	4	10,726	1,604	11,768	2,084	10,250	1,112	11,138	1,412
10	6	11,994	2,874	12,752	3,066	11,018	1,878	11,726	1,998

Table 5. Comparison of signing key (sk), verification key (vk), and signature size (sig) between [19], our improvement, and XMSS$^+$ [11] for common (H, K, w) parameter sets. All sizes are reported in bytes.

			MJH-256			MJH-160			[19] (MJH-256)			[19] (MJH-160)			XMSS$^+$ [11]		
H	K	w	sk	vk	sig	sk	vk	sig	sk	vk	sig	sk	vk	sig	sk	vk	sig
16	2	2	5,335	32	2,640	3,547	20	1,680	2,423	32	2,640	1,727	20	1,680	3,760	544	3,476
16	2	4	5,335	32	1,584	3,547	20	1,008	2,423	32	1,584	1,727	20	1,008	3,200	512	1,892
20	4	2	7,049	32	2,768	4,649	20	1,760	3,209	32	2,768	2,249	20	1,760	4,303	608	3,540
20	4	4	7,049	32	1,712	4,649	20	1,088	3,209	32	1,712	2,249	20	1,088	3,744	576	1,956

5   Conclusion

We presented a novel algorithmic improvement for authentication path computation in MSS that balances leaf computations. The proposed improvements have been implemented on two platforms and were compared to previous proposed algorithms showing significant improvements. Furthermore, we gave explicit formulas to quantify the number of leaf computations when using MSS.

We stated theoretically achievable performance gains and verified them practically. The algorithmic improvement decreases the required computation time for signature creation in theory as well as in practice. The performance figures show that Merkle signatures are not only practical, but also resource-friendly and fast. As such they are an advantageous choice for, e.g., digital signature smart cards.

Acknowledgments. This material is based in part upon work supported by the National Science Foundation under Grant No. 1261399 and by grant 01ME12025 SecMobil of the German Federal Ministry of Economics and Technology.

References

1. Atmel. ATxmega128A1 Data Sheet,
   http://www.atmel.com/dyn/resources/prod_documents/doc8067.pdf
2. Atmel. AVR XMEGA A Manual,
   http://www.atmel.com/dyn/resources/prod_documents/doc8077.pdf
3. Atmel. AVR XPLAIN board,
   http://www.atmel.com/dyn/resources/prod_documents/doc8203.pdf
4. Biryukov, A., Khovratovich, D., Nikolić, I.: Distinguisher and related-key attack on the full aes-256. In: Halevi, S. (ed.) CRYPTO 2009. LNCS, vol. 5677, pp. 231–249. Springer, Heidelberg (2009)
5. Buchmann, J., Dahmen, E., Ereth, S., Hülsing, A., Rückert, M.: On the Security of the Winternitz One-Time Signature Scheme. In: Nitaj, A., Pointcheval, D. (eds.) AFRICACRYPT 2011. LNCS, vol. 6737, pp. 363–378. Springer, Heidelberg (2011)
6. Buchmann, J., Dahmen, E., Hülsing, A.: XMSS - A Practical Forward Secure Signature Scheme Based on Minimal Security Assumptions. In: Yang, B.-Y. (ed.) PQCrypto 2011. LNCS, vol. 7071, pp. 117–129. Springer, Heidelberg (2011)
7. Buchmann, J., Dahmen, E., Szydlo, M.: Hash-based Digital Signature Schemes. In: Bernstein, D.J., Buchmann, J., Dahmen, E. (eds.) Post-Quantum Cryptography, pp. 35–93. Springer, Heidelberg (2009)
8. Dods, C., Smart, N.P., Stam, M.: Hash Based Digital Signature Schemes. In: Smart, N.P. (ed.) Cryptography and Coding 2005. LNCS, vol. 3796, pp. 96–115. Springer, Heidelberg (2005)
9. Eisenbarth, T., von Maurich, I., Ye, X.: Faster Hash-based Signatures with Bounded Leakage. In: Selected Areas in Cryptography, SAC 2013 (August 2013)
10. Hülsing, A.: W-OTS+ - Shorter Signatures for Hash-Based Signature Schemes. In: Youssef, A., Nitaj, A., Hassanien, A.E. (eds.) AFRICACRYPT 2013. LNCS, vol. 7918, pp. 173–188. Springer, Heidelberg (2013)
11. Hülsing, A., Busold, C., Buchmann, J.: Forward Secure Signatures on Smart Cards. In: Knudsen, L.R., Wu, H. (eds.) SAC 2012. LNCS, vol. 7707, pp. 66–80. Springer, Heidelberg (2013)
12. Intel. Intel Core i7 2620M Specifications,
    http://ark.intel.com/products/52231/
    Intel-Core-i7-2620M-Processor-4M-Cache-2_70-GHz
13. Intel. Whitepaper on the Intel AES Instructions Set,
    http://software.intel.com/file/24917
14. Lamport, L.: Constructing Digital Signatures from a One-Way Function. Technical report, CSL-98, SRI International (1979)
15. Lee, J., Stam, M.: MJH: A Faster Alternative to MDC-2. In: Kiayias, A. (ed.) CT-RSA 2011. LNCS, vol. 6558, pp. 213–236. Springer, Heidelberg (2011)
16. Matyas, S.M., Meyer, C.H., Oseas, J.: Generating strong one-way functions with cryptographic algorithm. IBM Technical Disclosure Bulletin 27(10A), 5658–5659 (1985)
17. Menezes, A., Van Oorschot, P., Vanstone, S.: Handbook of Applied Cryptography. CRC, Algorithm 9.41 (1997)
18. Merkle, R.C.: A Certified Digital Signature. In: Brassard, G. (ed.) CRYPTO 1989. LNCS, vol. 435, pp. 218–238. Springer, Heidelberg (1990)
19. Rohde, S., Eisenbarth, T., Dahmen, E., Buchmann, J., Paar, C.: Fast Hash-Based Signatures on Constrained Devices. In: Grimaud, G., Standaert, F.-X. (eds.) CARDIS 2008. LNCS, vol. 5189, pp. 104–117. Springer, Heidelberg (2008)

20. Standaert, F.-X., Pereira, O., Yu, Y., Quisquater, J.-J., Yung, M., Oswald, E.: Leakage Resilient Cryptography in Practice. In: Sadeghi, A.-R., Naccache, D., Basin, D., Maurer, U. (eds.) Towards Hardware-Intrinsic Security, Information Security and Cryptography, pp. 99–134. Springer, Heidelberg (2010)
21. Szydlo, M.: Merkle Tree Traversal in Log Space and Time. In: Cachin, C., Camenisch, J.L. (eds.) EUROCRYPT 2004. LNCS, vol. 3027, pp. 541–554. Springer, Heidelberg (2004)

A   Appendix

Algorithm 1. Improved treehash update

Input: Height h, current index s, RIGHTNODES array
Output: updated RIGHTNODES array, updated Treehash instance TREEHASH$_h$

    Compute the sth leaf: NODE$_1$ ←LEAFCALC(s)

    if $s \equiv 2^h - 1 \left(\bmod\, 2^h \right)$ **and** NODE$_1$.height() $< h$ **then**
        offset $= h\,(h - 1)\,/2$
        RIGHTNODES[offset] ←NODE$_1$
    end if
    while NODE$_1$ has the same height as the top node on TREEHASH$_h$ **do**
        Pop the top node from the stack: NODE$_2$ ←TREEHASH$_h$.pop()
        Computer their parent node: NODE$_1$ ← g(NODE$_2$‖NODE$_1$)
        if $s \equiv 2^h - 1 \left(\bmod\, 2^h \right)$ **then**
            offset $=$ offset $+ 1$
            RIGHTNODES[offset] ←NODE$_1$
        end if
    end while

    Push the parent node on the stack: TREEHASH$_h$.push(NODE$_1$)

Algorithm 2. Key generation and initial setup for the improved traversal algorithm.

Input: H, K
Output: Public key $\nu_H\,[0]$, Authentication path, RIGHTNODES array, TREEHASH stacks, RETAIN stacks

1: **Public Key**
   Calculate and publish tree root, $\nu_H\,[0]$.
2: **Initial Right Nodes**
   $i = 0$
   for $h = 1$ **to** $H - K - 1$ **do**
      for $j = 0$ **to** $h - 1$ **do**
         Set RIGHTNODES[i] $= \nu_j \left[2^{2+h-j} - 1 \right]$.
         $i = i + 1$
3: **Initial Authentication Nodes**
   for each $h \in \{0, 1, \ldots, H - 1\}$ **do**
      Set AUTH$_h = \nu_h\,[1]$.
4: **Initial Treehash Stacks**
   for each $h \in \{0, 1, \ldots, H - K - 1\}$ **do**
      Setup TREEHASH$_h$ stack with $\nu_h\,[3]$.
5: **Initial Retain Stacks**
   for each $h \in \{H - K, \ldots, H - 2\}$ **do**
      for each $j \in \left\{ 2^{H-h-1}, \ldots, 0 \right\}$ **do**
         RETAIN$_h$.push($\nu_h\,[2j + 3]$).

Algorithm 3. Algorithm for authentication path computation as presented in [7]

Input: $s \in \left\{0, \dots, 2^H - 2\right\}$, H, K, and the algorithm state.

Output: Authentication path A_{s+1} for leaf $s + 1$.

1: Let $\tau = 0$ if leaf s is a left node or let τ be the height of the first parent of leaf s which is a left node: $\tau \leftarrow \max\{h : 2^h | (s+1)\}$

2: If the parent of leaf s on height $\tau + 1$ is a left node, store the current authentication node on height τ in KEEP_τ:
    if $\lfloor s/2^{\tau+1} \rfloor$ is even **and** $\tau < H - 1$ **then** $\text{KEEP}_\tau \leftarrow \text{AUTH}_\tau$

3: If leaf s is a left node, it is required for the authentication path of leaf $s + 1$:
    if $\tau = 0$ **then** $\text{AUTH}_0 \leftarrow \text{LEAFCALC}(s)$

4: Otherwise, if leaf s is a right node, the auth. path for leaf $s + 1$ changes on heights $0, \dots, \tau$:
    if $\tau > 0$ **then**
    a) The authentication path for leaf $s + 1$ requires a new left node on height τ. It is computed using the current authentication node on height $\tau - 1$ and the node on height $\tau - 1$ previously stored in $\text{KEEP}_{\tau-1}$. The node stored in $\text{KEEP}_{\tau-1}$ can then be removed:
    $\text{AUTH}_\tau \leftarrow g\left(\text{AUTH}_{\tau-1} \| \text{KEEP}_{\tau-1}\right)$, remove $\text{KEEP}_{\tau-1}$
    b) The authentication path for leaf $s + 1$ requires new right nodes on heights $h = 0, \dots, \tau - 1$. For $h < H - K$ these nodes are stored in TREEHASH_h and for $h \geq H - K$ in RETAIN_h:
    for $h = 0$ **to** $\tau - 1$ **do**
        if $h < H - K$ **then** $\text{AUTH}_h \leftarrow \text{TREEHASH}_h.\text{pop}()$
        if $h \geq H - K$ **then** $\text{AUTH}_h \leftarrow \text{RETAIN}_h.\text{pop}()$
    c) For heights $0, \dots, \min\{\tau - 1, H - K - 1\}$ the Treehash instances must be initialized anew. The Treehash instance on height h is initialized with the start index $s + 1 + 3 \cdot 2^h < 2^H$:
    for $h = 0$ **to** $\min\{\tau - 1, H - K - 1\}$ **do**
        $\text{TREEHASH}_h.\text{initialize}(s + 1 + 3 \cdot 2^h)$

5: Next we spend the budget of $(H - K)/2$ updates on the Treehash instances to prepare upcoming authentication nodes:
    repeat $(H - K)/2$ **times**
    a) We consider only stacks which are initialized and not finished. Let k be the index of the Treehash instance whose lowest tail node has the lowest height. In case there is more than one such instance we choose the instance with the lowest index:
    $k \leftarrow \min\left\{h : \text{TREEHASH}_h.\text{height}() = \min_{j=0,\dots,H-K-1}\left\{\text{TREEHASH}_j.\text{height}()\right\}\right\}$
    b) The Treehash instance with index k receives one update: $\text{TREEHASH}_k.\text{update}()$

6: The last step is to output the authentication path for leaf $s + 1$: **return** $\text{AUTH}_0, \dots, \text{AUTH}_{H-1}$.

Privacy-Preserving Reconciliation Protocols: From Theory to Practice

Ulrike Meyer[1] and Susanne Wetzel[2]

[1] RWTH Aachen, Department of Computer Science
Mies-van-der-Rohe-Str. 15, 52074 Aachen, Germany
meyer@umic.rwth-aachen.de
[2] Stevens Institute of Technology, Department of Computer Science
Castle Point on Hudson, Hoboken NJ 07030, USA
swetzel@stevens.edu

Abstract. In this paper we provide a brief summary of our work on privacy-preserving reconciliation protocols. The main focus of the paper is to review two of our two-party protocols. We detail the protocols and provide a comprehensive theoretical performance analysis. Furthermore, we briefly describe some of our work on multi-party protocols. We also show how we have translated our theoretical results into practice—including the design and implementation of a library as well as developing iPhone and Android apps.

1 Introduction

Over the past few years, smart devices have become ubiquitous and much of our daily activities leave some form of a digital fingerprint online. We use services such as Doodle to find a suitable time for the next project meeting or ZocDoc to make a doctor's appointment; we use tools such as Google Calendar to efficiently keep track of appointments and commitments; we use eBay to buy and auction off goods; we rely on technology that enables us to be connected at all times by having our smart devices seamlessly roam between our own home wireless network, business networks, or various wireless service providers.

Today, in most of these cases not all parties involved have the possibility to define the parameters and conditions under which and how they would like to see these activities being carried out. For example, Doodle does not allow the user to specify that he prefers some meeting times over others. And even if such preferences could be specified, there typically is no transparency whether and to what extent these will be considered. Furthermore, the issue of privacy is generally neglected and more information is disclosed than is needed or desired. For example, at least the participant starting a Doodle poll can see the selections made by all the others.

It is in this context that our work on privacy-preserving reconciliation strives to develop solutions that allow for a reconciliation process that yields an optimal solution while recognizing the preferences of the involved parties and assuring suitable privacy guarantees.

M. Fischlin and S. Katzenbeisser (Eds.): Buchmann Festschrift, LNCS 8260, pp. 183–210, 2013.

Our work falls into the area of Secure Multi-Party Computation (SMPC). Loosely speaking, SMPC enables parties to compute a commonly known function of their private inputs where nothing else but the result of the computation becomes known to the parties. Since Yao's seminal paper [35], SMPC has evolved greatly both with regard to general-purpose protocols as well as specialized protocols. In fact, these developments do not only extend in theory but also there is an ongoing discussion as to whether or not special-purpose protocols typically outperform general-purpose solutions in practice (e.g., [15,7]). At the core of much of our work to date are Private Set Intersection (PSI) protocols. While our early work on two-party protocols directly builds on the work by Freedman *et al.* [9], our later contributions in the two-party context use PSI as a generic building block which allow the use of any PSI protocol (e.g., [8,14,5]). For multi-party settings, our work builds on multi-party protocols for specific privacy-preserving multiset operations. Such protocols were first proposed by Kissner *et al.* [16]. While there are some works on set intersection (e.g., [33,6,25,2]) and set union (e.g., [10]) operations in the multi-party setting, the work of Kissner *et al.* still is the only work that supports all operations required by our multi-party protocols. On the practical side there have been recent advances both with regard to implementations of generic constructions (e.g., [11,1,3]) as well as special-purpose protocols (e.g., [4,8]).

Outline. In the following, we first review important notations and definitions that are used in the remainder of this paper (Section 2). At the core of the paper, we then provide a detailed description and analysis of two of our two-party privacy-preserving reconciliation protocols (Section 3). This is followed by a high-level description of our work in the context of multi-party reconciliation protocols (Section 4). Then, we provide some insights on how we have taken our theoretical results and translated them into practice—focusing not only on the design and implementation of a library but also exploring real-world applications (Section 5). We close this paper with some remarks on future work.

2   Definitions and Tools

2.1   Definitions

Throughout this paper, we consider $n \geq 2$ parties $P_1, ..., P_n$ with private input sets $S_1, ..., S_n$ with exactly k distinct elements each, chosen from a common domain $\mathcal{D}$. Each party has certain preferences associated with its input set which allow a party to define a total order on its private inputs. The goal of reconciliation protocols is to find one (or more) common input element(s) while taking the preferences of all parties into account in an equal manner. In the following, we more formally define these notions, starting with the input sets of each party:

Definition 1 (Ordered sets & ranking). *Let $\mathcal{D}$ be the* domain. *$(S, <) \in 2^{\mathcal{D}} \times 2^{\mathcal{D} \times \mathcal{D}}$ is an ordered set if $<$ is a strict total order on S. We write $\{x_1 > ... > x_k\}$ for the ordered set $(\{x_1 ..., x_k\}, \{(x_j, x_i) \mid 1 \leq i < j \leq k\})$. The ranking function $rank_S : S \to \mathbb{N}$ is defined as $rank_S(x_i) := k - i + 1$.*

In order to allow for a more formal definition of reconciliation protocols that take the preferences of all parties into account in an equal manner, we first need to properly define the composition of multiple ordered sets into a single pre-ordered set:

Definition 2 (Composition schemes). *Let* $(S_1, <_1), ..., (S_n, <_n)$ *be ordered sets. The pre-ordered set* $(S_1 \cap ... \cap S_n, \leq_{\{P_1,..,P_n\}})$ *is called the* combined pre-ordered set *of* $(S_1, <_1), ..., (S_n, <_n)$ *w.r.t.* $\mathcal{F}_{comp}$ *if* $\leq_{\{P_1,..,P_n\}}$ *is the pre-order induced by the function* $\mathcal{F}_{comp} : (S_1 \cap ... \cap S_n) \to \mathbb{N}$. *The function* $\mathcal{F}_{comp}$ *is called the* composition scheme.

In the following, we consider two composition schemes that are reasonable choices in the sense that they consider the orders of all the individual parties in an equal manner—also referred to as unbiasedness. In particular, the level of unbiasedness that is achieved depends on how the composition scheme is defined, i.e., how and to what degree it recognizes the orders of the individual parties.

The first composition scheme is referred to as *minimum of ranks* scheme and determines the rank of an element in the intersection of all sets based on the minimum rank assigned to it by any party.

Definition 3 (Minimum of ranks). *The Minimum of Ranks (MR) composition scheme is defined by the function*

$$\mathcal{F}_{MR}(x) = min\{rank_{S_1}(x), ..., rank_{S_n}(x)\}.$$

The second composition scheme is referred to as the *sum of ranks* scheme. It determines the rank of an element in the intersection of the ordered sets as the sum of the ranks assigned to the element by the individual parties.

Definition 4 (Sum of ranks). *The Sum of Ranks (SR) composition scheme is defined by the function*

$$\mathcal{F}_{SR}(x) = rank_{S_1}(x) + ... + rank_{S_n}(x).$$

We can now formally define a reconciliation protocol on ordered sets:

Definition 5 (Reconciliation on ordered sets). *A privacy-preserving, preference-maximizing protocol for an order composition scheme* $\mathcal{F}_{comp}$ *is a multi-party protocol between* n *parties* $P_1, ..., P_n$ *each with an ordered input set* $(S_i, <_i)$ *drawn from the same domain* $\mathcal{D}$. *Upon completion of the protocol, each party learns* (χ, τ) *with*

$$\chi = \underset{x \in (S_1 \cap ... \cap S_n)}{arg\ max}\ \mathcal{F}_{comp}(x)$$

$$\tau = \underset{x \in (S_1 \cap ... \cap S_n)}{max}\ \mathcal{F}_{comp}(x)$$

where $arg\ max_{x \in (S_1 \cap ... \cap S_n)} \mathcal{F}_{comp}(x) = \{x | \forall y \in (S_1 \cap ... \cap S_n) : \mathcal{F}_{comp}(y) \leq \mathcal{F}_{comp}(x)\}$.

No party P_i *learns anything about the inputs and preferences of the other parties except what can be deduced from* χ, τ, *and its private input set* $(S_i, <_i)$.

The result of the protocol is the set of common elements with maximum rank in the combined preference order and the corresponding maximum rank. In the following, we will refer to the two-party reconciliation protocols as PROS (Privacy-preserving Reconciliation on Ordered Sets) and the multi-party reconciliation protocols as MPROS (Multi-party Privacy-preserving Reconciliation on Ordered Sets). The corresponding variants for the maximized minimum of ranks and sum of ranks composition schemes will be referred to as $PROS^{MR}$, $PROS^{SR}$, $MPROS^{MR}$, and $MPROS^{SR}$, respectively.

2.2 Adversary Models

In secure multi-party computation two or more parties compute the output of some function of their individual private inputs in a distributed fashion without revealing any more information about each others' inputs than what can be deduced from the intended output. Two adversary models are commonly used: the *semi-honest* and the *malicious* models. Both models assume the existence of pairwise encrypted and authenticated channels between the participating parties such that external attackers do not have to be considered.

Security in the Semi-honest Adversary Model. A semi-honest adversary is an inside attacker that tries to infer as much (secret) information as possible from its view of a protocol run, but strictly follows the prescribed actions of the protocol.

Security in the Malicious Adversary Model. A malicious adversary is an inside attacker that can almost arbitrarily deviate from the protocol. The only actions it cannot take are the ones that cannot be prevented, namely: refusal to participate in the protocol, manipulation of its own input, and protocol abortion [12].

For more formal definitions refer to [12].

2.3 Tools

Additively Homomorphic Cryptosystems. Our two-party protocols require a semantically secure, additively homomorphic, asymmetric cryptosystem. Similarly, our multi-party protocols require a threshold version of such a cryptosystem.

A cryptosystem which has *additively homomorphic* properties provides an operation in the ciphertext domain to compute the encrypted sum of two plaintexts given only the corresponding ciphertexts. More formally, let $E_{pk}(\cdot)$ be the encryption function with public key pk. In an additively homomorphic cryptosystem there is an operation $+_h$ such that $E_{pk}(a + b) = E_{pk}(a) +_h E_{pk}(b)$ which can be computed efficiently given only $E_{pk}(a), E_{pk}(b)$, and pk. It is also possible to perform an efficient scalar multiplication with a scalar q (chosen from the plaintext domain) by repeatedly applying the $+_h$ operation. Formally,

$$q \times_h E_{pk}(a) = \underbrace{E_{pk}(a) +_h \ldots +_h E_{pk}(a)}_{q \text{ times}} = E_{pk}(q \cdot a).$$

In the following, we use the notation $\tilde{\sum}$ to denote an iterative homomorphic sum, i.e., $\tilde{\sum}_{1 \le i \le \ell} c_i = c_1 +_h c_2 +_h \ldots +_h c_\ell$.

In addition, we require the cryptosystem to provide for semantic security [13]. That is, if given two ciphertexts, it is infeasible for an attacker to find any meaningful relationship between the plaintexts based only on the ciphertexts.

In a (t, n)-threshold version of an additively homomorphic cryptosystem, the private key sk is shared among the n parties with each party P_i holding a private share of the key. Using its private key share, a party P_i can compute a *partial decryption* of a ciphertext. To successfully decrypt a given ciphertext, t of the n key shares are required to compute the plaintext by combining t partial decryptions of the ciphertext.

The Paillier cryptosystem [31] is a prominent and widely-used example for a semantically secure additively homomorphic cryptosystem for which there also is a threshold variant.

Private Set Intersection in the Two-Party Case. Our PROS protocols make use of a PSI protocol. The first such protocol was proposed by Freedman *et al.*[9]. Further two-party protocols solving the PSI problem include [8,14,5]. In the following, we describe Freedman *et al.*'s protocol (FNP) in more detail.

The FNP construction requires a semantically secure additively homomorphic cryptosystem. It assumes a chooser C and a sender Y and allows these two parties to compute the intersection $I_C \cap I_Y$ of their private sets I_C and I_Y (chosen from the same domain). At the end of the protocol, C only learns which elements C and Y have in common and Y learns nothing. The technique used by FNP is oblivious polynomial evaluation. Oblivious means that Y only sees encrypted coefficients and evaluates the polynomial without having access to its actual coefficients. The intersection protocol then works as follows:

First, C generates a polynomial containing all her set elements c_i as roots:

$$p_C(X) = (X - c_1) \cdot (X - c_2) \ldots (X - c_\ell) = \sum_{i=0}^{\ell} \alpha_i \cdot X^i$$

Then, C encrypts all α_i using her public key pk_C and sends them to Y. Y chooses a random r_i for each y_i in his set and obliviously computes $p_C^{y_i} := E_{pk_C}(r_i \cdot p_C(y_i) + y_i)$. Y then sends all $p_C^{y_i}$ back to C. C uses her private key to decrypt the result. Note that $p_C^{y_i}$ decrypts to y_i if y_i is in I_C and to a random number otherwise. Thus, C can tell which elements of Y are also in her set. In order for Y to be able to also determine the intersection, the protocol is executed in the opposite direction. The protocol is proven to be privacy-preserving in the semi-honest model [9].

Privacy-Preserving Multiset Operations. Our MPROS protocols make use of multi-party protocols for privacy-preserving multiset operations, specifically intersection, union, and reduction.[1] Kissner *et al.* [16] were the first to introduce such protocols.

Representing Multisets as Polynomials. The essential idea of the operations proposed by Kissner *et al.* is to encode the elements of multisets as the roots of an encrypted polynomial and compute the result of the set operations using an additively homomorphic cryptosystem. A private input multiset $M_i = \{m_{i,1}, ..., m_{i,k}\}$ is encoded by defining a polynomial $f_i(X) = \prod_{j=1}^{k}(X - m_{i,j})$ and then encrypting the coefficients of the resulting polynomial.

The result of multiset operations can now be computed solely by manipulating encrypted polynomials. The sum of two polynomials can be computed by means of homomorphic addition of their coefficients. We denote this operation by $\phi +_h \gamma$, where ϕ and γ are encrypted polynomials.

Furthermore, the encrypted result π of a polynomial multiplication of an encrypted polynomial ϕ and a plaintext polynomial g can be computed using homomorphic addition and scalar multiplication. Specifically, the encrypted coefficients π_i of the polynomial π can be computed as

$$\pi_i = \sum_{j=0}^{i} \phi_j \times_h g_{i-j} \text{ for } 0 \leq i \leq deg(\phi) + deg(g).$$

We denote this operation by $\phi \times_h g$.

Finally, one can compute the derivative ϕ' of an encrypted polynomial ϕ using only homomorphic operations. A coefficient ϕ'_i of the derivative of ϕ can be computed as

$$\phi'_i = (i + 1) \times_h \phi_{i+1} \text{ for } 0 \leq i \leq deg(\phi) - 1.$$

Using polynomial addition, multiplication, and derivation, privacy-preserving multiset operations can be achieved as follows.

Union. For an encrypted polynomial ϕ and a plaintext polynomial g, representing the two multisets F and G of arbitrary size, a polynomial representation of the union $F \cup G$ can be computed by means of the homomorphic polynomial multiplication

$$\phi \times_h g.$$

The product contains all roots of f and g in the corresponding aggregated multiplicity (where f is the plaintext polynomial corresponding to the encrypted polynomial ϕ). It is important to note that it is not possible to learn more information from the decryption of $\phi \times_h g$ than one can derive from $F \cup G$.

---

[1] A multiset is a generalization of a set in that a specific set element may appear more than once. The multiplicity of an element is the number of times that the element occurs in the multiset.

Intersection. For two encrypted polynomials ϕ and γ representing two multisets F and G of equal size, the intersection can be determined by means of

$$\phi \times_h s +_h \gamma \times_h r.$$

Here, s and r are random plaintext polynomials of degree $deg(\phi)$. The roots of the resulting polynomial are those common to ϕ and γ (with minimum multiplicity) and thus represent the elements of $F \cap G$. Again, from the decryption of the resulting polynomial one cannot learn more than from $F \cap G$.

Set Reduction. For an encrypted polynomial γ representing a multiset G, one can compute the element reduction $Rd_t(G)$ as

$$\sum_{i=0}^{\tilde{t}} \gamma^{(i)} \times_h F_i \times_h r_i.$$

Each r_i is chosen uniformly and independently from the set of plaintext polynomials of degree $deg(\gamma^{(i)})$. The polynomials F_i are of degree i and must meet the additional requirement that these polynomials do not share any roots, i.e., $gcd(F_0, ..., F_t) = 1$. Unlike the r_i's, the F_i's may be fixed. For any element $g \in G$ with multiplicity b in G, $Rd_t(G)$ contains g with multiplicity $max\{b - t, 0\}$.

Other Protocols. While a variety of multi-party protocols solving the (multi)set intersection problem have been proposed following Kissner *et al.*'s work (e.g., [33,6,25,2]), there still is only a few privacy-preserving union protocols (e.g., [10]). And to date, Kissner *et al.*'s work remains to be the only one that solves the multiset reduction problem. In the following, we therefore use the protocols proposed by Kissner *et al.* to design our multi-party reconciliation protocols. Yet, it is important to note that our protocols could use other other solutions that support all three set operations (union, intersection, and reduction) and are composable in the manner that Kissner *et al.*'s protocols are. Specifically, the work by Kissner *et al.* allows the computation of any combination of set operations that can be expressed by the grammar

$$\Upsilon ::= M_i | Rd_t(\Upsilon) | \Upsilon \cap \Upsilon | \Upsilon \cup M_i | M_i \cup \Upsilon$$

where M_i are the input multisets of the parties $P_1, \ldots, P_n$.

3   Two-party Reconciliation Protocols

In this section we present some of our PROS protocols. We first introduce generic constructions that can use any PSI protocol as a building block. In particular, we show how any PSI protocol (e.g., [8,14,5,9]) can be used to construct generic $PROS^{MR}$ and $PROS^{SR}$ protocols. We then provide a detailed description and analysis of protocols that specifically use the PSI protocol introduced by Freedman *et al.* [9].

In order to improve readability of this section, we will use A and B to refer to the two parties (instead of P_1 and P_2 or even P_A and P_B). Also, we denote their input sets as $S_A = \{a_1, a_2, \ldots, a_k\}$ and $S_B = \{b_1, b_2, \ldots, b_k\}$ and assume that they are ordered, i.e., $\{a_1 >_A a_2 \ldots >_A a_k\}$ and $\{b_1 >_B b_2 \ldots >_B b_k\}$. As stated before, the input sets are drawn from the common domain $\mathcal{D}$. Furthermore, for the remainder of this section we assume $|\mathcal{D}| = N$.

3.1   Generic Construction from any PSI

We recall that the objective of a PROS protocol is to determine the common element(s) in the intersection of the input sets of the parties A and B that maximize the rank with regard to the combined preference order in a privacy-preserving manner.

Using a PSI as a building block, the PROS^{SR} protocol can generically be implemented as illustrated in Figure 1 on the example where both parties A and B have four distinct input elements each. In order to determine a common element

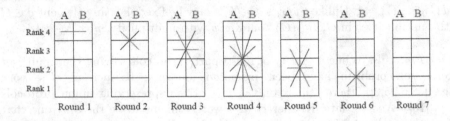

Fig. 1. PROS^{SR} : comparison of input elements of parties A and B for $k = 4$

of the input sets of the parties A and B that maximizes the SR composition scheme, the protocol progresses in a number of rounds. Each round consists of a number of PSIs (each indicated by a line in Figure 1). For example, in Round 1, parties A and B carry out one PSI to determine whether there is a match between A's highest ranked input element and the highest ranked one of Party B. If so, the protocol terminates as it found a common element in the intersection of the input sets that maximizes the MR composition scheme. Otherwise, the protocol moves on to the next round. All PSIs carried out in one round result in the same sum of ranks. In particular, the sum of ranks in Round 1 is eight (in the example, or $2k$ in case of both parties holding k distinct inputs). In each round, the sum decreases by one, i.e., Round i corresponds to a sum of ranks of $2k - i + 1$—thus guaranteeing that the protocol will in fact always maximize the sum of ranks. In the worst case, the protocol will carry out $2k - 1$ rounds of PSIs.

As illustrated in Figure 2, the order in which the PSIs are executed is somewhat different when using the MR composition scheme. Unlike before, the focus in each round is now on the common element with the smaller preference since

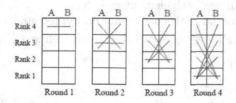

Fig. 2. PROS^{MR} : comparison of input elements of parties A and B for $k = 4$

the goal is to maximize that. Consequently, the respective PROS^{MR} protocol requires fewer rounds than the PROS^{SR} protocol. In fact, the number of protocol rounds never exceeds the number of inputs the parties hold. As before, both parties A and B each carry out a certain number of PSIs in each round of the PROS^{MR} protocol. For example, in Round 3 in Figure 2, parties A and B will carry out up to five PSIs. Obviously, the minimum preference of all elements considered in Round 3 is two. If no intersection is found, the protocol will move on to the next round which is characterized by the fact that the minimum of the preferences of the elements considered is decreased by one. This procedure ensures that the protocol will determine a common set element that maximizes the MR composition scheme.

The assumption that the PSI is privacy-preserving in the semi-honest model automatically implies that the PROS^{SR} and PROS^{MR} protocols are privacy-preserving in the semi-honest model—irrespective of the PSI protocol and composition scheme used. This is due to the fact that in the semi-honest model both parties will follow the protocol. Consequently, if the PSI protocol used is privacy-preserving in the semi-honest model, then a party will not learn anything about any elements of the other party's input set unless a common set element is determined through the respective PSI. Obviously, an analogous statement for the malicious model does not hold true as malicious behavior in the PROS protocols may come into play both within a specific round (including multiple PSIs) as well as in-between different rounds.

3.2 PROS Protocol for the Sum of Ranks Composition Scheme

We now detail the PROS^{SR} protocol using the PSI introduced by Freedman *et al.* Unlike in the generic construction above, we formally describe the protocol messages and introduce optimizations (both in the number of messages exchanged as well as their respective payloads). We also provide a detailed security and performance analysis.

We assume that prior to the execution of any of the protocols introduced in this section, the two parties A and B agree upon a semantically secure homomorphic encryption scheme. The parties choose their public and private key pairs for the encryption scheme and exchange their public keys. We denote the public encryption functions of A and B with E_A and E_B and their private decryption functions with D_A and D_B.

PROS^{SR} **Protocol:** For each element $a_i \in S_A$ $(i = 1, \ldots, k)$ we define $f_A^{(i)} := (X - a_i)$. The polynomial with coefficients encrypted under E_A is denoted by $E_A^{(i)}$. Similarly, for each $b_i \in S_B$ $(i = 1, \ldots, k)$ we define $f_B^{(i)}$ and $E_B^{(i)}$.

COMMITMENT PHASE:

ROUND 0: B sends $E_B^{(1)}$ to A.

ROUND 1: A chooses a random $r_{A,1}^{(1)}$, computes $c_{A,1}^{(1)} = E_B\left(r_{A,1}^{(1)} \cdot f_B^{(1)}(a_1) + a_1 \right)$, and sends it together with $E_A^{(1)}$ to B. B decrypts $c_{A,1}^{(1)}$ under D_B and compares the result to b_1. If $b_1 = D_B(c_{A,1}^{(1)})$, then $b_1 = a_1$, that is, B has found MAX $:= b_1 = a_1$ and continues with the *Match Confirmation Phase*. Otherwise, B sends $E_B^{(2)}$ to A and A continues with Round 2 of the *Commitment Phase*.

ROUND $2 \leq i \leq k$: For $j = 1, \ldots, \lceil i/2 \rceil$, A chooses random $r_{A,j}^{(i-j+1)}$ and computes the ciphertexts

$$c_{A,j}^{(i-j+1)} := E_B\left(r_{A,j}^{(i-j+1)} \cdot f_B^{(i-j+1)}(a_j) + a_j \right)$$

and sends them as well as $E_A^{(i)}$ to B. For $j = 1, \ldots, \lceil i/2 \rceil$, B decrypts $c_{A,j}^{(i-j+1)}$ with D_B and checks whether the decrypted value equals b_{i-j+1}. If for some $j =: m$ the ciphertext $c_{A,m}^{(i-m+1)}$ decrypts to b_{i-m+1}, B has found MAX $:= b_{i-m+1} = a_m$ and continues with the *Match Confirmation Phase*. Otherwise, for $1 \leq j \leq \lfloor i/2 \rfloor$, B chooses random $r_{B,j}^{(i-j+1)}$ and computes the ciphertexts

$$c_{B,j}^{(i-j+1)} := E_A\left(r_{B,j}^{(i-j+1)} \cdot f_A^{(i-j+1)}(b_j) + b_j \right)$$

and sends them to A. For $2 \leq i < k$, B additionally sends $E_B^{(i+1)}$ to A. For $1 \leq j \leq \lfloor i/2 \rfloor$, A decrypts $c_{B,j}^{(i-j+1)}$ with D_A and checks whether the decrypted value equals a_{i-j+1}. If $c_{B,m}^{(i-m+1)}$ decrypts to a_{i-m+1}, A has found MAX $:= a_{i-m+1} = b_m$ and continues with the *Match Confirmation Phase*. Otherwise, A continues with Round $i + 1$ of the *Commitment Phase*.

ROUND $k < i < 2k - 1$: For $j = 1, \ldots, k - \lfloor i/2 \rfloor$, A chooses random $r_{A,i-k+j}^{(k-j+1)}$ and computes the ciphertexts

$$c_{A,i-k+j}^{(k-j+1)} := E_B\left(r_{A,i-k+j}^{(k-j+1)} \cdot f_B^{(k-j+1)}(a_{i-k+j}) + a_{i-k+j} \right)$$

and sends them to B. For $j = 1, \ldots, k - \lfloor i/2 \rfloor$, B decrypts $c_{A,i-k+j}^{(k-j+1)}$ with D_B and checks whether the decrypted value equals b_{k-j+1}. If for some $j =: m$ the ciphertext $c_{A,i-k+m}^{(k-m+1)}$ decrypts to b_{k-m+1}, B has found MAX $:= b_{k-m+1} = a_{i-k+m}$ and continues with the *Match Confirmation Phase*. Otherwise, for $1 \leq j \leq k - \lceil i/2 \rceil$, B chooses random $r_{B,i-k+j}^{(k-j+1)}$ and computes the ciphertexts

$$c_{B,i-k+j}^{(k-j+1)} := E_A\left(r_{B,i-k+j}^{(k-j+1)} \cdot f_A^{(k-j+1)}(b_{i-k+j}) + b_{i-k+j} \right)$$

and sends them to A. For $1 \leq j \leq k - \lceil i/2 \rceil$, A decrypts $c_{B,i-k+j}^{(k-j+1)}$ with D_A and checks whether the decrypted value equals a_{k-j+1}. If $c_{B,i-k+m}^{(k-m+1)}$ decrypts to a_{k-m+1}, A has found MAX $:= a_{k-m+1} = b_{i-k+m}$ and continues with the *Match Confirmation Phase*. Otherwise, A continues with Round $i+1$ of the *Commitment Phase*.

ROUND $i = 2k - 1$: A chooses random $r_{A,k}^{(k)}$ and computes the ciphertext

$$c_{A,k}^{(k)} := E_B \left(r_{A,k}^{(k)} \cdot f_B^{(k)}(a_k) + a_k \right)$$

and sends it to B. B decrypts $c_{A,k}^{(k)}$ with D_B and checks whether the decrypted value equals b_k. If $c_{A,k}^{(k)}$ decrypts to b_k, B has found MAX $:= b_k = a_k$ and continues with the *Match Confirmation Phase*. Otherwise, B sends a message to A indicating that no match was found and aborts the protocol.

MATCH CONFIRMATION PHASE: If B found the match $b_1 = a_1 =$ MAX in the first round of the *Commitment Phase*, then B computes $c_{B,1}^{(1)}$ and sends it to A. A decrypts the received ciphertext with D_A to a_1. Thus, A and B both know that MAX $= a_1 = b_1$.

If B found MAX $= b_{i-m+1} = a_m$ in Round $2 \leq i \leq k$ of the *Commitment Phase*, then B computes $c_{B,i-m+1}^{(m)}$ and sends it to A. A decrypts the received value under D_A to a_m. Thus, A and B both know that $a_m = b_{i-m+1} =$ MAX.

If A found MAX $= a_{i-m+1} = b_m$ in Round $2 \leq i \leq k$ of the *Commitment Phase*, A computes $c_{A,i-m+1}^{(m)}$ and sends it to B. B decrypts $c_{A,i-m+1}^{(m)}$ with D_B and checks that $D_B(c_{A,i-m+1}^{(m)})$ decrypts to a value $b_m \in \{b_1, \ldots, b_{\lfloor i/2 \rfloor}\}$. Thus, both parties A and B know that $a_{i-m+1} = b_m =$ MAX.

If B found MAX $= b_{k-m+1} = a_{i-k+m}$ in Round $k < i < 2k - 1$ of the *Commitment Phase*, then B computes $c_{B,k-m+1}^{(i-k+m)}$ and sends it to A. A decrypts the received value under D_A to a_{i-k+m}. Thus, A and B both know that $a_{i-k+m} = b_{k-m+1} =$ MAX.

If A found MAX $= a_{k-m+1} = b_{i-k+m}$ in Round $k < i < 2k - 1$ of the *Commitment Phase*, A computes $c_{A,k-m+1}^{(i-k+m)}$ and sends it to B. B decrypts $c_{A,k-m+1}^{(i-k+m)}$ with D_B and checks that $D_B(c_{A,k-m+1}^{(i-k+m)})$ decrypts to a value $b_{i-k+m} \in \{b_{i-k+1}, \ldots, b_{\lfloor i/2 \rfloor}\}$. Thus, both parties A and B know that $a_{k-m+1} = b_{i-k+m} =$ MAX.

If B found MAX $= b_k = a_k$ in Round $i = 2k - 1$ of the *Commitment Phase*, then B computes $c_{B,k}^{(k)}$ and sends it to A. A decrypts the received value with D_A to a_k. Thus, A and B both know that $a_k = b_k =$ MAX.

Intuitively, the protocol works as follows: The homomorphic encryption function allows a Party A to encrypt a polynomial corresponding to one of its own inputs in a way that the encrypted polynomial does not reveal information to Party B. In addition, if B evaluates the encrypted polynomial on one of its own inputs, then its input is blinded and encrypted. From this blinded ciphertext A learns nothing about B's input, unless the input A used for the creation of the

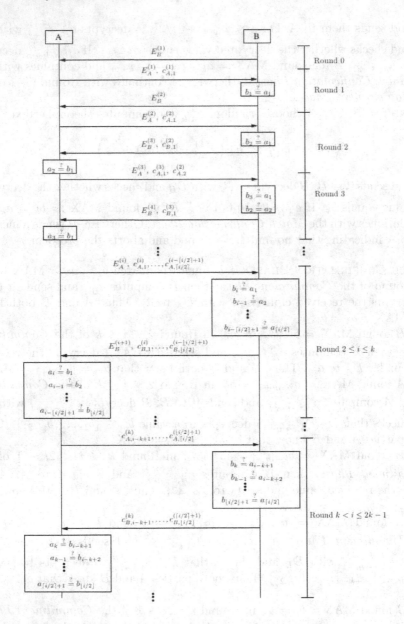

Fig. 3. Commitment Phase of the PROS^{SR} protocol. For example, in Round $2 \leq i \leq k$, A for $j = 1, \ldots, \lfloor i/2 \rfloor$ compares a_{i-j+1} with b_j and B for $j = 1, \ldots, \lceil i/2 \rceil$ compares b_{i-j+1} with a_j without obtaining knowledge of these values (unless they are equal), that is, in a privacy-preserving manner. Note that in each Round $1 \leq i \leq 2k - 1$ all inputs are compared that result in the same sum of ranks $i + 1$.

polynomial and the input on which B evaluated the polynomial coincide. This is due to the fact that the blinded ciphertext decrypts to A's input if and only if the inputs coincide. A and B can thus determine a common input without revealing inputs to each other that they do not have in common. Our protocols make use of this basic procedure in multiple rounds.

In each round, parties A and B exchange polynomials and blinded ciphertexts corresponding to previously received polynomials. They each decrypt the blinded ciphertext and compare the result to the inputs corresponding to the previously sent polynomials. The order in which the information is exchanged guarantees that the protocol terminates when a common input is found that maximizes the sum of ranks of the common inputs of A and B. For example, in the first round of the protocol, B can check whether the two most preferred inputs of both parties are the same ($a_1 \overset{?}{=} b_1$). If this is not the case, B does not learn anything about a_1. As a general rule, in Round i of the protocol, A and B compare all inputs that result in the same sum of ranks in the combined preference order of A and B.

Security Analysis: For $S_A \cap S_B \neq \emptyset$, the *Commitment Phase* of the PROS^{SR} protocol always terminates with one of the parties finding a match. This is due to the fact that for each pair (l, s) $1 \leq l, s \leq k$, a_l is compared with b_s in Round $l + s - 1$. To be precise, for $l < s$, a_l is compared with b_s by A and for $s \leq l$, b_s is compared with a_l by B in Round $l + s - 1$. Let MAX $= a_l = b_s$ be a match found in Round $l + s - 1$. Then, no match was found in any previous round of the *Commitment Phase*. We claim that then MAX maximizes the sum of the ranks of all elements in the intersection $S_A \cap S_B$. If this was not the case, then there would be a pair (a_r, b_v) with $a_r = b_v \in S_A \cap S_B$ and $r + v < s + l$. As a consequence, a_r would have been compared to b_v already in Round $r + v - 1$ of the *Commitment Phase*. This obviously contradicts the assumption.

In order to prove that the PROS^{SR} protocol is privacy-preserving in the semi-honest model it is necessary to show that the information that either party can deduce from the PROS^{SR} protocol is equivalent to either party knowing nothing else but what can be deduced from knowing the matching input MAX (maximizing the SR composition scheme) and its rank $\mathcal{F}_{SR}(\text{MAX})$. As discussed before, finding a match with the PROS^{SR} protocol implies that both parties know the result MAX which maximizes the combined preference order as well as the round in which it was found. The latter corresponds to the rank of the matching input under the SR composition scheme.[2] In turn, knowing MAX itself and its rank $\mathcal{F}_{SR}(\text{MAX})$ allows Party A (B) to determine $rank_B(\text{MAX})$ ($rank_A(\text{MAX})$). Furthermore, since MAX maximizes $\mathcal{F}_{SR}(x)$ with $x \in S_A \cap S_B$, it holds that $\nexists\, a \in S_A, b \in S_B$ such that $a = b$ and $\mathcal{F}_{SR}(a) = \mathcal{F}_{SR}(b) > \mathcal{F}_{SR}(\text{MAX})$. Since

---

[2] It is important to note that if a match is found by A (B), A (B) may learn more than one preference maximizing common input. This is due to the fact that more than one pair of inputs may be evaluated by A (B) in each round. This asymmetry can be prevented by adding subrounds to the *Commitment Phase* such that each party commits to only one input in each subround of each round in the *Commitment Phase*. This, however, would result in a higher communication overhead.

$\mathcal{F}_{SR}(\text{MAX})$ directly corresponds to the round in which the PROS^{SR} protocol would find the match, this implies that the operations in Rounds i with
$i < f_{SR}(\text{MAX})$ in the PROS^{SR} protocol should yield no information other
than that the respective inputs do not match. Using Freedman's PSIs in each
round, the PROS^{SR} protocol achieves this by construction as Freedman's PSI
is privacy-preserving in the semi-honest model.[3]

Performance Analysis: We evaluate the worst case performance of the protocol by counting (i) the number of times A and B have to compare a received
decrypted input of the other party with one of their inputs, (ii) the number of
decryptions A and B have to perform, (iii) the number of encryptions of coefficients of polynomials, (iv) the number of ciphertexts A and B have to compute
to commit to their inputs, (v) the number of messages exchanged between A
and B, as well as (vi) the overall size of the payload of all messages exchanged,
counted in the number of ciphertexts included in all messages.

We count the above numbers for three different relations between the size N
the domain $\mathcal{D}$ of possible inputs to choose from and the number k of inputs A
and B choose from the $\mathcal{D}$.

In the first case, we assume that N equals k. In this case, A and B share all
inputs and differ only in their preferences. The worst case occurs if the inputs
of A and B are in exactly the opposite order. In this case, the match is found in
Round k of the protocol. Table 1 shows the results.

Table 1. PROS^{SR} : worst case performance for $N = k$

$k = N$	
# Comparisons	$\frac{k(k-1)}{2} + 2$
# Encryption of coefficients	$4k - 3$
# Decryptions	$\frac{k(k-1)}{2} + 2$
# Commitment of inputs	$\frac{k^2+2}{2}$ (k even); $\frac{k^2+3}{2}$ (k odd)
# Messages	$2k + 1$
Size of payload in # ciphertexts	$k^2/2 + 4k - 2$

For example, the number of comparisons is counted as follows: In the worst
case, the match is found in Round k by one comparison and confirmed in the
confirmation phase by yet another comparison. The preceding $k - 1$ rounds are
unsuccessful. In the ith round ($i = 1, \ldots, k-1$), i comparisons take place. Therefore, the overall number of comparisons is $1+2+\cdots+k-1+2 = (\sum_{r=1}^{k-1} r) + 2 = \frac{k(k-1)}{2} + 2$.

---

[3] While deviations from the protocol always imply privacy violations, a malicious party
will not necessarily profit from deviations in terms of maximizing its preferences. This
is due to the fact that the combined preference order depends on the preference order
of the honest party. In order to profit from deviations a malicious party would need
to know the honest party's preference order at the time of initiation of reconciliation.

In the second case, we assume that $k \leq N \leq 2k - 1$. That is A and B differ in at least one input and have at least one input in common. In this case, the worst case occurs if A and B share only exactly one input and this input is the least preferred input of both parties. In this case, the match is found in Round $2k - 1$ of the protocol. The results are shown in Table 2.

Table 2. PROS^{SR} : worst case performance for $k \leq N \leq 2k - 1$

$k \leq N \leq 2k - 1$	
# Comparisons	$k^2 + 1$
# Encryption of coefficients	$4k$
# Decryptions	$k^2 + 1$
# Commitment of inputs	$k^2 + 1$
# Messages	$4k - 1$
Size of payload in # ciphertexts	$k^2 + 4k + 1$

Finally, in the third case, we assume that $N > 2k - 1$. In this case, the inputs of A and B may be disjoint, which is also the worst case for this relation between N and k. A and B realize that they do not have an input in common when no match is found after $2k - 1$ rounds. Table 3 shows the results for this case.

Table 3. PROS^{SR} : worst case performance for $N > 2k - 1$

$N > 2k - 1$	
# Comparisons	k^2
# Encryption of coefficients	$4k$
# Decryptions	k^2
# Commitment of inputs	k^2
# Messages	$4k - 1$
Size of payload in # ciphertexts	$k^2 + 4k$

Overall we can conclude that the computational overhead as well as the communication overhead of the protocol are bounded by $O(k^2)$.

3.3   PROS Protocol for the Minimum of Ranks Scheme

The focus of this section is on a detailed description of the PROS^{MR} protocol using the PSI protocol introduced by Freedman *et al.* As discussed previously, the main difference to the PROS^{SR} protocol is the order in which PSIs are executed (see Figure 2). We once again formally describe all protocol messages as well as the operations carried out by the two parties A and B.

PROS^{MR} **Protocol:** Within this protocol description, we reuse the definitions of $f_A^{(i)}, f_B^{(i)}, E_A^{(i)}, E_B^{(i)}, c_{A,j}^{(i)}$, and $c_{B,j}^{(i)}$ introduced in the previous section.

COMMITMENT PHASE:

ROUND 0: B sends $E_B^{(1)}$ to A.

ROUND 1: A sends $E_A^{(1)}$ and $c_{A,1}^{(1)}$ to B. B decrypts $c_{A,1}^{(1)}$ under D_B and compares the result to b_1. If $b_1 = D_B(c_{A,1}^{(1)})$, then $b_1 = a_1$, that is, B has found MAX $:= a_1 = b_1$ and continues with the *Match Confirmation Phase*. Otherwise B sends $E_B^{(2)}$ to A and A continues with Round 2 of the *Commitment Phase*.

ROUND $2 \leq i \leq k$: For $j = 1, \ldots, i$, A computes the ciphertexts $c_{A,j}^{(i)}$ and sends them as well as $E_A^{(i)}$ to B. B decrypts $c_{A,1}^{(i)}, \ldots, c_{A,i}^{(i)}$ with D_B and checks whether any decrypted value equals b_i. If for some $j =: m$ the ciphertext $c_{A,m}^{(i)}$ decrypts to b_i, B has found MAX $:= b_i = a_m$ and continues with the *Match Confirmation Phase*. Otherwise, B computes the ciphertexts $c_{B,j}^{(i)}$ for $1 \leq j \leq i-1$ and sends them to A. For $1 \leq i < k$, B additionally sends $E_B^{(i+1)}$ to A. A decrypts $c_{B,1}^{(i)}, \ldots, c_{B,i-1}^{(i)}$ with D_A and checks whether any decrypted value equals a_i. If for some $j =: m$ the ciphertext $c_{B,m}^{(i)}$ decrypts to a_i, A has found MAX $:= a_i = b_l$ and continues with the *Match Confirmation Phase*. Otherwise, if $i + 1 \leq k$, A continues with Round $i + 1$ of the *Commitment Phase*. If $i + 1 > k$, the protocol terminates without a match being found.

MATCH CONFIRMATION PHASE: If it was A who found MAX in Round i of the *Commitment Phase*, A computes $c_{A,i}^{(m)}$ and sends it to B. B decrypts $c_{A,i}^{(m)}$ with D_B and checks that $D_B(c_{A,i}^{(m)})$ decrypts to a value $b_m \in \{b_1, \ldots, b_i\}$. Thus, both parties A and B know that $a_i = b_m = $ MAX.

If it was B who found MAX in Round i of the *Commitment Phase*, then B computes $c_{B,i}^{(m)}$ and sends it to A. A decrypts the received value under D_A to a_m. Thus, A and B both know that $a_m = b_i = $ MAX.

Intuitively speaking, the difference between this and the previously described protocol is that the order in which A and B exchange information and compare inputs is changed. As a general rule, in Round i of the protocol A and B compare all inputs that lead to the same maximum of minimum of ranks.

Security Analysis: For $S_A \cap S_B \neq \emptyset$, the *Commitment Phase* of the PROS^{MR} protocol always terminates with one of the parties finding a match. This is due to the fact that for each $1 \leq i \leq k$, a_i is compared with $b_1, \ldots, b_{i-1}$ by A in Round i. Similarly, for each $1 \leq i \leq k$, b_i is compared with $a_1, \ldots a_i$ by B.

Let MAX be the first match found in Round i of the *Commitment Phase* and let without loss of generality $max(min(rank_A(\text{MAX}), rank_B(\text{MAX}))) = rank_A(\text{MAX})$. Then, MAX maximizes the minimum of the ranks of all inputs in the intersection $S_A \cap S_B$. If this was not the case, then there would be an input $p \in S_A \cap S_B$ with $min(rank_A(p), rank_B(p)) > rank_A(\text{MAX})$.

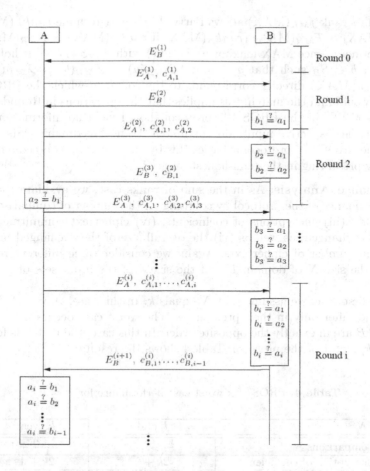

Fig. 4. Commitment Phase of the PROS^{MR} protocol. In Round i, A compares a_i with $b_1, \ldots, b_{i-1}$ and B compares b_i with $a_1, \ldots, a_i$ without obtaining knowledge of these values, that is, in a privacy-preserving manner.

This implies that $rank_A(p) > rank_A(\text{MAX})$ and $rank_B(p) > rank_A(\text{MAX})$ and A as well as B would have committed to p in a round prior to Round i. As a consequence, p would have been found in a prior round. This obviously contradicts the assumption.

In order to prove that the PROS^{MR} protocol is privacy-preserving in the semi-honest model, it is necessary to show that the information that either party can deduce from the PROS^{MR} protocol is equivalent to either party knowing nothing else but what can be deduced from knowing the matching input MAX (maximizing the MR composition scheme) and its rank $\mathcal{F}_{MR}(\text{MAX})$. As discussed before, finding a match with the PROS^{MR} protocol implies that both parties know the common input MAX which maximizes the combined preference order as well as the round in which it was found. The latter corresponds to the rank of the matching input under the MR composition scheme.[2] In turn, knowing MAX itself and its rank $f_{MR}(\text{MAX})$ allows Party A (B) to determine $rank_B(\text{MAX})$ if $rank_A(\text{MAX}) > \mathcal{F}_{MR}(\text{MAX})$ ($rank_A(\text{MAX})$ if $rank_B(\text{MAX}) > \mathcal{F}_{MR}(\text{MAX})$).

Furthermore, since MAX maximizes $\mathcal{F}_{MR}(x)$ with $x \in S_A \cap S_B$, it holds that $\not\exists\ a \in S_A, b \in S_B$ such that $a = b$ and $\mathcal{F}_{MR}(a) = \mathcal{F}_{MR}(b) > \mathcal{F}_{MR}(\text{MAX})$. Since $\mathcal{F}_{MR}(MAX)$ directly corresponds to the round in which the PROS^{MR} protocol would find the match, this implies that the operations in Rounds i with $i < \mathcal{F}_{MR}(\text{MAX})$ in the PROS^{MR} protocol should yield no information other than that the respective inputs do not match. Using Freedman's PSIs in each round, the PROS^{MR} protocol achieves this by construction as Freedman's PSI is privacy-preserving in the semi-honest model.[3]

Performance Analysis: As in the sum of ranks case, we evaluate the worst case performance of the protocol by counting (i) the number of comparisons, (ii) decryptions, (iii) encryptions of coefficients, (iv) ciphertext commitments, (v) messages exchanged, as well as (vi) the overall size of the exchanged messages counted in number of ciphertexts. Again, we consider three different relations between the size N of domain $\mathcal{D}$ and the size k of the input sets of parties A and B.

In the first case, we assume that N equals k. In this case, A and B share all inputs and differ only in their preferences. The worst case occurs if the inputs of A and B are in exactly the opposite order. In this case, the match is found in Round $\lfloor k/2 \rfloor + 1$ of the protocol. Table 4 shows the results.

Table 4. PROS^{MR} : worst case performance for $N = k$

$k = N$	k odd	k even
# Comparisons	$\frac{k^2+1}{4} + 2$	$\frac{k(k+2)}{4} + 1$
# Encryption of coefficients	$2k + 2$	$2k + 4$
# Decryptions	$\frac{k^2+1}{4} + 2$	$\frac{k(k+2)}{4} + 1$
# Commitment of inputs	$\frac{k-1}{2}(\frac{k-1}{2} + 1) + 2$	$\frac{k}{2}(\frac{k}{2} + 1) + 1$
# Messages	$k + 2$	$k + 3$
Size of payload in # ciphertexts	$2k + 4 + \frac{k-1}{2}(\frac{k-1}{2} + 1)$	$2k + 5 + \frac{k}{2}(\frac{k}{2} + 1)$

In the second case, we assume that $k \leq N \leq 2k - 1$. In this case, A and B have at least one input in common but may differ greatly in their preferences. In this case, the worst case occurs if A and B share only exactly one input and this input is the least preferred one of both parties. In this case, the match is found in Round k of the protocol. The results are shown in Table 5.

Finally, in the third case, we assume that $N > 2k - 1$. In this case, the input sets of A and B may be disjoint, which is also the worst case for this relation between N and k. A and B realize that they do not have an input in common when no match is found after k rounds. Table 6 shows the results for this case.

Overall we can conclude that the computational overhead as well as the communication overhead of the protocol are bounded by $O(k^2)$.

Table 5. PROS^{MR} : worst case performance for $k \le N \le 2k - 1$

$k \le N \le 2k - 1$	
# Comparisons	k^2
# Encryption of coefficients	$4k$
# Decryptions	k^2
# Commitment of inputs	k^2
# Messages	$2k + 1$
Size of payload in # ciphertexts	$k^2 + 4k$

Table 6. PROS^{MR} : worst case performance for $N > 2k - 1$

$N > 2k - 1$	
# Comparisons	k^2
# Encryption of coefficients	$4k$
# Decryptions	k^2
# Commitment of inputs	k^2
# Messages	$2k + 1$
Size of payload in # ciphertexts	$4k + k^2$

3.4   Further Work in Theory on Two-party Reconciliation Protocols

Our recent work on PROS protocols includes the development of protocols that are secure in the malicious model [21]. Furthermore, we have proposed privacy-preserving protocols for interval computations that have the promise to yield improved reconciliation protocols based on the use of intervals.

4   Multi-party Protocols

In [29,30], we propose multi-party reconciliation protocols for the sum of ranks as well as the maximized minimum of ranks composition scheme, which are secure in the semi-honest model. It is important to note that while it is possible to generalize the two-protocols described in the previous section in a straight-forward fashion, this approach is highly inefficient requiring k^n rounds of multi-party set intersection operations (see [29]). Instead, the main idea of our multi-party protocols proposed in [29] is to represent the ordered input sets of the parties as multisets by encoding the rank of an element in the multiplicity of the element in the multiset. For example, an ordered set $\{A > B > C\}$ is encoded as the multiset $\{A, A, A, B, B, C\}$. We call this the *rank encoding* which is formally defined as

$$ renc\left(S_i\right) = \left\{ s_{ij}^{\mathrm{rank}_{S_i}(s_{ij})}, \; j = 1, ..., k \right\}. $$

By carefully designing a sequence of multiset operations that are performed on the multisets, we can now solve the reconciliation problem for the MR and SR composition schemes.

More specifically, given the parties P_i ($i = 1, ..., n$) and their private input sets S_i (each with k distinct elements $s_{i1}, ..., s_{ik}$), the MPROS^{MR} protocol computes

$$Rd_t(renc(S_1) \cap ... \cap renc(S_n)).$$

for $t = k-1, k-2, ..., 0$ until the resulting set is non-empty for the first time. All elements in this non-empty set then maximize the MR composition scheme. The non-empty check is done by means of a threshold decryption of the resulting set and a computation of the roots of the decrypted polynomial. Intuitively speaking, the MPROS^{MR} protocol entails the following steps:

1. Each party determines the polynomial representation of its multiset $renc(S_i)$.
2. All parties compute the set intersection which is based on oblivious polynomial evaluation on input sets $renc(S_1), ..., renc(S_n)$. After this step, all parties hold an encrypted polynomial that represents $renc(S_1) \cap ... \cap renc(S_n)$. This intersection encodes not only the elements which the parties have in common but also the respective minimum preference for each of these elements across all the participants.
3. All parties iteratively compute the element reduction by t. The first time this step is executed, the reduction value $t = k - 1$ is used. The reduction operation is applied to the result of the set intersection in Step 2. The goal of the reduction step is to determine the element(s) in the intersection $renc(S_1) \cap ... \cap renc(S_n)$ for which the minimum preference is maximized, i.e., to perform element reduction using the largest possible t. A reduction by t eliminates up to t occurrences of each unique element in the multiset $renc(S_1) \cap ... \cap renc(S_n)$. If t is too large, the reduction will yield the encryption of a polynomial representing the empty set.
4. All parties participate in the threshold decryption of the result of Step 3. Specifically, each party checks whether at least one of its input elements is a root of the polynomial computed in Step 3. This will not be the case until Step 3 results in a non-empty set which in turn corresponds to the maximum of the minimum of ranks. As long as Step 3 yields an empty set, the parties iterate Steps 3 and 4 with a decreasing value of t.

Similarly, the protocol MPROS^{SR} computes

$$Rd_t((renc(S_1) \cup ... \cup renc(S_n)) \cap (menc(S_1) \cap ... \cap menc(S_n)))$$

using the so-called multiplicity encoding

$$menc(S_i) = \left\{ s_{ij}^{n \cdot k}, \ \forall s_{ij} \in S_i \right\}.$$

The protocol continues for $t = kn - 1, kn - 2, ..., n - 1$ until the resulting set is non-empty for the first time. All elements in this non-empty set then maximize the sum of ranks assigned by each party. The auxiliary multisets $menc(S_i)$ ensure that only inputs that are common to all parties are in the resulting set. Intuitively speaking, the MPROS^{SR} protocol entails the following steps:

1. Each party computes the polynomial representations of its multisets $renc(S_i)$ and $menc(S_i)$.
2. All parties compute the set union on their input multisets $renc(S_1), \ldots, renc(S_n)$. After completing this step, all parties hold an encrypted polynomial that represents $renc(S_1) \cup \ldots \cup renc(S_n)$. This union encodes not only all the input elements that are held by the parties but it also represents the sum of the preferences for each element across all parties.
3. All parties then compute the set intersection on input sets $menc(S_1), \ldots, menc(S_n)$ and the result of Step 2. This step is necessary in order to ensure that those elements are eliminated from $renc(S_1) \cup \ldots \cup renc(S_n)$ which are held by some but not all parties.
4. All parties iteratively participate in the element reduction by t with $t = nk - 1, \ldots, n - 1$ on the result of Step 3. As in the case of the $MPROS^{MR}$ protocol, the purpose of this step is to maximize the composition scheme. This is ensured by possibly repeating Steps 4 and 5 for decreasing t.
5. All parties participate in the threshold decryption of the result of Step 4 and check whether at least one of its input elements is a root of the polynomial. If this is the case, the common elements with the maximum sum of ranks are found. If this is not the case (i.e., t was too large and the polynomial computed as part of Step 4 corresponds to the empty set), the parties repeat Steps 4 and 5 with a value t decreased by one.

The computational complexity of $MPROS^{MR}$ is $O(k^6 + n \cdot k^4)$ and $O(n^4 \cdot k^6)$ for $MPROS^{SR}$. For the communication we have $O(n \cdot k^3)$ in the case of $MPROS^{MR}$ and $O(n^3 \cdot k^3)$ for $MPROS^{SR}$ where n is the number of parties and k the number of inputs. The protocols are secure in the semi-honest model with up to $c < n$ colluding attackers [29,30].

4.1 Further Work in Theory on Multi-party Reconciliation Protocols

In more recent work, we further optimized the above protocols and analyzed their performance in detail [26]. Also, by using novel zero-knowledge proofs we were able to construct multi-party protocols that are secure in the malicious model [27]. Furthermore, we explored how the use of a fully-homomorphic cryptosystem would improve the performance of our multi-party protocols [34].

5 Practice

As mentioned earlier, our work on privacy-preserving reconciliation protocols focuses not only on advances in theory but also strives to have practical impact. In particular, the practical aspects include the implementation of our newly-developed protocols, the comprehensive testing and performance analysis, the consideration of real-world applications, as well as the development of respective apps and comprehensive frameworks.

In the following, we provide a high-level description of our library for privacy-preserving operations and reconciliation protocols (Section 5.1) and briefly review our other practical contributions (Section 5.2).

5.1   Library

Figure 5 shows the simplified stack of our library. The main goal in designing the library was to provide for a modular structure while enabling an efficient implementation of privacy-preserving operations and protocols.

The library consists of four main layers. The first layer implements basic data types (e.g., sets, multisets, polynomials, intervals) and includes a network interface which facilitates efficient communication between parties. It builds on the multi-precision integer arithmetic GMP.

The second layer implements cryptographic primitives that are needed for the various privacy-preserving operations and protocols. The functionality includes zero-knowledge proofs, commitment schemes, hash functions, and various (homomorphic) encryption schemes (e.g., (threshold) Paillier, ElGamal).

The third layer of the library includes fundamental secure multi-party computation protocols for privacy-preserving (multi)set operations (e.g., union, intersection, reduction), secure comparison, oblivious transfer, etc.

The fourth layer of our library provides the implementations of our newly-developed protocols. These include the two-party/multi-party reconciliation protocols described previously which were proven secure in the semi-honest model. Furthermore, this layer implements various privacy-preserving interval operations which were also proven secure in the semi-honest model. Also included are various newly-developed protocols which were proven secure in the malicious model (e.g., a two-party protocol for verifiable private equality testing and multi-party reconciliation protocols).

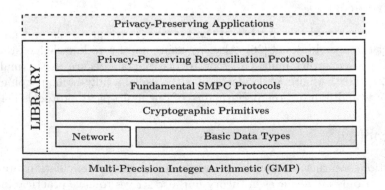

Fig. 5. Simplified logical layers of the library

There are two implementations of the library—one in C++ and one in Java. While both implementations share the basic structure as outlined in Figure 5, the two currently differ in the functionality they provide. While the C++ version provides more two-party functionality, the focus of the Java implementation was on the multi-party protocols. For the future, we plan to further complete the implementations with the respective functionalities.

For a detailed description of the library, its design, its functionality, and implementations please refer to [17,20,26,27,28,32].

5.2   Real-World Applications, Apps, and Frameworks

Mobile apps. Building on the library stack and its implementation described above, we have developed an app *appoint* for both iOS and Android which allows for the privacy-preserving negotiation of a common meeting time while recognizing the preferences of the users. Figure 6 and Figure 7 show some screenshots of the apps. The network component implemented for the app uses peer-to-peer networking via Bluetooth. As such the app assumes that the parties striving to negotiate a meeting time are in the same physical location [20]. Aside from scheduling, we have explored applying our protocols to other real-world contexts, including electronic voting and auctions [18].

Fig. 6. iPhone app   **Fig. 7.** Android app

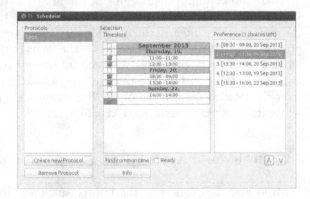

Fig. 8. SMC-MuSe: An-
droid client

Fig. 9. SMC-MuSe: Desktop client

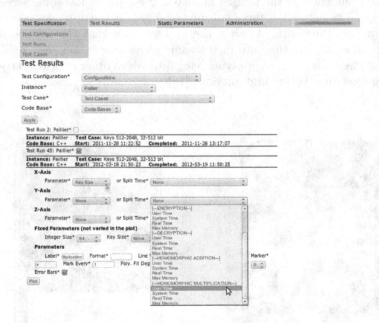

Fig. 10. Testing framework

Framework for SMC. Extending on the system model used for the apps, we
have designed and implemented SMC-MuSE—a framework for Secure Multi-
Party Computation for MultiSets. Specifically, SMC-MuSe allows for asyn-
chronous communication and provides for a fully-automated key management.
Most importantly, SMC-MuSe addresses the challenge of properly implementing
assumptions that are typically underlying the theoretical solutions and protocols
in SMPC [28]. Figures 8 and 9 show some GUI components of the multi-party
scheduling application implemented on top of the framework.

Testing Framework. An orthogonal component of our practical work is the development of a suitable framework that allows for the efficient performance testing of the various components of the library. While CaPTIF (Comprehensive Performance TestIng Framework) was designed and implemented in the context of our work on privacy-preserving reconciliation protocols, CaPTIF is a framework that allows for the definition of test inputs, the scheduling and execution, as well as the plotting, browsing and analysis of comprehensive performance tests in arbitrary contexts. Figure 10 shows a screenshot of the framework's web interface. For more details refer to [19].

6   Conclusion and Future Work

The paper at hand provides some initial insights into a research project on privacy-preserving reconciliation protocols. The results obtained so far and the ongoing research agenda for this project include both theoretical and practical components. More details on this work can be found in our publications [17,18,19,20,21,23,24,26,27,28,29,30,34] and on our project website [32].

Ongoing and future efforts for the project include the extension of the functionality of the library as well as the frameworks CaPTIF and SMC-MuSe. Also, we currently are pursuing other fields of application beyond scheduling, voting, and auctions. On the theoretical side, the focus is on extending the privacy-preserving interval operations and exploring game-theoretic approaches. Furthermore, we are in the process of conducting user studies with the goal to better understand and eventually provide for alternative notions of unbiasedness to the ones used to date.

Acknowledgments. In part, this work has been supported through the DFG grant ME 3704/1-1 and the NSF Award CCF 1018616.

We particularly thank our PhD students Georg Neugebauer and Daniel Mayer who greatly advanced our initial ideas leading to many publications—as briefly sketched in this paper. We also thank Post-Doc Sotiris Ioannidis and the many students who have worked in the context of this project contributing to its results and success to date: Dominik Teubert, Christian Bien, Lucas Brutschy, Florian Weingarten, Samuel Schüppen, Jonathan Voris, Kenneth Bodzak, and Orie Steele.

7   Context of this Work

We have specifically chosen to report on this work in the Festschrift for Johannes as this work is connected to him, his efforts, and his accomplishments in a number of ways.

During his career, Johannes has repeatedly spent time abroad working with others in the community and building international collaborations. As a mentor, Johannes has continuously provided his mentees with opportunities to travel abroad and experience the importance and benefits of extending their horizons internationally.

As a PhD student, Ulrike visited Stevens Institute of Technology many times to collaborate with Susanne. The work described in this paper was started during one of the visits. It was motivated by Ulrike's PhD work on secure roaming and handover procedures in wireless access networks [22]. Specifically, the starting point for this work was the fact that the decision on which mechanisms are used to protect mobile communication is made by the network components and forced onto the mobile device (and thus the user). We first explored whether it was possible and, if so, how to recognize the preferences of both the network and the user in determining the most suitable mechanism. Subsequently, this led to considering privacy issues and extending the focus to other fields of application.

We continued to meet in Darmstadt to further work on the project even after Ulrike had finished her PhD. This was possible because we were always welcome at Johannes' institute.

However, foremost, this project follows in the footsteps of Johannes' work in providing not only for solid theoretical advances in the field but also translating them into practice—a central theme of Johannes' accomplishments. Examples include his contributions in algorithmic number theory and the development of the LiDIA library as well as his advancing of public-key cryptography and the founding of the spin-off company FlexSecure GmbH.

References

1. Ben-David, A., Nisan, N., Pinkas, B.: FairplayMP: A System for Secure Multi-Party Computation. In: Conference on Computer and Communications Security (CCS). ACM (2008)
2. Blanton, M., Aguiar, E.: Private and Oblivious Set and Multiset Operations. In: Symposium on Information, Computer and Communications Security (ASIACCS). ACM (2012)
3. Bogdanov, D., Laur, S., Willemson, J.: Sharemind: A Framework for Fast Privacy-Preserving Computations. In: Jajodia, S., Lopez, J. (eds.) ESORICS 2008. LNCS, vol. 5283, pp. 192–206. Springer, Heidelberg (2008)
4. Burkhart, M., Strasser, M., Many, D., Dimitropoulos, X.: SEPIA: Privacy-Preserving Aggregation of Multi-Domain Network Events and Statistics. In: USENIX Security Symposium. USENIX (2010)
5. Camenisch, J., Zaverucha, G.M.: Private Intersection of Certified Sets. In: Dingledine, R., Golle, P. (eds.) FC 2009. LNCS, vol. 5628, pp. 108–127. Springer, Heidelberg (2009)
6. Dachman-Soled, D., Malkin, T., Raykova, M., Yung, M.: Efficient Robust Private Set Intersection. In: Abdalla, M., Pointcheval, D., Fouque, P.-A., Vergnaud, D. (eds.) ACNS 2009. LNCS, vol. 5536, pp. 125–142. Springer, Heidelberg (2009)
7. De Cristofaro, E., Tsudik, G.: Experimenting with Fast Private Set Intersection. In: Katzenbeisser, S., Weippl, E., Camp, L.J., Volkamer, M., Reiter, M., Zhang, X. (eds.) Trust 2012. LNCS, vol. 7344, pp. 55–73. Springer, Heidelberg (2012)
8. De Cristofaro, E., Tsudik, G.: Practical Private Set Intersection Protocols with Linear Complexity. In: Sion, R. (ed.) FC 2010. LNCS, vol. 6052, pp. 143–159. Springer, Heidelberg (2010)

9. Freedman, M.J., Nissim, K., Pinkas, B.: Efficient Private Matching and Set Intersection. In: Cachin, C., Camenisch, J.L. (eds.) EUROCRYPT 2004. LNCS, vol. 3027, pp. 1–19. Springer, Heidelberg (2004)
10. Frikken, K.B.: Privacy-Preserving Set Union. In: Katz, J., Yung, M. (eds.) ACNS 2007. LNCS, vol. 4521, pp. 237–252. Springer, Heidelberg (2007)
11. Geisler, M.: Cryptographic Protocols: Theory and Implementation. PhD Thesis, Aarhus University (2010)
12. Goldreich, O.: Foundations of Cryptography, vol. 1. Press Syndicate of the University of Cambridge (2004)
13. Goldwasser, S., Micali, S.: Probabilistic Encryption & How to Play Mental Poker Keeping Secret all Partial Information. In: Symposium on Theory of Computing (STOC). ACM (1984)
14. Hazay, C., Nissim, K.: Efficient Set Operations in the Presence of Malicious Adversaries. In: Nguyen, P.Q., Pointcheval, D. (eds.) PKC 2010. LNCS, vol. 6056, pp. 312–331. Springer, Heidelberg (2010)
15. Huang, Y., Evans, D., Katz, J.: Private Set Intersection: Are Garbled Circuits Better than Custom Protocols. In: Network and Distributed System Security Symposium (NDSS). Internet Societey (2012)
16. Kissner, L., Song, D.: Privacy-Preserving Set Operations. In: Shoup, V. (ed.) CRYPTO 2005. LNCS, vol. 3621, pp. 241–257. Springer, Heidelberg (2005)
17. Mayer, D.: Design and Implementation of Efficient Privacy-Preserving and Unbiased Reconciliation Protocols. PhD Thesis, Stevens Institute of Technology (2012)
18. Mayer, D., Neugebauer, G., Meyer, U., Wetzel, S.: Enabling Fair and Privacy-Preserving Applications Using Reconciliation Protocols on Ordered Sets. In: Sarnoff Symposium. IEEE (2011)
19. Mayer, D.A., Steele, O., Wetzel, S., Meyer, U.: CaPTIF: Comprehensive Performance TestIng Framework. In: Nielsen, B., Weise, C. (eds.) ICTSS 2012. LNCS, vol. 7641, pp. 55–70. Springer, Heidelberg (2012)
20. Mayer, D.A., Teubert, D., Wetzel, S., Meyer, U.: Implementation and Performance Evaluation of Privacy-Preserving Fair Reconciliation Protocols on Ordered Sets. In: Conference on Data and Application Security and Privacy (CODASPY). ACM (2011)
21. Mayer, D.A., Wetzel, S.: Verifiable Private Equality Test: Enabling Unbiased 2-Party Reconciliation on Ordered Sets in the Malicious Model. In: Symposium on Information, Computer and Communications Security (ASIACCS). ACM (2012)
22. Meyer, U.: Secure Roaming and Handover Procedures in Wireless Access Networks. PhD Thesis, Darmstadt University of Technology (2005)
23. Meyer, U., Wetzel, S., Ioannidis, S.: Distributed Privacy-Preserving Policy Reconciliation. In: International Conference on Communications (ICC). IEEE (2007)
24. Meyer, U., Wetzel, S., Ioannidis, S.: New Advances on Privacy-preserving Policy Reconciliation. Cryptology ePrint Archive, 2010/064 (2010)
25. Narayanan, G.S., Aishwarya, T., Agrawal, A., Patra, A., Choudhary, A., Rangan, C.P.: Multi Party Distributed Private Matching, Set Disjointness and Cardinality of Set Intersection with Information Theoretic Security. In: Garay, J.A., Miyaji, A., Otsuka, A. (eds.) CANS 2009. LNCS, vol. 5888, pp. 21–40. Springer, Heidelberg (2009)
26. Neugebauer, G., Brutschy, L., Meyer, U., Wetzel, S.: Design and Implementation of Privacy-Preserving Reconciliation Protocols. In: International Workshop on Privacy and Anonymity in the Information Society (PAIS). ACM (2013)

27. Neugebauer, G., Brutschy, L., Meyer, U., Wetzel, S.: Privacy-Preserving Multi-Party Reconciliation Secure in the Malicious Model. In: International Workshop on Data Privacy Management (DPM). ACM (2013)
28. Neugebauer, G., Meyer, U.: SMC-MuSe: A Framework for Secure Multi-Party Computation on MultiSets, RWTH Aachen University Technical Report AIB-2012-16 (2012)
29. Neugebauer, G., Meyer, U., Wetzel, S.: Fair and Privacy-Preserving Multi-Party Protocols for Reconciling Ordered Input Sets. In: Burmester, M., Tsudik, G., Magliveras, S., Ilić, I. (eds.) ISC 2010. LNCS, vol. 6531, pp. 136–151. Springer, Heidelberg (2011)
30. Neugebauer, G., Meyer, U., Wetzel, S.: Fair and Privacy-Preserving Multi-Party Protocols for Reconciling Ordered Input Sets (Extended Version) (2011), http://eprint.iacr.org/2011/200
31. Paillier, P.: Public-Key Cryptosystems Based on Composite Degree Residuosity Classes. In: Stern, J. (ed.) EUROCRYPT 1999. LNCS, vol. 1592, pp. 223–238. Springer, Heidelberg (1999)
32. PreFairAppl—Private and Fair Applications (2013), http://www.prefairappl.info
33. Sang, Y., Shen, H.: Privacy Preserving Set Intersection Based on Bilinear Groups. In: Proc. of the Thirty-First AACCS. Australian Computer Science Conference (ACSC), vol. 74, Australian Computer Society, Inc. (2008)
34. Weingarten, F., Neugebauer, G., Meyer, U., Wetzel, S.: Privacy-Preserving Multi-Party Reconciliation using Fully Homomorphic Encryption. In: Lopez, J., Huang, X., Sandhu, R. (eds.) NSS 2013. LNCS, vol. 7873, pp. 493–506. Springer, Heidelberg (2013)
35. Yao, A.C.: Protocols for Secure Computations. In: Symposium on Foundations of Computer Science (SFCS). IEEE (1982)

Defining Privacy Based on Distributions
of Privacy Breaches

Matthias Huber[1], Jörn Müller-Quade[2], and Tobias Nilges[2]

[1] FZI Forschungszentrum Informatik
[2] Institut für Kryptographie und Sicherheit
Fakultät für Informatik
Karlsruher Institut für Technologie (KIT)
huber@fzi.de, {mueller-quade,tobias.nilges}@kit.edu

Abstract. In contrast to classical cryptography, the challenge of privacy in the context of databases is to find a trade-off between a security guarantee and utility. Individuals in a database have to be protected while preseving the usefullnes of the data. In this paper, we provide an overview over the results in the field of database privacy with focus on privacy notions. On the basis of these notions, we provide a framework that allows for the definition meaningful guarantees based on the distribution on privacy breaches and sesitive predicates. Interestingly, these notions do not fulfill the privacy axioms defined by Kifer et al. in [1,2].

1 Introduction - What Is Privacy?

In classical cryptography, privacy is often used as a synonym for confidentiality. For example a voting scheme that provides privacy for the voters [3,4] guarantees the confidentiality of single votes. A mix network that provides privacy guarantees the confidentiality of the origin of a message [5].

Recently, privacy is also considered in the context of database disclosure. Here, a database containing sensitive data has to be anonymized in order to allow disclosure, for example to third parties that want to analyze the disclosed database [6].

Consider an institution (e.g. the census bureau or a hospital organization) wanting to disclose a database in order to allow for data mining. Since the privacy of the individuals of the database has to be preserved, the database is anonymized prior to disclosure (cf. Figure 1). In contrast to confidentiality, leakage of some data is desired – even necessary [7] – while the privacy of individuals has to be protected. An adversary should not be able to learn sensitive information given the anonymized database. On the other hand it should still have utility in the sense that the anonymized data is useful, e. g. for data mining. This trade-off between utility and privacy is described by formal privacy notions.

Another application of privacy notions is secure database outsourcing. Here, a client wants to outsource a database to a not necessarily trustworthy party. This party should be able to execute queries on the database for the client while

M. Fischlin and S. Katzenbeisser (Eds.): Buchmann Festschrift, LNCS 8260, pp. 211–225, 2013.

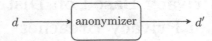

Fig. 1. Privacy perserving database disclosure. An anonymizer transforms a database d, and discloses the result d'.

learning as little as possible about the data [8,9]. Again, there is a trade off between privacy and utility. But here, the utility is needed in order to support efficient execution of queries.

In this paper, we present an overview of the developments in the field of database privacy with a focus on privacy notions. Furthermore, we present a framework for defining meaningful privacy notions. It extends the Pufferfish [10] framework in a natural way and allows for explicitly modelling what predicates the resulting notion should hide as well as how effective it should be. It introduces privacy breaches of different magnitudes and assigns probabilities to each breach, thereby enabling detailed and precise definitions of privacy notions. Based on this definition classical notions such as (ϵ, δ)-differential privacy [11] can be derived. We provide example notions as well as methods and argue that they are incompatible with the privacy axioms defined in [2].

This paper is structured as follows: In Section 2, we present and discuss previous work. In Section 3 we present the framework. In Section 4, we provide examples and discuss them in the context of the privacy axioms [1,2]. Section 5 concludes.

2   Previous Work

The first attempt to formally treat anonymity resulted in the concept of *k*-*anonymity* by Samarati and Sweeney [12,6]. Their idea was to achieve anonymity by letting each individual blend into a crowd of k individuals. Since this notion has several shortcomings, it was iteratively improved to remove the inherent problems, e.g. by Machanavajjhala et al. [13] in the form of *l-diversity* and by Li et al. [14] in the form of *t-closeness*. A lot of variations and improvements followed [15,16,17,18]. One main problem, however, was not solved: the combination of anonymized datasets can lead to a deanonymization of at least a part of the database, as was shown by Ganta et al. [19]. Additionally, all of these notions have another problem: they do not give a guarantee, but describe the result of an process to achieve some form of anonymity. This problem does not occur in [20], where it is guaranteed that the correlation between the secret value and individuals in the database is hidden.

As it turns out, in general the actual goal should not be to achieve anonymity in a crowd but rather to achieve privacy, i.e. the association of an individual to a certain value or even the value itself has to be kept secret. In her seminal work Dwork [7] introduced the notion of *Differential Privacy*, which is the first privacy notion defined with a cryptographic flavour. In contrast to previous work, differential privacy gives a guarantee that for any two databases that differ in a single entry the result of the anonymization process cannot be distinguished except by a small factor ϵ. This can be achieved by adding noise from a Laplacian distribution to the actual result. Here it is important to point out that the works based on k-anonymity all expect a release of the complete sanitized database, while differential privacy assumes a curator algorithm that answers queries about the data without releasing a completely anonymized database. Many follow-up works have been presented, see [21,22,23,24].

Differential privacy is the current gold standard in privacy research, although there are some drawbacks. Kifer and Machanavajjhala [25] show that the definition of differential privacy is too strict and thereby diminishes the utility of the obtained data severely. The term utility in this context means the degree of distortion of the actual result. The main reason for this is that only worst case scenarios are considered and thus the amount of noise needed to achieve the desired privacy is very large. Two different solutions were proposed to circumvent this problem. One proposition is to use the noise that is implicit in the data such that the utility is not reduced by the addition of noise. The other approach is to weaken the definition such that a breach is not impossible, but unlikely.

Concerning the first approach, the assumption is made that an adversary does not have the complete knowledge of a database except for a single entry. Instead, it is assumed that the adversary's knowledge has a given uncertainty which can be described as a distribution on the attribute value [26,27]. From an adversary's point of view this means that the data has implicit noise. It was shown by Duan [26] that this uncertainty can be enough to achieve differential privacy for sum queries. Additionally, privacy measures similar to differential privacy based on the implicit noise of the data were proposed [27]. Albeit, to maintain privacy for several queries it is necessary that some fraction of the database is updated between the queries, otherwise the adversary can learn the complete database over time. The aforementioned work also incorporates the second idea, namely a weakening of the privacy guarantee by allowing a privacy breach with small probability.

One of the first relaxations of differential privacy was put forward by Dwork et al. [21] and is called (ε, δ)-*differential privacy*, where δ signifies the probability that the factor between the two distributions is greater than ε. A lot of notions [28,29,30,23,31,32] also seize the aforementioned idea, although in somewhat different contexts. Before the upcoming of differential privacy Chaudhuri and Mishra [29] use an (ε, δ)-notion of privacy, where they show that a random subset of a database can be released privately if no values with low occurence are in this subset. The work by Blum et al. [31,32] is based on a concept that has proven to be of increasing interest. The knowledge of the adversary is depicted as a distribution of possible databases which is a generalization of uncertainty of database entries. They release information for a class of queries where the utility of the information has only been reduced by a (fixed) small amount. This concept was taken a step further by the recent results of Kifer and Machanavajjhala [33], who introduced the *Pufferfish* framework. Instead of a strict and static guarantee, they allow to specify certain predicates for which the privacy is enforced. This was already previously used by Blum et al. [28], where a priori and a posteriori beliefs over binary predicates were considered, and by Evfimievski et al. [30] for the case where a change in the probability of a predicate indicates a privacy breach. The idea of [33] is to specify two mutually exclusive predicates: one that describes the value to be hidden, the other serves as an indistinguishable value. It is e. g. possible to choose the predicates *value x is in the database* and *value x is not in the database*. With apropriate predicates, Differential privacy can be modeled as a special case of this framework. Based on these results, Kifer et al. extract a set of axioms for privacy notions [1,2] which should be fulfilled by any anonymity notion. They define a privacy definition as a set of randomized algorithms. The first axiom says that any privacy definition must be invariant to transformations, i. e. the application of an arbitrary algorithm to the result does not affect the privacy of the result. The second axiom says that any two algorithms that satisfy a privacy definition can be applied interchangeably while still satisfying the privacy definition. Based on these axioms a mathematical analysis is possible, such that semantic guarantees of a privacy definition can be made explicit [10].

In the following we will show that these axioms can be questioned to be essential for all anonymity notions, because there exist natural notions of anonymity that contradict these axioms. Additionally, by weakening the Pufferfish definition to an (ε, δ)-notion we derive a notion that maintains a high utility, because the implicit noise of the data makes an addition of noise unnecessary in most cases. This new notion might enable additional methods to achieve privacy, like subset sampling, which were not possible before.

3   Notions Based on Distributions of Privacy Breaches

In this section, we present a way to define privacy notions by explicitly declaring the sensitive predicates it should hide, and its effectivity. This framework extends the Pufferfish [10] framework. Notions in this framework are defined by a probability distribution of breaches. A breach is the change of the probability of a sensitive predicate given the release. As in similar frameworks, we model the background knowledge of the adversary as a probability distribution over possible datasets.

As mentioned above, we define sensitive predicates that an adversary should not learn. This means that, given the anonymized database, the belief of the adversary about the predicates should not change by much. We argue that in order to allow for methods that provide good utility, we need notions that allow to fail with a small probability. Previous notions based on this concept only considered the worst case, which made for large amounts of noise. In the following, we describe the notation the framework uses and provide definitions. An overview over the notation is depicted in Figure 2. Figure 3 depicts an overview of how the elements of the framework interact.

U	Universe of all possible datasets
$\mathcal{X}$	Set of possible distributions X of U
E	Mechanism $E : U \rightarrow D$ that generates a databaset d from a datase $u \in U$
L	Set of sensitive predicates, $l \in L, l : U \rightarrow true, false, l$ not a constant
F	Anonymization mechanism, $F : D \rightarrow D$
$\mathcal{A}$	the adversary

Fig. 2. Notation used by the framework

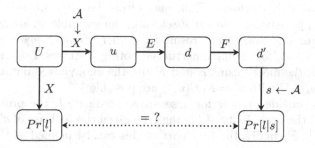

Fig. 3. The notion in a nutshell. A dataset u is drawn from the universe of all possible datasets U according to a distribution X. A mechanism E allows for transformation of the dataset prior to anonymization. The mechanism F anonymizes the resulting database. Given the anonymized database d', the adversary $\mathcal{A}$ has to guess a sensitive predicate l.

The set of all possible datasets is called the Universe U. It is distributed according to a distribution X. Please note, that an element of U may describe more than one individual since it is possible that the probabilities of two individuals occuring in the same dataset are not independet of each other. An example for U is the power set of all possible residents of a city. The distribution X models the background knowledge of the adversary. The set of possible distributions $\mathcal{X}$ allows to limit the background knowledge of the adversary. For example an adversary may know a number of individuals that unlikely live in the city U describes. She also may know that for example for a survey, a reporter selects a number of residents close to 1000. This makes some elements of U more probable to be selected that others.

The database d that is to be anonymized is generated from a dataset $u \in U$ with a mechanism E known to the adversary. If E involes probabilistic processes, the random coins used are unknown to the adversary. This introduces entropy into the database in a natural way. An example for E that involves a probabilistic process is randomized response [34]. The reporter may use this technique in order to get more honest answers. Another reason to introduce this selection mechanism E into the framework are the sensible predicates. We want to be able to define predicates about individuals that are not in the database d. The predicates, however, have to have a basic set even if they are not defined on d. Therefore, we define predicates on U and introduce a mechanism that generates the database d from a $u \in U$.

As mentioned above, we introduce a set of sensitive predicates L. With a predicate $l \in L$, $l : U \to$ {true, false}, we can model sensitive information the adversary should not be able to learn given the anonymized database. We require for every $l \in L$, that the probability of $l(u) = true$ for a $u \in U$ drawn according to X is between 0 and 1. Examples for such predicates are "individual a is in the dataset" or "individual b will vote republican". It is pointless to define a sensitive predicate that is true (or false) for *every* possible $u \in U$.

The anonymization mechanism F anonymizes the database d. This anonymized database d' is disclosed. This may either be an anonymized database (*offline privacy preserving database disclosure*), for example an anonymized voting database, or an anonymized result to a query (*online privacy preserving database disclosure*), for example perturbed voting statistics. Depending on the distribution X, the mechanisms E and F and the anonymized database d', several datasets $u_i \in U$ with $d' = F(E(u_i))$ are possible.

In order to provide privacy for a sensitive predicate l, the probability of l being true and the probability of l being true given the realease of d' should not differ too much. Formally, the first part of this can be defined as the following game:

Definition 1. *Let U be the universe of possible datasets, L be a set of sensitive predicates defined on U, and $\mathcal{X}$ a set of distributions of U. We define the experiment $Priv_\mathcal{A}$ as follows:*

$$Priv_\mathcal{A}(U, L, \mathcal{X}):$$
$$\mathcal{A} \leftarrow U, L$$
$$X \leftarrow \mathcal{A}, X \in \mathcal{X}$$
$$u \leftarrow X$$
$$d \leftarrow E(u)$$
$$l, s \leftarrow \mathcal{A}^{F(d)}, l \in L$$

In this game, the adversary $\mathcal{A}$ gets the universe of all possible datasets and a set of sensitive predicates. Then, she may choose a distribution X for U. According to X, a dataset u is drawn. This dataset is transformed by the mechanism E to the actual database d. Then, the adversary $\mathcal{A}$ gets oracle acces to $F(d)$ and has to output a sensitive predicate l and a string s. The string s can for example be a transcript of the interaction of $\mathcal{A}$ and $F(d)$

If the probability of a sensitive predicate l being true differs from the probability of l being true given s, we say there is a breach. Instead of $Pr[l = true]$ we write $Pr[l]$.

Definition 2. *For a given universe of datasets U with distribution X, a sensitive predicate l and a string s, a breach b is defiend as*

$$b = |\frac{Pr[l|s] - Pr[l]}{Pr[l]}|$$

With these two definitions, we can define the probability of a breach. Since the Privacy-Game involves probabilistic processes, we can consider probabilities for specific breaches:

Definition 3. *For a given universe of datasets U with distribution X, a set of sensitive predicates L, two mechanisms E and F, and a string s the probability of a specific breach b is called* probability of breach b *and defined as*

$$Pr[b = \frac{Pr[l|s] - Pr[l]}{Pr[l]}]$$

For example a breach of 1 may have probability 0, while a breach of ϵ may have a probability of 1. An anonymization mechanism and therefore its anonymity notion may also have different breaches with different probabilities. This depends on the one hand on the number of sensitive predicates and the distribution X, and on the other hand on the probabilistic processes involved in the mechanis E and as well in the anonymization mechanism F. Ideally, all breaches greater than 0 have probability 0. This, however, may not always be the case. Consider for example Figure 4. There are three breaches with a probability > 0. With probability 0.25 there is a breach of 0.5; with probability 0.5 a breach of 0.5 and with probabiliy 0.25 a breach of 1. An anonymization mechanism may for example be well suited for a specific predicate l, but not very well for another predicate l'. Consequently, an anonymity notion can be defined as follows:

Definition 4. *For a given universe of possible datasets U, an* anonymity notion *is a tuple* $(L, p, \mathcal{X})$*, where L is a set of sensitive predicates, $p : [0, 1] \rightarrow [0, 1]$, and $\mathcal{X}$ is a set of distributions X. Each distribution models possible background knowledge an adversary can have. An anonymization mechanism F fulfills a privacy notion $(L, p, \mathcal{X})$ with respect to a mechanism E iff for all strings s and all distributions X that can be generated by an adversary $\mathcal{A}$ in the experiment in Definition 1 the the breach probability distribution of the mechanism F is dominated by p.*

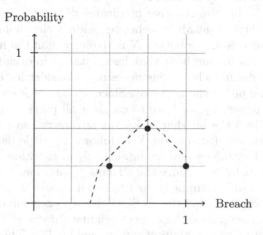

Fig. 4. Example for a probability distribution of breaches B (dots) and a dominating function p (dashed line)

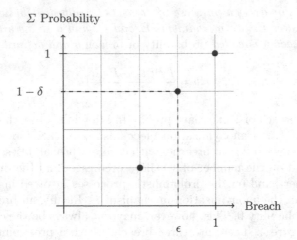

Fig. 5. Cumulative distribution of the distribution in Figure 4. The probability of a breach equal or less than ϵ is $1 - \delta$.

For two anonymity notions $A = (L, p, \mathcal{X})$ and $B = (L', p', \mathcal{X}')$ we say that notion B is stronger than A $(A \subseteq B)$ iff $L \subseteq L'$, $p' \leq p$, and $\mathcal{X} \subseteq \mathcal{X}'$.

In order to get more classical anonymity notions such as (ϵ, δ)-differential privacy we can consider the cumulative distribution of the breach probability distribution B, as seen in Figure 5 for the distribution of Figure 4. With such a cumulative distribution, we can easily give guarantees that with probability $p = 1 - \delta$ all breaches are equal or less than ϵ. Depending on the sensitive predicates the breach is based on, this translates to notions such as (ϵ', δ)-differential privacy.

Introducing the thresholds ϵ and δ also enables us to compare different privacy notions. Let for two privacy notions, $N = (P, B)$ and $N' = (P', B')$, the sensitive predicates P' of N' be also sensitive predicates of N $(P' \subseteq P)$, and for both notions, the probability that all breaches be below a threshold ϵ is $1 - \delta$. Then we can say that for this δ, the notion N is stronger than the notion N'.

Such notions allow for methods that leak sensitive information with a small probability δ. We deem such notions necessary. Consider for example a notion that only leaks one bit to the adversay. Since we do not know the background knowledge of the adversary, we have to consider all possible distributions X of the universe U. Given the bit that leaks, the adversary can partion U into two partitions and choose a distribution X as follows: A single database d_1 in partition 1 has probability 0.5 as well as database d_2 in partition 2. The remaining databases have probability 0. Since the adversary partitioned U given the single bit that leaks, she can determine if the original database to a given anonymization is from partition 1 or partition 2. Since in each partition only one database is possible, she can determine the correct original database. Therefore, the privacy notion fails. A more formal proof can be found in [25]. This shows that there are distributions for which a notion cannot hold and notions need to account for such cases. Therefore, we need to allow privacy notions to fail with a low probability.

4   Examples

In this sections, we will provide and discuss example instanciations of our framework. We will show, that the framework allows for meaningful notions that are not compatible to the privacy axioms postulated by Kifer and Lin [2,1]. Additionaly, we show that the framework allows to describe the privacy of aggregates such as average and that (ϵ, δ)-differential privacy can be modeld in the framework.

(ϵ, δ)-Privacy and the Privacy Axoims

Kifer and Lin [2,1] presented two axioms which they deem are essential for every privacy definition. They aim to achieve a consistent basis that allows a more stringent mathematical evaluation of privacy notions. These axioms are known to hold for differential privacy [7] and some notions of similar structure like the Pufferfish framework [33].

Axiom 1. (Transformation Invariance). *Let $\mathcal{M}$ be a privacy mechanism for a particular privacy definition and let $\mathcal{A}$ be a randomized algorithm whose input space contains the output space of $\mathcal{M}$ and whose randomness is independent of both the data and the randomness in $\mathcal{M}$. Then $\mathcal{M}' \equiv \mathcal{A} \circ \mathcal{M}$ must also be a privacy mechanism satisfying the privacy definition.*

This axiom states that any computation on the sanitized data should not compromise the privacy, but even stronger the privacy definition must still hold, i. e. the combined algorithms satisfy the same definition.

Axiom 2. (Convexity). *Let $\mathcal{M}_1$ and $\mathcal{M}_2$ be privacy mechanisms that satisfy a particular privacy definition (and such that the randomness in $\mathcal{M}_1$ is independent of the randomness in $\mathcal{M}_2$). For any $p \in \{0, 1\}$, let $\mathcal{M}_p$ be a randomized algorithm that on input i outputs $\mathcal{M}_1(i)$ with probability p (independent of the data and the randomness in $\mathcal{M}_1$ and $\mathcal{M}_2$) and outputs $\mathcal{M}_2(i)$ with probability $1 - p$. Then $\mathcal{M}_p$ is a privacy mechanism that satisfies the privacy definition.*

The second axiom says that two mechanisms for a privacy definition are supposed to be interchangeable and the fact which algorithm was used cannot be guessed.

In the following we will provide two examples of meaningful privacy definitions that contradict the above mentioned axioms. For the first axiom, the transformation algorithm $\mathcal{A}$ can change the parameters of an (ε, δ)-notion in such a way that it is not comparable to the original notion. Suppose an intelligence service $\mathcal{M}$ has a release mechanism that tells the truth to a query half of the time, while answering "no comment" the rest of the time. Further assume a politician $\mathcal{A}$ has to relay the information to a reporter, but does not want to give fishy answers. So instead of the "no comment" the politician answers these queries randomly. While this second mechanism always leaks some information, it also gives the politician the possibility to deny some of his statements. Lets formalize this example considering binary predicates.

$$\Pr[\mathcal{M}(b) = b] = \Pr[\mathcal{M}(b) = \bot] = 0.5, \qquad \mathcal{A}(\mathcal{M}(b)) = \begin{cases} b & \mathcal{M}(b) = b \\ rand\{0,1\} & \mathcal{M}(b) = \bot \end{cases}$$

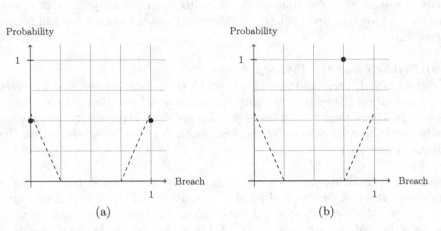

Fig. 6. (a) Breach distribution for the Mechanism $\mathcal{M}$ of the first example. With probability 0.5, there is a breach of 0 and with the same probability there is a breach of 1. (b) Breach distribution for the Mechanism $\mathcal{A} \circ \mathcal{M}$ of the first example. With probability 1, there is a breach of 0.75. The dominating function of M in (a) does not dominate the breach distribution of $\mathcal{A} \circ \mathcal{M}$.

After application of $\mathcal{M}$ it holds that $\Pr[b_{in} = 0 \mid b_{out} = 0] = 1$ and $\Pr[b_{in} = 1 \mid b_{out} = 0] = 0$, i. e. with probability $\delta = 0.5$ the breach is 1 if we consider the release of the true value to be a security breach. The second mechanism $\mathcal{A}$ changes these probabilities: it holds that $\Pr[b_{in} = 0 \mid b_{out} = 0] = 0.75$, while the probability that this information is released is 1, i. e. $\delta = 1$. Thus the severity of the breach is smaller, while the probability of an occurence is increased. Obviously, after application of $\mathcal{A}$ the security definition of the first algorithm is not met anymore (cf. Figure 6). Still, as described in the example above, this definition is meaningful.

Now consider the following scenario. Mechanism $\mathcal{M}_1$ hides a predicate with high probability v, say $v = 0.99$. Another mechanism that also hides the predicate (perfectly) is a mechanism $\mathcal{M}_2$ that returns $r \in \{0,1\}$ uniformly at random. Seperately, both mechanisms satisfy the privacy definition that the predicate has to be hidden with probability greater than 98 percent. If we combine them as required per axiom 2 suddenly the random values are correlated to the real predicate, thus decreasing the probability that the predicate is hidden to $p \cdot (1 - v) + (1 - p) * 0.5$. This does not hide the predicate with probability greater 0.98 for any p. To summarize our results, there are meaningfull privacy guarantees that do not fulfill the privacy axioms mentioned above.

Averages

Publishing the average of for example salaries intuitively preservers the privacy of individuals if the uncertainty of the adversary is high enough. If a privacy notion, however, does not allow for high breaches with a small probability, averaging can not be considered to be a privacy preserving mechanism. In the following example, we consider that every salary in the range from 1 to 3 and has identical independent probability of $p = \frac{1}{3}$. Assuming our company employs 3 individuals, every database has a probability of $(\frac{1}{3})^3$. We consider the salary of employee A sensitive, but want to publish the average salary. There are 27 databases and 7 different possible average salaries. The table in Figure 7 shows the probabilities of each average salary, and the breach induced by its publication.

Publishing an average salary of 2 for example results in a breach of $\frac{6}{21}$. Publishing an average salary of 1 results in fully disclosing the salary of A. In this example, publishing the average is a mechanism, that fulfills privacy with $\epsilon = \frac{2}{3}$ and $\delta = \frac{6}{27}$. Figure 8 depicts the breach distribution and the cumulative breach distribution as well as ϵ and δ for this example. This is a very simple example and larger databases with a bias to an average salary different to the expected value are even less probable. This will result in a smaller ϵ and δ. Consequently, high breaches are even less probable in larger examples.

Differential Privacy

As the Pufferfish framework, our framework is also able to model differential privacy, even (ϵ, δ)-differential privacy. In the following, we present an outline of a proof.

Lemma 1. *Set the universe U to the set of all possible databases of a given domain, the mechanism E to the identity function, and the set of allowed*

average salary	probability	breach
1	$\frac{1}{27}$	2
$\frac{4}{3}$	$\frac{3}{27}$	1
$\frac{5}{3}$	$\frac{6}{27}$	$\frac{14}{21}$
2	$\frac{7}{27}$	$\frac{6}{21}$
$\frac{7}{3}$	$\frac{6}{27}$	$\frac{14}{21}$
$\frac{8}{3}$	$\frac{3}{27}$	1
3	$\frac{1}{27}$	2

Fig. 7. Average salaries and their probabilities in the averaging mechanism example. Figure 8 depicts the breach distribution and the cumulative breach distribution.

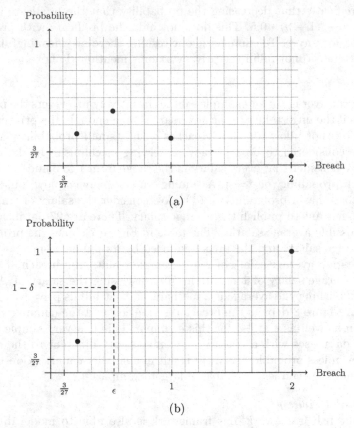

Fig. 8. Breach distribution (a) and cumulative breach distribution (b) for the averaging mechanism example. Even in this small example, the thresholds δ and ϵ are relatively small.

distributions $\mathcal{X}$ to all distributions X with the following property: There are database d_1 and d_2 with probability $\frac{1}{2}$, with $d_1 \Delta d_2 = 1$. All other databases have probability 0. If the set of sensitive predicates L contains all predicates of the form "Individual i is in the database d" for all Individuals i, then, a mechanism F that fulfills $(\frac{1}{e^\epsilon - 1}, \delta)$-privacy fulfills (ϵ, δ)-differential privacy.

Proof outline:
With probability $1 - \delta$, for all breaches b holds:

$$b = \frac{Pr[l|F(d)] - Pr[l]}{Pr[l]} \leq \frac{1}{e^\epsilon - 1}$$

After applicaton of the bayes rule, this is equal to:

$$Pr[F(d)|l] \leq \frac{e^\epsilon}{e^\epsilon - 1} Pr[F(d)]$$

$$\Leftrightarrow Pr[F(d)|l] \leq \frac{e^\epsilon}{e^\epsilon - 1}(Pr[F(d)|l] + Pr[F(d)|\neg l])$$

$$\Leftrightarrow Pr[F(d)|l] \leq e^\epsilon Pr[F(d)|\neg l]$$

$$\Leftrightarrow Pr[F(d_1) = x] \leq e^\epsilon Pr[F(d_2) = x]$$

Since the last equation holds with probability $1 - \delta$ for all l and d_1, d_2, it implies (ϵ, δ)-differential privacy.

5   Conclusion and Outlook

In this paper, we presented an overview of the development in the field of database privacy and a framework that allows for modeling privacy notions based on allowed breaches for sensitive predicates. This framework allows for notions that fail with a small probability δ.

Furthermore, we provided meningfull examples and argued, that this framework allows more general notions than allowed by the privacy axioms of Kifer and Lin [2,1]. We showed that this framework allows to describe the privacy of aggregates such as average. Additionally, we showed that (ϵ, δ)-differential privacy can be modeled in the framework.

For future work, we want to also consider infeasibility and computational assumptions. This will enable methods like subset sampling or computing averages to be used as an anonymization mechanism without the addition of noise. This also will allow to use privacy notions defined in this framework to be used for secure database outsourcing, where encryption is used as a way to hide sensitive data.

References

1. Kifer, D., Lin, B.R.: An axiomatic view of statistical privacy and utility. Journal of Privacy and Confidentiality 4(1), Article 2
2. Kifer, D., Lin, B.R.: Towards an axiomatization of statistical privacy and utility. In: Proceedings of the Twenty-Ninth ACM SIGMOD-SIGACT-SIGART Symposium on Principles of Database Systems, PODS 2010, pp. 147–158. ACM, New York (2010)
3. Demirel, D., Henning, M., van de Graaf, J., Ryan, P.Y.A., Buchmann, J.: Prêt à voter providing everlasting privacy. In: Heather, J., Schneider, S., Teague, V. (eds.) Vote-ID 2013. LNCS, vol. 7985, pp. 156–175. Springer, Heidelberg (2013)
4. Langer, L., Schmidt, A., Volkamer, M., Buchmann, J.: Classifying privacy and verifiability requirements for electronic voting. In: Fischer, S., Maehle, E., Reischuk, R. (eds.) GI Jahrestagung. LNI, vol. 154, pp. 1837–1846. GI (2009)
5. Buchmann, J., Demirel, D., van de Graaf, J.: Towards a publicly-verifiable mix-net providing everlasting privacy. In: Sadeghi, A.-R. (ed.) FC 2013. LNCS, vol. 7859, pp. 197–204. Springer, Heidelberg (2013)
6. Sweeney, L.: k-anonymity: a model for protecting privacy. Int. J. Uncertain. Fuzziness Knowl. -Based Syst. 10, 557–570 (2002)

7. Dwork, C.: Differential privacy. In: Bugliesi, M., Preneel, B., Sassone, V., Wegener, I. (eds.) ICALP 2006. LNCS, vol. 4052, pp. 1–12. Springer, Heidelberg (2006)

8. Hacigümüş, H., Iyer, B., Li, C., Mehrotra, S.: Executing sql over encrypted data in the database-service-provider model. In: SIGMOD 2002: Proceedings of the 2002 ACM SIGMOD International Conference on Management of Data, pp. 216–227. ACM, New York (2002)

9. Popa, R.A., Redfield, C., Zeldovich, N., Balakrishnan, H.: CryptDB: Protecting Confidentiality with Encrypted Query Processing. In: Symposium on Operating Systems Principles (SOSP), Cascais, Portugal (October 2011)

10. Lin, B.R., Kifer, D.: A framework for extracting semantic guarantees from privacy. CoRR abs/1208.5443 (2012)

11. Dwork, C., Kenthapadi, K., McSherry, F., Mironov, I., Naor, M.: Our data, ourselves: privacy via distributed noise generation. In: Vaudenay, S. (ed.) EUROCRYPT 2006. LNCS, vol. 4004, pp. 486–503. Springer, Heidelberg (2006)

12. Samarati, P., Sweeney, L.: Protecting privacy when disclosing information: k-anonymity and its enforcement through generalization and suppression. Technical report, CMU SRI (1998)

13. Machanavajjhala, A., Gehrke, J., Kifer, D., Venkitasubramaniam, M.: l-diversity: Privacy beyond k-anonymity. In: ICDE, vol. 24 (2006)

14. Li, N., Li, T., Venkatasubramanian, S.: t-closeness: Privacy beyond k-anonymity and l-diversity. In: ICDE, pp. 106–115 (2007)

15. LeFevre, K., DeWitt, D.J., Ramakrishnan, R.: Incognito: Efficient full-domain k-anonymity. In: SIGMOD Conference, pp. 49–60 (2005)

16. Meyerson, A., Williams, R.: On the complexity of optimal k-anonymity. In: PODS, pp. 223–228 (2004)

17. Bayardo Jr., R.J., Agrawal, R.: Data privacy through optimal k-anonymization. In: ICDE, pp. 217–228 (2005)

18. Tian, H., Zhang, W.: Extending l-diversity to generalize sensitive data. Data Knowl. Eng. 70(1), 101–126 (2011)

19. Ganta, S.R., Kasiviswanathan, S.P., Smith, A.: Composition attacks and auxiliary information in data privacy. In: KDD, pp. 265–273 (2008)

20. Heidinger, C., Buchmann, E., Huber, M., Böhm, K., Müller-Quade, J.: Privacy-aware folksonomies. In: Lalmas, M., Jose, J., Rauber, A., Sebastiani, F., Frommholz, I. (eds.) ECDL 2010. LNCS, vol. 6273, pp. 156–167. Springer, Heidelberg (2010)

21. Dwork, C., Kenthapadi, K., McSherry, F., Mironov, I., Naor, M.: Our data, ourselves: Privacy via distributed noise generation. In: Vaudenay, S. (ed.) EUROCRYPT 2006. LNCS, vol. 4004, pp. 486–503. Springer, Heidelberg (2006)

22. Dwork, C., McSherry, F., Nissim, K., Smith, A.: Calibrating noise to sensitivity in private data analysis. In: Halevi, S., Rabin, T. (eds.) TCC 2006. LNCS, vol. 3876, pp. 265–284. Springer, Heidelberg (2006)

23. Gehrke, J., Lui, E., Pass, R.: Towards privacy for social networks: A zero-knowledge based definition of privacy. In: Ishai, Y. (ed.) TCC 2011. LNCS, vol. 6597, pp. 432–449. Springer, Heidelberg (2011)

24. Gehrke, J., Hay, M., Lui, E., Pass, R.: Crowd-blending privacy. In: Safavi-Naini, R., Canetti, R. (eds.) CRYPTO 2012. LNCS, vol. 7417, pp. 479–496. Springer, Heidelberg (2012)

25. Kifer, D., Machanavajjhala, A.: No free lunch in data privacy. In: Proceedings of the 2011 ACM SIGMOD International Conference on Management of data, SIGMOD 2011, pp. 193–204 (2011)

26. Duan, Y.: Privacy without noise. In: CIKM, pp. 1517–1520 (2009)
27. Bhaskar, R., Bhowmick, A., Goyal, V., Laxman, S., Thakurta, A.: Noiseless database privacy. In: Lee, D.H., Wang, X. (eds.) ASIACRYPT 2011. LNCS, vol. 7073, pp. 215–232. Springer, Heidelberg (2011)
28. Blum, A., Dwork, C., McSherry, F., Nissim, K.: Practical privacy: the sulq framework. In: Proceedings of the Twenty-Fourth ACM SIGMOD-SIGACT-SIGART Symposium on Principles of Database Systems, PODS 2005, pp. 128–138 (2005)
29. Chaudhuri, K., Mishra, N.: When random sampling preserves privacy. In: Dwork, C. (ed.) CRYPTO 2006. LNCS, vol. 4117, pp. 198–213. Springer, Heidelberg (2006)
30. Evfimievski, A., Fagin, R., Woodruff, D.P.: Epistemic privacy. In: Proceedings of the Twenty-Seventh ACM SIGMOD-SIGACT-SIGART Symposium on Principles of Database Systems, PODS 2008, pp. 171–180 (2008)
31. Blum, A., Ligett, K., Roth, A.: A learning theory approach to non-interactive database privacy. In: STOC, pp. 609–618 (2008)
32. Blum, A., Ligett, K., Roth, A.: A learning theory approach to noninteractive database privacy. J. ACM 60(2), 12 (2013)
33. Kifer, D., Machanavajjhala, A.: A rigorous and customizable framework for privacy. In: Proceedings of the 31st Symposium on Principles of Database Systems, PODS 2012, pp. 77–88 (2012)
34. Warner, S.L.: Randomized response: A survey technique for eliminating evasive answer bias. Journal of the American Statistical Association 60(309), 63–69 (1965)

A Constructive Perspective
on Key Encapsulation

Sandro Coretti, Ueli Maurer, and Björn Tackmann

Department of Computer Science, ETH Zürich, Switzerland
{corettis,maurer,bjoernt}@inf.ethz.ch

Abstract. A *key-encapsulation mechanism (KEM)* is a cryptographic
primitive that allows anyone in possession of some party's public key
to securely transmit a key to that party. A KEM can be viewed as a
key-exchange protocol in which only a single message is transmitted; the
main application is in combination with symmetric encryption to achieve
public-key encryption of messages of arbitrary length.

The security of KEMs is usually defined in terms of a certain game
that no efficient adversary can win with non-negligible advantage. A
main drawback of game-based definitions is that they often do not have
clear semantics, and that the security of each higher-level protocol that
makes use of KEMs needs to be proved by showing a tailor-made security
reduction from breaking the security of the KEM to breaking the security
of the combined protocol.

We propose a novel approach to the security and applications of
KEMs, following the constructive cryptography paradigm by Maurer and
Renner (ICS 2011). The goal of a KEM is to *construct* a resource that
models a shared key available to the honest parties. This resource can
be used in designing and proving higher-level protocols; the composition
theorem guarantees the security of the combined protocol without the
need for a specific reduction.

1 Introduction

Key establishment is a cryptographic primitive that allows two parties to obtain
a shared secret key, which can subsequently be used in cryptographic mechanisms such as encryption schemes or message authentication codes (MACs).
The most important application of key-establishment protocols is in the setup
phases of protocols for secure communication, such as TLS or IPSec, but, furthermore, their unidirectional variant—*key-encapsulation mechanisms (KEMs)*—are
an important building block in most practical public-key encryption schemes.

This paper is dedicated to Johannes Buchmann on the occasion of his 60th
birthday. The topic of the paper, the key-establishment problem, a fundamental problem in cryptography, is one of the areas to which he has contributed
significantly (e.g., [3–5, 21]).

In this paper, we focus on the particular case where keys are established using
KEMs and only unidirectional communication. We build on [8], where public-key
encryption is treated in constructive cryptography, and some parts of this work
are taken from that paper.

M. Fischlin and S. Katzenbeisser (Eds.): Buchmann Festschrift, LNCS 8260, pp. 226–239, 2013.

1.1   Security Notions for Key-Encapsulation Mechanisms

An important question for the application of KEMs is which level of KEM security is required in order for a higher-level protocol that makes use of a KEM to be secure. To define KEM security, game-based security notions for public-key encryption have been adapted to work with KEMs. A game-based definition is usually characterized by a security property that is to be maintained in the presence of an adversary launching a certain attack against the scheme in question. Both security property and attack are encoded only implicitly into the security game. As a consequence, the traditional answer to the above question is that for each protocol one needs to identify the appropriate security notion and provide a reduction proof to show that a KEM satisfying this notion yields a secure protocol.

An alternative approach is to capture the semantics of a security notion by characterizing directly what it achieves, making explicit in which applications it can be used securely. The constructive cryptography framework [15, 16] was proposed with this general goal in mind. Resources such as different types of communication channels and keys are modeled explicitly, and the goal of a cryptographic protocol or scheme π is to *construct* a stronger or more useful resource S from an assumed resource R, denoted as $R \overset{\pi}{\longmapsto} S$. Two such construction steps can then be composed, i.e., if we additionally consider a protocol ψ that assumes the resource S and constructs a resource T, the composition theorem states that

$$R \overset{\pi}{\longmapsto} S \ \wedge \ S \overset{\psi}{\longmapsto} T \quad \Longrightarrow \quad R \overset{\psi \circ \pi}{\longmapsto} T,$$

where $\psi \circ \pi$ denotes the composed protocol.

Following the constructive paradigm, a protocol is built in a modular fashion from isolated construction steps. A security proof guarantees the soundness of one such step, and each proof is independent of the remaining steps. The composition theorem then guarantees that several such steps can be composed. While the general approach to protocol design based on reduction proofs is in principle sound, it is substantially more complex, more error-prone, and not suitable for re-use.

In this spirit, we treat the use of KEMs as such a construction step. We show how one can construct, using KEMs, shared keys from authenticated and insecure channels.

1.2   Constructing Keys Using KEMs

From the perspective of constructive cryptography [15, 16], the purpose of a KEM is to construct a shared key from non-confidential communication, which is modeled as channels. Channels and keys are *resources* that involve a sender, a receiver, and—to model different levels of security—an attacker. The security properties of a particular resource are captured by the capabilities available to the attacker. In the case of a channel they might, e.g., include reading or modifying the messages in transmission.

The parties access resources through the interfaces they provide; these are specific to each party. For example, the sender's interface of a channel allows to input messages, and the receiver's interface allows to receive them. We refer to the interfaces by labels A, B, and E, where A and B are the sender's and the receiver's interfaces, respectively, and E is the adversary's interface. In this work, we consider two types of channels and two types of keys:[1]

- An *insecure channel*, denoted $--\twoheadrightarrow$, allows the adversary to read, deliver, and to delete all messages input at A, as well as to inject her own messages.
- An *authenticated channel*, denoted $\bullet\!\!-\!\!\diamond\!\!-\!\!\twoheadrightarrow$, still allows to read all messages, but the adversary is limited to forwarding or deleting messages input at interface A.
- A *unilateral key*, denoted $=\!\!\diamond\!\!=\!\!\bullet$, allows A to request keys. The adversary at interface E may choose to deliver or delete keys requested at interface A to B or to inject her own keys.
- A *bilateral key*, denoted $\bullet\!\!=\!\!\diamond\!\!=\!\!\bullet$, also allows A to request keys, but the adversary may not inject any keys.

In the straightforward application of a KEM, the receiver initially generates a key pair and transmits the public key to the sender. The sender needs to obtain the correct public key, which corresponds to assuming that the channel from B to A is authenticated ($\leftarrow\!\!-\!\!\bullet$[2]). The sender, using the public key, generates a key and corresponding ciphertext, which he sends to the receiver over a channel that could either be authenticated or completely insecure.

The type of key that is constructed depends on the type of assumed channel used to transmit the ciphertext to the receiver: We show that if the assumed channel is authenticated ($\bullet\!\!-\!\!\diamond\!\!-\!\!\twoheadrightarrow$) and the KEM scheme is ind-cpa-secure, the constructed key is bilateral ($\bullet\!\!=\!\!\diamond\!\!=\!\!\bullet$). If the assumed channel is insecure ($--\twoheadrightarrow$) and the PKE scheme is ind-cca-secure, the constructed key is unilateral ($=\!\!\diamond\!\!=\!\!\bullet$).

Using the above notation, for protocols π and π' based on ind-cpa and ind-cca KEMs, respectively, these constructions can be written as

$$[\leftarrow\!\!-\!\!\bullet, \bullet\!\!-\!\!\diamond\!\!-\!\!\twoheadrightarrow] \quad \xmapsto{\ \pi\ } \quad \bullet\!\!=\!\!\diamond\!\!=\!\!\bullet$$

and

$$[\leftarrow\!\!-\!\!\bullet, --\twoheadrightarrow] \quad \xmapsto{\ \pi'\ } \quad =\!\!\diamond\!\!=\!\!\bullet,$$

where the bracket notation means that both resources in the brackets are available.

---

[1] The notation is taken from [18]: The "$\bullet$" in the notation signifies that the capabilities at the marked interface, i.e., sending or receiving, are exclusive to the respective party. If the "$\bullet$" is missing, the adversary also has these capabilities. The $\diamond$-symbol is explained in Section 2.4.

[2] The simple arrow indicates that $\leftarrow\!\!-\!\!\bullet$ is a single-use channel, i.e., only one message can be transmitted.

The notion of constructing the bilateral or unilateral keys from the two assumed non-confidential channels is made precise in a simulation-based sense [15, 16], where the simulator can be interpreted as translating all attacks on the protocol into attacks on the constructed (ideal) key. As the constructed key is secure by definition, there are no attacks on the protocol.

The composability of the construction notion then means that the constructed key can again be used as an assumed resource (possibly along with additional assumed or constructed resources) in other protocols. For instance, if a higher-level protocol uses a unilateral key in symmetric encryption to confidentially transmit a message, then the proof of this protocol is based on the "idealized" unilateral key and does not (need to) include a reduction to the security of the KEM. In the same spirit, the authenticated channel from B to A could be a physically authenticated channel, but it could also be constructed by using, for instance, a digital signature scheme to authenticate the transmission of the public key (which is done by certificates in practice).

1.3   Related Work

Game-Based Security. The notion of KEMs was introduced by Cramer and Shoup [9, 22], when they proposed an ISO standard for public-key encryption (PKE). They adapted the standard security notions for PKE schemes to work with KEMs. Today's standard security notions for PKE are indistinguishability under chosen-plaintext attack (ind-cpa) and (the stronger) indistinguishability under chosen-ciphertext attack (ind-cca). Bellare et al. [2] compare them to various other PKE security notions. Herranz et al. [12] analyze the security of PKE schemes obtained from KEMs in conjunction with symmetric encryption schemes with various security levels.

Real-World/Ideal-World Security. The idea of defining protocol security with respect to an ideal execution was first proposed by Goldreich et al. [10]; the concept of a simulator can be traced back to the seminal work by Goldwasser et al. [11] on zero-knowledge proofs. Canetti [6] introduced the universal composability framework (UC), which allows the formalization of arbitrary functionalities to be realized by cryptographic protocols. The framework is designed from a bottom-up perspective (starting from a selected machine model), whereas we follow the top-down approach of [16], which leads to simpler, more abstract definitions and statements. Nagao et al. [20] give a treatment of KEMs in the UC framework. They show that combining a cca-secure KEM with cca-secure symmetric encryption realizes an "ideal KEM" functionality, which can be used to implement the secure channel functionality of [7].

The constructive cryptography paradigm has been introduced by Maurer and Renner [15, 16]. Several primitives have been modeled and analyzed following this paradigm, including symmetric encryption [17] and public-key encryption [8]. Similarly to this work, both papers model the primitives as constructions and compare the resulting security definitions to (game-based) notions from the literature. The unilateral key resource we use here is described in [19], where it is shown to be achievable by an interactive protocol based on (signatures and) a cpa-secure KEM.

2 Preliminaries

2.1 Systems: Resources, Converters, Distinguishers, and Reductions

Resources and converters are modeled as systems. At the highest level of abstraction (following the hierarchy in [16]), systems are objects with interfaces by which they connect to (interfaces of) other systems; each interface is labeled with an element of a label set and connects to only a single other interface. This concept, which we refer to as *abstract systems*, captures the topological structures that result when multiple systems are connected in this manner.

The abstract systems concept, however, does not model the behavior of systems, i.e., *how* the systems interact via their interfaces. Consequently, statements about cryptographic protocols are statements at the next (lower) abstraction level. In this work, we describe all systems in terms of (probabilistic) discrete systems, which we explain in Section 2.2.

Resources and Converters. *Resources* in this work are systems with three interfaces labeled by A, B, and E. A protocol is modeled as a pair of two so-called *converters* (one for each honest party), which are directed in that they have an *inside* and an *outside* interface, denoted by in and out, respectively. As a notational convention, we generally use upper-case, bold-face letters (e.g., $\mathbf{R}$, $\mathbf{S}$) or channel symbols (e.g., $\bullet\!-\!\diamond\!-\!\twoheadrightarrow$) to denote resources and lower-case Greek letters (e.g., α, β) or sans-serif fonts (e.g., snd, rcv) for converters. We denote by Φ the set of all resources and by Σ the set of all converters.

The topology of a composite system is described using a term algebra, where each expression starts from one (or more) resources on the right-hand side and is subsequently extended with further terms on the left-hand side. An expression is interpreted in the way that all interfaces of the system it describes can be connected to interfaces of systems which are appended on the left. For instance, for a single resource $\mathbf{R} \in \Phi$, all its interfaces A, B, and E are accessible.

For $I \in \{A, B, E\}$, a resource $\mathbf{R} \in \Phi$, and a converter $\alpha \in \Sigma$, the expression $\alpha^I \mathbf{R}$ denotes the composite system obtained by connecting the inside interface of α to interface I of $\mathbf{R}$; the outside interface of α becomes the I-interface of the composite system. The system $\alpha^I \mathbf{R}$ is again a resource (cf. Figure 1 on page 235).

For two resources $\mathbf{R}$ and $\mathbf{S}$, $[\mathbf{R}, \mathbf{S}]$ denotes the parallel composition of $\mathbf{R}$ and $\mathbf{S}$. For each $I \in \{A, B, E\}$, the I-interfaces of $\mathbf{R}$ and $\mathbf{S}$ are merged and become the *sub-interfaces* of the I-interface of $[\mathbf{R}, \mathbf{S}]$, which we denote by $I.1$ and $I.2$. A converter α that connects to the I-interface of $[\mathbf{R}, \mathbf{S}]$ has two inside sub-interfaces, denoted by in.1 and in.2, where the first one connects to $I.1$ of $\mathbf{R}$ and the second one connects to $I.2$ of $\mathbf{S}$.

Any two converters α and β can be composed sequentially by connecting the inside interface of β to the outside interface of α, written $\beta \circ \alpha$, with the effect that $(\beta \circ \alpha)^I \mathbf{R} = \beta^I \alpha^I \mathbf{R}$. Moreover, converters can also be taken in parallel, denoted by $[\alpha, \beta]$, with the effect that $[\alpha, \beta]^I [\mathbf{R}, \mathbf{S}] = [\alpha^I \mathbf{R}, \beta^I \mathbf{S}]$.

We assume the existence of an identity converter $\mathrm{id} \in \Sigma$ with $\mathrm{id}^I \mathbf{R} = \mathbf{R}$ for all resources $\mathbf{R} \in \Phi$ and interfaces $I \in \{A, B, E\}$ and of a special converter $\perp \in \Sigma$ with an inactive outside interface.

Distinguishers. A *distinguisher* is a special type of system $\mathbf{D}$ that connects to all interfaces of a resource $\mathbf{U}$ and outputs a single bit at the end of its interaction with $\mathbf{U}$. In the term algebra, this appears as the expression $\mathbf{DU}$, which defines a binary random variable. The *distinguishing advantage of a distinguisher* $\mathbf{D}$ *on two systems* $\mathbf{U}$ *and* $\mathbf{V}$ is defined as

$$\Delta^{\mathbf{D}}(\mathbf{U}, \mathbf{V}) := |\mathrm{P}[\mathbf{DU} = 1] - \mathrm{P}[\mathbf{DV} = 1]|.$$

The advantage of a class $\mathcal{D}$ of distinguishers is defined as

$$\Delta^{\mathcal{D}}(\mathbf{U}, \mathbf{V}) := \sup_{\mathbf{D} \in \mathcal{D}} \Delta^{\mathbf{D}}(\mathbf{U}, \mathbf{V}).$$

The distinguishing advantage measures how much the output distribution of $\mathbf{D}$ differs when it is connected to either $\mathbf{U}$ or $\mathbf{V}$. There is an equivalence notion on systems (which is defined on the discrete systems level), denoted by $\mathbf{U} \equiv \mathbf{V}$, which implies that $\Delta^{\mathbf{D}}(\mathbf{U}, \mathbf{V}) = 0$ for all distinguishers $\mathbf{D}$. The distinguishing advantage satisfies the triangle inequality, i.e., $\Delta^{\mathbf{D}}(\mathbf{U}, \mathbf{W}) \le \Delta^{\mathbf{D}}(\mathbf{U}, \mathbf{V}) + \Delta^{\mathbf{D}}(\mathbf{V}, \mathbf{W})$ for all resources $\mathbf{U}$, $\mathbf{V}$, and $\mathbf{W}$ and distinguishers $\mathbf{D}$.

Games. We capture games defining security properties as distinguishing problems in which an adversary $\mathbf{A}$ tries to distinguish between two *game systems* $\mathbf{G}_0$ and $\mathbf{G}_1$. Game systems (or simply *games*) are single-interface systems, which appear, similarly to resources, on the right-hand side of the expressions in the term algebra. The adversary is similar to a distinguisher, except that it connects to a game instead of a resource.

Reductions. When relating two distinguishing problems, it is convenient to use a special type of system $\mathbf{C}$ that translates one setting into the other. Formally, $\mathbf{C}$ is a converter that has an *inside* and an *outside* interface. When it is connected to a system $\mathbf{S}$, which is denoted by $\mathbf{CS}$, the inside interface of $\mathbf{C}$ connects to the (merged) interface(s) of $\mathbf{S}$ and the outside interface of $\mathbf{C}$ is the interface of the composed system. $\mathbf{C}$ is called a *reduction system* (or simply *reduction*).

To reduce distinguishing two systems $\mathbf{S}, \mathbf{T}$ to distinguishing two systems $\mathbf{U}, \mathbf{V}$, one exhibits a reduction $\mathbf{C}$ such that $\mathbf{CS} \equiv \mathbf{U}$ and $\mathbf{CT} \equiv \mathbf{V}$.[3] Then, for all distinguishers $\mathbf{D}$, we have $\Delta^{\mathbf{D}}(\mathbf{U}, \mathbf{V}) = \Delta^{\mathbf{D}}(\mathbf{CS}, \mathbf{CT}) = \Delta^{\mathbf{DC}}(\mathbf{S}, \mathbf{T})$. The last equality follows from the fact that $\mathbf{C}$ can also be thought of as being part of the distinguisher.

---

[3] For instance, we consider reductions from distinguishing game systems to distinguishing resources. Then, $\mathbf{C}$ connects to a game on the inside and provides interfaces A, B, and E on the outside.

2.2   Discrete Systems

Protocols that communicate by passing messages and the respective resources are described as (probabilistic) discrete systems. Their behavior can be formalized by random systems as in [14], i.e., as families of conditional probability distributions of the outputs (as random variables) given all previous inputs and outputs of the system. For systems with multiple interfaces, the interface to which an input or output is associated is explicitly specified as part of the input or output. For the restricted (but here sufficient) class of systems that for each input provide (at most) a single output, an execution of a collection of systems is defined as the consecutive evaluation of the respective random systems (this is similar to the models in [6, 13]).

2.3   The Notion of Construction

Recall that we consider resources with interfaces A, B, and E, where A and B are interfaces of honest parties and E is the interface of the adversary. We formalize the security of protocols via the following notion of *construction* (which is a special case of the abstraction notion from [16]):

Definition 1. *Let Φ and Σ be as in Section 2.1. A protocol $\pi = (\pi_1, \pi_2) \in \Sigma^2$ constructs resource $\mathbf{S} \in \Phi$ from resource $\mathbf{R} \in \Phi$ within ε and with respect to distinguisher class $\mathcal{D}$, denoted*

$$\mathbf{R} \stackrel{(\pi, \varepsilon)}{\Longmapsto} \mathbf{S},$$

if

$$\begin{cases} \Delta^{\mathcal{D}}(\pi_1^A \pi_2^B \bot^E \mathbf{R}, \bot^E \mathbf{S}) \le \varepsilon & (availability) \\ \exists \sigma \in \Sigma : \quad \Delta^{\mathcal{D}}(\pi_1^A \pi_2^B \mathbf{R}, \sigma^E \mathbf{S}) \le \varepsilon & (security). \end{cases}$$

The availability condition captures that a protocol must correctly implement the functionality of the constructed resource in the absence of the adversary. The security condition models the requirement that everything the adversary can achieve in the *real-world system* (i.e., the assumed resource with the protocol) he can also accomplish in the *ideal-world system* (i.e., the constructed resource with the simulator). More details can be found in [15].

An important property of Definition 1 is its composability. Intuitively, if a resource $\mathbf{S}$ is used in the construction of a larger system, then the composability implies that $\mathbf{S}$ can be replaced by a construction $\pi_1^A \pi_2^B \mathbf{R}$ without affecting the security of the composed system. Security and availability are preserved under composition. More formally, if for some resources $\mathbf{R}$, $\mathbf{S}$, and $\mathbf{T}$ and protocols π and ϕ,

$$\mathbf{R} \stackrel{(\pi, \varepsilon)}{\Longmapsto} \mathbf{S} \quad \text{and} \quad \mathbf{S} \stackrel{(\phi, \varepsilon')}{\Longmapsto} \mathbf{T},$$

then

$$\mathbf{R} \stackrel{(\phi \circ \pi, \varepsilon + \varepsilon')}{\Longmapsto} \mathbf{T},$$

as well as

$$[\mathbf{R}, \mathbf{U}] \stackrel{([\pi,\mathrm{id}],\varepsilon)}{\Longmapsto} [\mathbf{S}, \mathbf{U}] \quad \text{and} \quad [\mathbf{U}, \mathbf{R}] \stackrel{([\mathrm{id},\pi],\varepsilon)}{\Longmapsto} [\mathbf{U}, \mathbf{S}]$$

for any resource $\mathbf{U}$. More details can be found in [15].

2.4   Important Resources

In this work, we consider two types of resources: channels and keys. Both types initially expect a special cheating bit $b \in \{0, 1\}$ at interface E, indicating whether the adversary is present and intends to interfere with the transmission of the messages. The special converter $\perp$ (cf. Section 2.1) always sets $b = 0$, in which case all messages/keys input at the sender's interface are delivered immediately to the receiver. For simplicity, we will assume that whenever $\perp$ is not present, all cheating bits are set to 1.

An *authenticated channel* from A to B with message space $\mathcal{M}$ is a resource •–◇–» with interfaces A, B, and E and behaves as follows: When the ith message $m \in \mathcal{M}$ is input at interface A, it is recorded as (i, m) and (msg, i, m) is output at interface E. When (dlv, i') is input at interface E, if (i', m') has been recorded, m' is delivered at interface B. An *insecure channel* –→ behaves as •–◇–», but, additionally, when (inj, m') is input at interface E, m' is output at interface B.

A *bilateral key* between A and B with key space $\mathcal{K}$ is a resource •–◇–• with interfaces A, B, and E and behaves as follows: When req is input for the ith time at interface A, a key $k \leftarrow_R \mathcal{K}$ is chosen uniformly at random and is recorded as (i, k). Then, k is output at interface A and (key, i) at interface E. When (dlv, i') is input at interface E, if (i', k') has been recorded, k' is output at interface B. A *unilateral key* =◇=• behaves as •=◇=•, but, additionally, when (inj, k') is input at interface E, k' is output at interface B.[4]

For $\mathbf{R} \in \{-\twoheadrightarrow, \bullet\!-\!\diamond\!-\!\twoheadrightarrow, =\!\diamond\!=\!\bullet, \bullet\!=\!\diamond\!=\!\bullet\}$, denote by $\overset{n}{\mathbf{R}}$ the resource that processes only the first n queries at A and E (e.g., $\overset{n}{\bullet\!-\!\diamond\!-\!\twoheadrightarrow}$ only processes the first n messages input at A and the first n queries at E).

2.5   Key-Encapsulation Mechanisms

A key-encapsulation mechanism (KEM) with key space $\mathcal{K}$ is a triple of algorithms $\Pi = (K, E, D)$. The key generation algorithm K outputs a key pair $(\mathsf{pk}, \mathsf{sk}) \leftarrow K$, the (probabilistic) encryption algorithm takes a public key pk and outputs a pair $(k, c) \leftarrow E_{\mathsf{pk}}$, where $k \in \mathcal{K}$ and c is the corresponding ciphertext, and the decryption algorithm takes a secret key sk and a ciphertext c and outputs $k' \leftarrow D_{\mathsf{sk}}(c')$, where $k' = \diamond$ indicates an invalid ciphertext.

A KEM is correct if for $(k, c) \leftarrow E_{\mathsf{pk}}$, $D_{\mathsf{sk}}(c) = k$ (with probability 1 over the randomness of E) for all key pairs $(\mathsf{pk}, \mathsf{sk})$ generated by K. For security

---

[4] Note that neither the channels nor the keys prevent the adversary from reordering or replaying messages/keys sent by A. The ◇-symbol suggests the "internal buffer" in which a channel/key stores messages input at A.

properties of KEM schemes which are defined via a bit-guessing game, it will be more convenient to phrase the game as a distinguishing problem between two game systems (cf. Section 2.1). We consider the following games, which correspond to the (standard) notions of ind-cpa (cpa for short) and ind-cca2 (cca).[5]

CPA Game. Consider systems $\mathbf{G}_0^{\mathsf{cpa}}$ and $\mathbf{G}_1^{\mathsf{cpa}}$: For a KEM scheme Π, both initially run the key-generation algorithm to obtain $(\mathsf{pk}, \mathsf{sk})$ and output pk. When the query chall is input, they compute $(k, c) \leftarrow E_{\mathsf{pk}}$. Then, $\mathbf{G}_0^{\mathsf{cpa}}$ outputs (k, c) and $\mathbf{G}_1^{\mathsf{cpa}}$ outputs $(\bar{k}, c)$ for a randomly chosen $\bar{k} \in \mathcal{K}$. Ciphertext c is called the *challenge*.

CCA Game. Consider systems $\mathbf{G}_0^{\mathsf{cca}}$ and $\mathbf{G}_1^{\mathsf{cca}}$: They proceed as in the CPA case but also answer decryption queries (dec, c') by returning $k' \leftarrow D_{\mathsf{sk}}(c')$ unless c' equals the challenge c (if defined), in which case the answer is test.

3   Bilateral and Unilateral Keys from KEMs

In the spirit of constructive cryptography, we explicitly consider the purpose of KEMs, namely to achieve a shared key between A and B over an untrusted network. As a constructive statement this means that we view the KEM as a protocol, a pair of converters $(\mathsf{snd}, \mathsf{rcv})$, whose goal is to construct a shared key from non-confidential channels. Differentiating between the two cases where the communication from the sender A to the receiver B is authenticated and unauthenticated, this is written as

$$[\longleftarrow\!\bullet, \bullet\!\!-\!\diamond\!\!\rightarrow] \quad \overset{(\mathsf{snd},\mathsf{rcv})}{\Longrightarrow} \quad \bullet\!\!-\!\diamond\!\!-\!\bullet \tag{1}$$

and

$$[\longleftarrow\!\bullet, -\ -\rightarrow] \quad \overset{(\mathsf{snd},\mathsf{rcv})}{\Longrightarrow} \quad \Rightarrow\!\diamond\!\!-\!\bullet, \tag{2}$$

respectively.

In both cases, the channel $\longleftarrow\!\bullet$ captures the ability of the sender to obtain the receiver's public key in an authenticated fashion. In construction (1), the communication from the sender A to the receiver B is authenticated, which is modeled by the channel $\bullet\!\!-\!\diamond\!\!\rightarrow$. The goal is to achieve a bilateral key $\bullet\!\!-\!\diamond\!\!-\!\bullet$. In construction (2), the communication from A to B is completely insecure, which is captured by the insecure channel $-\ -\rightarrow$. Here, the goal is to achieve a unilateral key $\Rightarrow\!\diamond\!\!-\!\bullet$, which does not prevent the adversary from sending its own keys to B.

In the following we first show how a KEM Π can be seen as a converter pair $(\mathsf{snd}, \mathsf{rcv})$. We then prove that $(\mathsf{snd}, \mathsf{rcv})$ achieves construction (1) if Π is cpa-secure, and construction (2) if Π is cca-secure.

---

[5] We consider the so-called real-or-random versions of these games, which are equivalent to the more popular left-or-right formulations (as shown in [1] for symmetric encryption).

3.1   KEMs as Protocols

Let $\Pi = (K, E, D)$ be a KEM. Based on Π, we define a pair of protocol converters (snd, rcv) for constructions (1) and (2). Both converters have two sub-interfaces, denoted by in.1 and in.2, on the inside, as we connect them to a resource that is a parallel composition of two other resources (cf. Section 2.1).

Converter snd works as follows: It initially expects a public key pk at in.1. When req is input at the outside interface, snd computes $(k, c) \leftarrow E_{\mathsf{pk}}$ and outputs k at the outside interface and c at the inside interface in.2. Converter rcv initially generates a key pair (pk, sk) using key-generation algorithm K and outputs pk at in.1. When rcv receives c' at in.2, it computes $k' \leftarrow D_{\mathsf{sk}}(c')$ and, if $k' \neq \diamond$, outputs k' at the outside interface.

3.2   Bilateral Keys from Authenticated Channels

Towards proving that the protocol (snd, rcv) indeed achieves construction (1), note first that the correctness of the KEM immediately implies that the *availability* condition of Definition 1 is satisfied. To prove *security*, we need to exhibit a simulator σ such that the assumed resource $[\longleftarrow\!\bullet, \bullet\!\!-\!\!\diamond\!\!-\!\!\twoheadrightarrow]$ with the protocol converters is indistinguishable from the constructed resource $\bullet\!\!=\!\!\diamond\!\!=\!\!\bullet$ with the simulator (cf. Figure 1). Theorem 1 implies that (snd, rcv) realizes (1) if the underlying KEM is cpa-secure.

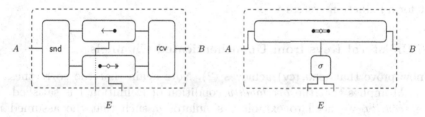

Fig. 1. Left: The assumed resource (two authenticated channels) with protocol converters snd and rcv attached to interfaces A and B, denoted $\mathsf{snd}^A \mathsf{rcv}^B [\longleftarrow\!\bullet, \bullet\!\!-\!\!\diamond\!\!-\!\!\twoheadrightarrow]$. Right: The constructed resource (a bilateral key) with simulator σ attached to the E-interface, denoted $\sigma^E \bullet\!\!=\!\!\diamond\!\!=\!\!\bullet$. In particular, σ must simulate the E-interfaces of the two authenticated channels. The protocol is secure if the two systems are indistinguishable.

Theorem 1. *There exists a simulator σ and for any $n \in \mathbb{N}$ there exists a (efficient) reduction $\mathbf{C}$ such that for every $\mathbf{D}$,*

$$\Delta^{\mathbf{D}}(\mathsf{snd}^A \mathsf{rcv}^B [\longleftarrow\!\bullet, \bullet\!\!-\!\!\diamond\!\!-\!\!\twoheadrightarrow], \sigma^E \overset{n}{\bullet\!\!=\!\!\diamond\!\!=\!\!\bullet}) \leq n \cdot \Delta^{\mathbf{DC}}(\mathbf{G}_0^{\mathsf{cpa}}, \mathbf{G}_1^{\mathsf{cpa}}).$$

Proof. Consider the following simulator σ for interface E of $\bullet\!\!=\!\!\diamond\!\!=\!\!\bullet$: Initially, it generates a key pair (pk, sk) and outputs (msg, 1, pk) at out.1. On (key, i) at the inside interface in, it generates $(k, c) \leftarrow E_{\mathsf{pk}}$ and outputs (msg, i, c) at out.2. On (dlv, i') at out.2, it outputs (dlv, i') at in. Consider the two systems

$$\mathsf{snd}^A \mathsf{rcv}^B [\leftarrow\!\!\bullet, \bullet\!\!\overset{1}{\diamond}\!\!\rightarrow\!\!\!\!\rightarrow] \quad \text{and} \quad \sigma^E \; \bullet\!\!\overset{1}{\diamond}\!\!\rightarrow\!\!\!\!\rightarrow\!\!\bullet \; .$$

Distinguishing $\mathbf{G}_0^{\mathsf{cpa}}$ from $\mathbf{G}_1^{\mathsf{cpa}}$ can be reduced to distinguishing these two systems via the following reduction system $\mathbf{C}'$, which connects to a game on the inside and provides interfaces A, B, and E on the outside (cf. Section 2.1 for details on reduction systems): Initially, it takes a value pk on the inside and outputs $(\mathsf{msg}, 1, \mathsf{pk})$ at the (outside) $E.1$-interface. On req at the A-interface, it outputs chall at in. On subsequent key and challenge (k, c) at in, it outputs k at A and $(\mathsf{msg}, 1, c)$ at $E.2$. On $(\mathsf{dlv}, 1)$ at the $E.2$-interface, it outputs k at interface B. We have

$$\mathbf{C}'\mathbf{G}_0^{\mathsf{cpa}} \equiv \mathsf{snd}^A \mathsf{rcv}^B [\leftarrow\!\!\bullet, \bullet\!\!\overset{1}{\diamond}\!\!\rightarrow\!\!\!\!\rightarrow] \quad \text{and} \quad \mathbf{C}'\mathbf{G}_1^{\mathsf{cpa}} \equiv \sigma^E \; \bullet\!\!\overset{1}{\diamond}\!\!\rightarrow\!\!\bullet,$$

and thus

$$\Delta^{\mathbf{D}}(\mathsf{snd}^A \mathsf{rcv}^B [\leftarrow\!\!\bullet, \bullet\!\!\overset{n}{\diamond}\!\!\rightarrow\!\!\!\!\rightarrow], \sigma^E \; \bullet\!\!\overset{n}{\diamond}\!\!\rightarrow\!\!\bullet)$$

$$\leq n\cdot\Delta^{\mathbf{D}\mathbf{C}''}(\mathsf{snd}^A \mathsf{rcv}^B [\leftarrow\!\!\bullet, \bullet\!\!\overset{1}{\diamond}\!\!\rightarrow\!\!\!\!\rightarrow], \sigma^E \; \bullet\!\!\overset{1}{\diamond}\!\!\rightarrow\!\!\bullet)$$

$$= n\cdot\Delta^{\mathbf{D}\mathbf{C}''}(\mathbf{C}'\mathbf{G}_0^{\mathsf{cpa}}, \mathbf{C}'\mathbf{G}_1^{\mathsf{cpa}})$$

$$= n\cdot\Delta^{\mathbf{D}\mathbf{C}}(\mathbf{G}_0^{\mathsf{cpa}}, \mathbf{G}_1^{\mathsf{cpa}}),$$

where $\mathbf{C} := \mathbf{C}''\mathbf{C}'$ and the first inequality follows from a standard hybrid argument for a reduction system $\mathbf{C}''$. $\square$

3.3   Unilateral Keys from Unauthenticated Channels

We now prove that $(\mathsf{snd}, \mathsf{rcv})$ achieves (2). Note again that the correctness of the KEM implies that the *availability* condition of Definition 1 is satisfied. To prove *security*, we need to exhibit a simulator σ such that the assumed resource $[\leftarrow\!\!\bullet, -\!\rightarrow\!\!\!\!\rightarrow]$ with the protocol converters is indistinguishable from the constructed resource $\Rightarrow\!\!\diamond\!\!\bullet$ with the simulator. Theorem 1 implies that $(\mathsf{snd}, \mathsf{rcv})$ realizes (2) if the underlying KEM is cca-secure.

Theorem 2. *There exists a simulator σ and for any $n \in \mathbb{N}$ there exists a (efficient) reduction $\mathbf{C}$ such that for every $\mathbf{D}$,*

$$\Delta^{\mathbf{D}}(\mathsf{snd}^A \mathsf{rcv}^B [\leftarrow\!\!\bullet, -\!\!\overset{n}{\rightarrow}\!\!\!\!\rightarrow], \sigma^E \Rightarrow\!\!\overset{n}{\diamond}\!\!\bullet) \leq n \cdot \Delta^{\mathbf{D}\mathbf{C}}(\mathbf{G}_0^{\mathsf{cca}}, \mathbf{G}_1^{\mathsf{cca}}).$$

Proof. Simulator σ for interface E of $\Rightarrow\!\!\diamond\!\!\bullet$ again has two outside sub-interfaces out.1 and out.2 and works as follows: Initially, it generates a key pair $(\mathsf{pk}, \mathsf{sk})$ and outputs $(\mathsf{msg}, 1, \mathsf{pk})$ at out.1. When it receives (key, i) at the inside interface in, it generates $(k, c) \leftarrow E_{\mathsf{pk}}$, outputs (msg, i, c) at out.1, and records (c, i). When (inj, c') is input at out.2, σ proceeds as follows: If (c', i') has been recorded for some i', it outputs (dlv, i') at in. Otherwise, it computes $k' \leftarrow D_{\mathsf{sk}}(c')$ and, if $k' \neq \diamond$, outputs (inj, k') at in.

Denote by $\xrightarrow{n,q}$ the insecure channel that processes the first n inputs at interface A and the first q inputs at interface E (and similarly for $\xrightarrow{n,q}$). Consider now the problem of distinguishing the two systems

$$\mathbf{U} := \mathsf{snd}^A \mathsf{rcv}^B [\longleftrightarrow\!\bullet, -\xrightarrow{1,n}] \qquad \text{and} \qquad \mathbf{V} := \sigma^E \xrightarrow{1,n}\!\bullet,$$

which are depicted in Figure 2. A distinguisher $\mathbf{D}$ connected to the real-world system $\mathbf{U}$ initially sees a public key at interface $E.1$. If $\mathbf{D}$ inputs req at interface A, a key k is output at A and its corresponding encryption is output at interface $E.2$ (both created by snd). When $\mathbf{D}$ inputs a ciphertext c' at $E.2$, it sees a decryption of c' (by rcv) at B. The ideal-world system $\mathbf{V}$ behaves differently: Initially, $\mathbf{D}$ also sees a public key. But when it inputs req at A, an encryption c of a key k unrelated to the key $\bar{k}$ output at A is output at interface $E.2$ (by simulator σ). When c is input at interface $E.2$, $\bar{k}$ is output at B (as σ issues a dlv-instruction to the key). When $c' \neq c$ is input at E, the decryption k' of c' (injected by σ) is output at B.

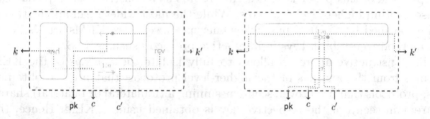

Fig. 2. The systems $\mathbf{U}$ and $\mathbf{V}$ with the "message flow" from the perspective of a distinguisher: Initially, a public-key pk is output at interface $E.1$. Requesting a key k at interface A causes a ciphertext c to be output at the $E.2$-interface. Note that c is the challenge in the cca-game. Inputting a ciphertext c' at interface $E.2$ results in a key k' being output at B. This corresponds to the decryption oracle in the cca-game.

The translation between the channel setting and the game setting is achieved by the following reduction system $\mathbf{C}'$: Initially, $\mathbf{C}'$ takes a value pk from the game and outputs $(\mathsf{msg}, 1, \mathsf{pk})$ at the $E.1$-interface. When req is input at interface A of $\mathbf{C}'$, chall is output to the game, which returns (k, c). Key k is output at A and challenge c is output as $(\mathsf{msg}, 1, c)$ at interface $E.2$. When (inj, c) is input at interface $E.2$, $\mathbf{C}'$ outputs k at interface B. When (inj, c') with $c' \neq c$ is input at interface $E.2$, $\mathbf{C}'$ passes c' to the game's decryption oracle and outputs the answer k' at interface B, provided $k' \neq \diamond$. We have

$$\mathbf{C}'\mathbf{G}_0^{\mathsf{cca}} \equiv \mathsf{snd}^A \mathsf{rcv}^B [\longleftrightarrow\!\bullet, -\xrightarrow{1,n}] \qquad \text{and} \qquad \mathbf{C}'\mathbf{G}_1^{\mathsf{cca}} \equiv \sigma^E \xrightarrow{1,n}\!\bullet,$$

and thus

$$\Delta^{\mathbf{D}}(\mathsf{snd}^{A}\mathsf{rcv}^{B}[\longleftarrow\!\bullet, -\overset{n}{\longrightarrow}\!], \sigma^{E}\overset{n}{\Longleftrightarrow}\!\bullet) \leq n \cdot \Delta^{\mathbf{DC}''}(\mathsf{snd}^{A}\mathsf{rcv}^{B}[\longleftarrow\!\bullet, -\overset{1,n}{\longrightarrow}\!], \sigma^{E}\overset{1,n}{\Longleftrightarrow}\!\bullet)$$
$$= n \cdot \Delta^{\mathbf{DC}''}(\mathbf{C}'\mathbf{G}_{0}^{\mathsf{cca}}, \mathbf{C}'\mathbf{G}_{1}^{\mathsf{cca}})$$
$$= n \cdot \Delta^{\mathbf{DC}}(\mathbf{G}_{0}^{\mathsf{cca}}, \mathbf{G}_{1}^{\mathsf{cca}}),$$

where $\mathbf{C} := \mathbf{C}''\mathbf{C}'$ and the first inequality follows from a standard hybrid argument for a reduction system $\mathbf{C}''$.   □

The unilateral key $\!-\!\diamond\!-\!\!\bullet$ is the best key one can construct from the two assumed channels. As the E-interface has the same capabilities as the A-interface at both the authenticated (from B to A) and the insecure channels, it will necessarily also be possible to inject keys to the receiver via the E-interface by simply applying the sender's protocol converter.

4   Conclusion

The purpose of this paper is to describe KEMs as a construction step that can be used within higher-level protocols. While the most widespread application is in the so-called KEM-DEM public-key encryption schemes, KEMs are also used as a method to establish keys in interactive protocols such as TLS.

The constructive approach allows to fully isolate the analysis of the KEM schemes from the analysis of the higher-level protocols that use the obtained keys; protocols that are proven secure assuming a (unilateral or bilateral) shared key remain secure if the respective key is obtained using a KEM. Hence, the approach supports a fully modular protocol design where each cryptographic mechanism is proven in isolation to achieve one construction step, and multiple such steps are composed by the general composition theorem.

Compared with the traditional approach of explicitly proving tailor-made security reductions from breaking the security of the underlying schemes to breaking the security of a complex protocol, this modular approach leads to security proofs which are simpler and consequently less prone to errors.

Acknowledgments. The work was supported by the Swiss National Science Foundation (SNF), project no. 200020-132794.

References

1. Bellare, M., Desai, A., Jokipii, E., Rogaway, P.: A Concrete Security Treatment of Symmetric Encryption. In: 38th FOCS, pp. 394–403 (1997)
2. Bellare, M., Desai, A., Pointcheval, D., Rogaway, P.: Relations Among Notions of Security for Public-Key Encryption Schemes. In: Krawczyk, H. (ed.) CRYPTO 1998. LNCS, vol. 1462, pp. 26–45. Springer, Heidelberg (1998)
3. Buchmann, J., Düllmann, S., Williams, H.C.: On the Complexity and Efficiency of a New Key Exchange System. In: Quisquater, J.-J., Vandewalle, J. (eds.) EUROCRYPT 1989. LNCS, vol. 434, pp. 597–616. Springer, Heidelberg (1990)

4. Buchmann, J., Williams, H.C.: A Key-Exchange System Based on Imaginary Quadratic Fields. Journal of Cryptology 1(2), 107–118 (1988)
5. Buchmann, J., Williams, H.C.: A Key Exchange System Based on Real Quadratic Fields. In: Brassard, G. (ed.) CRYPTO 1989. LNCS, vol. 435, pp. 335–343. Springer, Heidelberg (1990)
6. Canetti, R.: Universally Composable Security: A New Paradigm for Cryptographic Protocols. Cryptology ePrint Archive, Report 2000/067 (2000)
7. Canetti, R., Krawczyk, H.: Universally Composable Notions of Key Exchange and Secure Channels. In: Knudsen, L.R. (ed.) EUROCRYPT 2002. LNCS, vol. 2332, pp. 337–351. Springer, Heidelberg (2002)
8. Coretti, S., Maurer, U., Tackmann, B.: Constructing Confidential Channels from Authenticated Channels—Public-Key Encryption Revisited. In: Sako, K., Sarkar, P. (eds.) ASIACRYPT 2013. Springer, Heidelberg (2013)
9. Cramer, R., Shoup, V.: Design and Analysis of Practical Public-Key Encryption Schemes Secure against Adaptive Chosen Ciphertext Attack. SIAM Journal on Computing 33, 167–226 (2001)
10. Goldreich, O., Micali, S., Wigderson, A.: How to Play any Mental Game or A Completeness Theorem for Protocols with Honest Majority. In: 19th ACM STOC, pp. 218–229 (1987)
11. Goldwasser, S., Micali, S., Rackoff, C.: The Knowledge Complexity of Interactive Proof-Systems (Extended Abstract). In: 17th ACM STOC, pp. 291–304 (1985)
12. Herranz, J., Hofheinz, D., Kiltz, E.: Some (in)sufficient conditions for secure hybrid encryption. Cryptology ePrint Archive, Report 2006/265 (2006), http://eprint.iacr.org/
13. Hofheinz, D., Shoup, V.: GNUC: A New Universal Composability Framework. Cryptology ePrint Archive, Report 2011/303 (2011)
14. Maurer, U.: Indistinguishability of Random Systems. In: Knudsen, L.R. (ed.) EUROCRYPT 2002. LNCS, vol. 2332, pp. 110–132. Springer, Heidelberg (2002)
15. Maurer, U.: Constructive Cryptography—A New Paradigm for Security Definitions and Proofs. In: Mödersheim, S., Palamidessi, C. (eds.) TOSCA 2011. LNCS, vol. 6993, pp. 33–56. Springer, Heidelberg (2012)
16. Maurer, U., Renner, R.: Abstract Cryptography. In: Chazelle, B. (ed.) The Second Symposium in Innovations in Computer Science, ICS 2011, pp. 1–21. Tsinghua University Press (January 2011)
17. Maurer, U., Rüedlinger, A., Tackmann, B.: Confidentiality and Integrity: A Constructive Perspective. In: Cramer, R. (ed.) TCC 2012. LNCS, vol. 7194, pp. 209–229. Springer, Heidelberg (2012)
18. Maurer, U., Schmid, P.E.: A Calculus for Security Bootstrapping in Distributed Systems. Journal of Computer Security 4(1), 55–80 (1996)
19. Maurer, U., Tackmann, B., Coretti, S.: Key Exchange with Unilateral Authentication: Composable Security Definition and Modular Protocol Design. Cryptology ePrint Archive, Report 2013/555 (2013)
20. Nagao, W., Manabe, Y., Okamoto, T.: A Universally Composable Secure Channel Based on the KEM-DEM Framework. In: Kilian, J. (ed.) TCC 2005. LNCS, vol. 3378, pp. 426–444. Springer, Heidelberg (2005)
21. Scheidler, R., Buchmann, J., Williams, H.C.: Implementation of a Key Exchange Protocol Using Some Real Quadratic Fields. In: Damgård, I.B. (ed.) EUROCRYPT 1990. LNCS, vol. 473, pp. 98–109. Springer, Heidelberg (1991)
22. Shoup, V.: A Proposal for an ISO Standard for Public Key Encryption. IACR Cryptology ePrint Archive 2001, 112 (2001)

Why Are Business Processes Not Secure?

Günter Müller and Rafael Accorsi

Department of Telematics
University of Freiburg, Germany
{mueller,accorsi}@iig.uni-freiburg.de

Abstract. Security is simple to understand but hard to ensure. In the times of Internet, this task has been becoming harder every day. To date, computer science has not solved how to prevent the misuse of business processes. While data objects can be protected, a process cannot. The reason is the security of a process depends not only on its individual accesses and can only be accessed upon the process' termination or when cast into the context of other processes. Many unbelievable scandals encompassing sophisticated and powerful players, from Microsoft to Sony and credit card operators, from leakages in governments to cyber crime and war attacks could not be prevented despite heavy investment in security. The claim here is that the way in which computer science deals with security does not apply to processes. The key discipline in security is "cryptography", where the "laureate" Prof. Buchmann got his distinction from. This paper is about how cryptography can be applied as a basis to automate security and give participants in a market an equal position and prevent fraud. To complicate the issue, the goal is security in business processes. The reason is obvious. If one makes mistakes or vulnerabilities are left uncovered, huge fraud incidents might happen, the stockowners rebel, the government complains and employees are, in the worst case, deprived from their pension. This is a real, sensitive issue, with unclear solutions, ambivalent in nature, but rigorous in punishment. The issue is not just to protect, but also to deter "bad things", such as criminal intents. The option to judge people's intentions is not an option for mankind; it is not an option though for computer science. We need to automate security and establish procedures that, upon the event of misuse, ascertain accountability.

The main goal and challenge of security in business processes is, on one hand, to provide well-founded guarantees regarding the adherence to security, privacy and regulatory compliance requirements and, on the other hand, to integrate the corresponding mechanisms into the business process management lifecycle. This paper introduces this research area, its current status and upcoming practical challenges.

1   Insufficient Business Process Security

About 70% of the business processes in IT-based management systems are modeled and at least partially executed and managed by business process management systems (BPMS) [1]. This allows the flexible, adequate adaptation of processes to business

M. Fischlin and S. Katzenbeisser (Eds.): Buchmann Festschrift, LNCS 8260, pp. 240–254, 2013.
© Springer-Verlag Berlin Heidelberg 2013

changes and integration external resources on demand. However, the benefits of automating business processes are accompanied by the need to pre-design business events while changes amongst business interaction cannot be stopped. This built-in inflexibility is accompanied by significant risks concerning the adherence to security guidelines. If companies want to, e.g., automate their processes or outsource them into the cloud, they must have guarantees that corresponding security promises or properties are not violated and that mechanisms exist to timely detect such violations.

However, designing secure business process is as tricky as programming "Satan's computer". Recent fraud disasters show how subtle secure process engineering and control can be. The Swiss bank UBS suffered from a rogue trader scandal in 2011, which led to a loss of a then-estimated US$2 billion, was possible because the risk of trades could be disguised by using "forward-settling" Exchange-traded Funds (ETF) cash positions. Specifically, processes that implemented ETF transactions in Europe do not issue confirmations until after settlement has taken place. The exploitation of this process allows a party in a transaction to receive payment for a trade before the transaction has been confirmed [2]. While the cash proceeds in this scheme cannot be simply retrieved, the seller may still show the cash on their books and possibly use it in further transactions. Eventually, the mechanics of this attack allowed for a carrousel of transactions, thereby creating an ever growing snowball. Similar constellations, usually based upon insider threats, can also be made up in the fraud cases involving the well-documented Société Générale case [3], but also the WorldCom and Parmalat cases.

These cases evidence the fact that security cannot just partially be pre-planned, and that the need for control and audit of security: it must be ensured that processed information such as customer data, financial transactions and business' secrets remain protected and its own processes are encapsulated from foreign influence and internal attackers. The confidence in existing solutions so far is very low. Lack of security and compliance guarantees are currently considered as the main obstacle to the operational use of automated processes in the industry [1]. Here, despite the growing expenses in tool design for the compliant deployment and process certification, e.g. according to Control Objectives for Information and Related Technology (Cobit) and Committee of Sponsoring Organizations of the Treadway Commission (COSO), non-compliance and the monetary damage they incur are still soaring.

The trade-off to be solved occurs on many dimensions: Firstly, automating security costs time and performance and reduces response to customer demands, secondly security needs specified properties to design mechanisms to counter the violations, and thirdly, automating security need flexibility to even face the unplanned challenges.

2   Process: Ubiquitous Processing Structures

Processes are structured specifications of personal and business data usage. Traditionally, they are associated to business processes, which capture the operational aspects of an enterprise's value-chain, such as the purchase-to-pay (P2P) and supply chain tasks. Figure 1(a) depicts a P2P process definition using the Business Process Modeling and Notation process specification language. Process descriptions are, however, not restricted to business processes. In fact, processes ubiquitous entities in modern information systems. Figure 1(b) depicts, for example, the data analysis and visualization process in social networks as described in a Google patent. Further examples include the so-called "scientific workflows" which capture and process sensor data for weather forecasting or Tsunami and earthquake warning [4], ads customization in e-retailing [5] and the collection and processing of smart meter data in a smart grid [6]. While this paper focuses on business processes for the sake of concreteness, and the techniques put forward below equally apply to all these types of processes and, more generally, to structured workflow specifications.

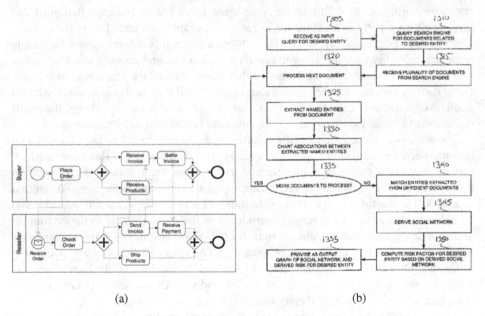

(a)   (b)

Fig. 1. Examples of processes

3   Secure Enterprise Business Model

The research area of "Enterprise Security" can be considered at different levels of abstraction. The so-called "Enterprise Architecture Metamodel" [7] provides a suitable abstraction to classify related works and depict the novelty of business process security. This model divides the support of IT in modern enterprise system architectures in three inter-layers.

- The *business layer* defines the business processes, -objects and -targets, the assets of the company, as well as its organizational structure and guidelines to be followed.
- The *application layer* contains the services and data schemas which are required for the execution of the processes. In this context service-oriented architectures (SOA) or higher service modes of Cloud and Grid computing are relevant.
- The *infrastructure layer* provides the software and hardware needed to automate the execution of services and, hence business processes. This includes, e.g. the required databases, the operating system and virtual machines.

Mechanisms of the enterprise security have to consider security and compliance requirements at all layers and compositionally enforce them across all the layers. These mechanisms, and more generally the state of the art, can themselves be split according to the timepoint (and entity) they act upon in "design-time" (process models), "runtime" (process instances) and "audit-time" (event logs). Figure 2 depicts the array containing both the abstraction layer and the time perspective. This is the security extended enterprise metamodel, as we have been proposing in Freiburg [56].

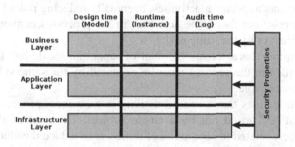

Fig. 2. The Freiburg Security Extended Enterprise Metamodel

While procedures for the application layer and infrastructure layer have been thoroughly investigated and may easily be adapted and carried over to new architectures, taking into account the business layer is a relatively new, challenging and, practically, indispensable area to be investigated. The aim of "Security in Business Process Management" is to provide well-founded guarantees regarding the adherence to security, privacy and regulatory compliance requirements. Furthermore, it must ensure that security is seamlessly integrated into the business process management lifecycle.

4   Requirements of Security in Business Processes

The relevant requirements in Security in BPM encompass the traditional safety and security properties, regulatory requirements and privacy and data protection requirements. Generally, they are called, depending on the specific application setting, "policies" or "controls" and can be classified according the following general types.

- *Authorization* means enforcing access control to ensure that only authorized individuals/roles are allowed to execute activities within a process [8]. In enterprise computing settings, this is usually achieved with role-based access control. The resulting authorization models can even in small enterprises easily exhibit more than 500 roles and a complex role hierarchy.
- *Usage control* expresses conditions that must hold after the access to a resource [9]. That way one can express, e.g., the maximal number of access to a resource (cardinality) or the obligation to delete local copies of a data item after accessing and using these data. In general, these requirements consider the interplay of different activities in the process (so-called processes' control-flow), which can be captured as patterns to facilitate re-use [10, 11]. Usage control policies can be used to capture the vast majority of regulatory compliance and privacy and data protection requirements.
- *Conflict of interest* aims to prevent the flow of sensitive information among competing organizations/departments taking part of in a process execution. These requirements are captured using the "Chinese-Wall" model [12] as an allusion to the metaphor in which, access restrictions to objects are built as walls around them.
- *Separation of duties* expresses constraints associated with the process to limit the abilities of agents to perform activities, eventually reducing risk of fraud [13]. Specifically, it prescribes that some activities in the processes cannot be executed by the same subject or by the same role.
- *Binding of duties* characterizes the dual of separation of duties. It prescribes that certain activities in the process must be executed by the same subject or subjects acting in the same role.
- *Four-eye rule* defines that a critical activity in a business process (e.g., final loan grant) cannot be carried out by a single subject. The goal of this requirement, which is closely related to separation of duties, is to reduce the chance of fraud, but also to avoid mistakes and inappropriate use of rights.
- *Isolation* generally says that information must remain confidential during the execution of a process. Put another way, there is no information or data leak [14]. This ranges from prescribing that executions of the process do not interfere one to the other (inter-instance leak, e.g. via covert-channels), but also that the roles involved in a process either "fully isolated" from each other or can only communicate over a "declassification channel" (intra-instance leak, e.g. via direct authorized access to data items).

Specifications of these requirements can be extensional, focusing on what has to be achieved, or intensional, focusing on how a certain property is guaranteed [15]. The corresponding formalisms to encode the requirements – for example, Petri nets, temporal logics or process calculi – are chosen accordingly. Irrespective of the concrete way in which specifications happen, there is a multitude of enforcement mechanisms to ensure/detect their adherence or diagnose their violation. These mechanisms are embedded into the business process management (BPM) lifecycle and act upon different process "entities".

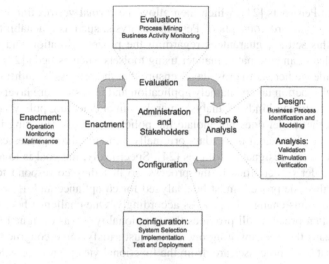

Fig. 3. BPM Lifecycle [16]

5   BPM Lifecycle and Control Mechanisms

Figure 3 depicts the BPM lifecycle according to [16]. When cast into the three process analysis timepoints depicted in Fig. 2, the following control mechanisms and their corresponding phases in the BPM lifecycle arise.

Design-Time Analysis

The design-time analysis takes place prior to the actual process execution and related to the "Design/Analysis" phase of the BPM lifecycle. The process entity to be considered is the "process models", which are specified using industrial modeling languages, such as BPMN and BPEL.

In this setting, there are mainly three active research directions and corresponding mechanisms: they focus on (a) the policies, (b) the process structure and, (c) the corrective interplay between policies and the process' structure. Focusing on the policies, specifically the optimal assignment of subjects, roles and activities in a role-based access control setting, Basin et al. [17], but also Wang and Li [18] attempt get as close as possible to the "least privilege". This principle defines that roles – and, hence, subjects acting in these roles – should only have the rights necessary to execute the process, and prohibits all the assignments that lead to more rights than necessary. The challenge here is to carry out this analysis in a context where processes or role models are flexible, or where role models are not know a priori (e.g. in Cloud settings).

Regarding the structure of the process, it has been shown that structural process vulnerabilities can lead to information leaks and the violation of usage control requirements [19, 20]. A fairly novel research strand investigates methods to detecting structural vulnerabilities in processes. For this, the specification of processes is translated into Petri nets [21], which then allow the formal verification of processes' structure using standard concepts of Petri net analysis, such as reachability and trace analysis. In this setting, guarantees regarding the process isolation and compliance can be provided in an automated manner using toolsets, such as Seda [22] and SWAT [23]. The challenge here is to cross-layer ensure that the process is faithfully executed at the lower abstraction layers, namely application and infrastructure layers.

The aforementioned kinds of analysis merely aim at detecting policy violations, or asserting that the model does not violate the policies. Process rewriting is a new research direction whose goal is to adapt originally non-compliant processes to enforce the adherence to the designated policies [24]. Specifically, the goal is to add, remove or swap the order of activities in the process, so that they correspond to fulfill the policies. For this, the process must be analyzed for compliance and, if necessary, the rewriting algorithms change the process accordingly. One challenge here is to ensure that the rewritten process still provides the functionality it was designed for. Technically, this means that the rewriting operators must satisfy some congruence relations that ensure that these processes are, from the functional viewpoint, equivalent.

Run-Time Analysis

Monitoring approaches trace the process execution and are responsible for run-time analysis of processes. The process entity considered is the process instance, i.e. an instantiated model with concrete subject and data assignments. In the BPM lifecycle, this corresponds to the "Enactment" phase, in which the processes are executed by the Workflow or Business Process Management System.

Monitors can exhibit a preventive or detective character. The former – and most traditional – stops the process execution before a policy violation; the latter detects the violation of a policy and signals it. The modality "preventive vs. detective" depends, on one hand, on the type of requirement to be monitored and, on the other hand, on the architectural deployment of the system that eventually characterizes the scope of a monitor's activity.

Preventive monitors stop the execution of a process whenever its continuation violates a policy. This is, however, not always reasonable in industrial processes: stopping and aborting a process means that it does not generate value for the enterprise. Technically, it might further lead to deadlocks in situations where other processes depend on the termination of processes just aborted. This chain reaction is known as "ripple effect" [25]. To avoid these situations, two sets of approaches exist to relax the "yes/no" decision. Risk-based [26, 27] approaches compute the risk, expressed as the probability that a policy will eventually be violated, and based the decisions upon this measurement. In these approaches, only "too risky" accesses are fully denied. Another approach is based upon the concept of "break glass" [28] originally developed to handle emergency situations. In break glass settings, the monitor stops the execution on the event of a policy violation and explicitly asks whether it should be resumed. If so, the transgressions of policies are recorded, as well as the further process execution, thereby providing a basis for a posteriori process analysis, traceability and accountability.

In contrast to preventive approaches, detective monitors merely observe the process execution without interfering with it. Detected violations are recorded in an operational log file, or communicated to the responsible subject. Detective monitors allow the timely reaction to policy violations, but they cannot prevent violations. This feature is particularly interesting in the context of flexible business process.

The general problem of monitoring approach is the performance overhead inherent to their activity. With figures ranging from 200 percent to 400 percent additional overhead, monitoring ads a prohibitive lag in, e.g., high frequency trading applications and event-intensive process management.

Audit-Time Analysis

The a posteriori analysis takes place at the audit-time using to this end the event logs generated during the business process execution. The audit-time analysis is related to the "Evaluation" phase. In this phase, process executions are analyzed as to whether (and how) the designated business goals are achieved and whether policies were adhered to. The corresponding process entities are the recorded process executions, so-called cases.

Approaches based upon Process Mining [29, 30] provide a set of tools for the event log analysis. These approaches can be roughly distinguished into two classes:

- *Process Discovery* approaches reconstruct the structure (control flow), the data exchange (data flow) or role model of the process that generated the event log, thereby generating models (e.g. Petri nets or directed graphs) or declarative process specifications. Focusing on process' structure, the reconstructed models denote the "de-facto" process executed in the system, which may then deviate from the "de-jure", designed process specifications using BPEL and BPMN. The so-called "delta-analysis" can be used to pinpoint these deviations and analysis methods "design-time analysis" used to detect policy violations [31, 32, 33, 34].
- *Conformance Checking* aims to detect discrepancies between the original process specification and the recorded traces without reconstructing the model. Further, the can be used to carry out a log-level root-cause analysis of the process based on each execution trace. That way it is not possible to measure the fitness between a process specification and its event log (i.e. how many traces can be "replayed" in the original specification), but also to analyze traces for adherence to policies, such as separation of duties and usage control [35, 36, 37].

The results of an audit-time analysis can serve as input for two further phases of the BPM lifecycle: the "Redesign" phase provides for incremental improvements on business process design, e.g. using process rewriting; the "Reengineering" phase, in contrast, carries out a fundamental change on the whole business process architecture. The focus of process reengineering, as task and process, lies outside the strictly technical scope.

6   Example

Information leaks can be a consequence of covert channels induced by the faulty structure of business process specifications. The detection of covert channels (and, hence, potential leaks) is required in various certification standards. For example, the TCSEC's "Part B: Mandatory Protection" demands the "analysis of covert channels and their bandwidth" [38]. Similarly, ISO/IEC 27001 Section 12.5 "Security in development and support processes" prescribes that "checks should be made for information leakage for example via covert channels" [39]. Concrete audit and certification procedures for SAS 70 and SAS 117 equally demand security [40, Chap. 24]. Several papers expose the risks of covert channels in fields related to business process management and deployment, e.g. Cloud Computing environments [41], virtualization [42] and web services [43]. The threat of leaks is considered imminent [44] and real [45, Chap. 8], thereby motivating the need for well-founded process analysis and covert channel detection.

Figure 4 depicts, schematically, the steps of an audit-time analysis. The left hand side of the picture depicts a simplified log file where only the case number and the execution traces are depicted. (Here, timestamps, involved subjects, their roles and the data types involved are omitted.) For readability, each capital letter denotes an abbreviation of an activity executed during the process. Specifically, A stands for "Enter Patient ID", B for "Open Patient Record", C for "Warn Invalid ID", D for "Update Record" and E for "Close Record". Process discovery can be used to the reconstruct the Petri net on the right hand side of the picture. This process is the so-called "Update Patient Record" and is, in different shapes, ubiquitous in process architectures in the e-health setting.

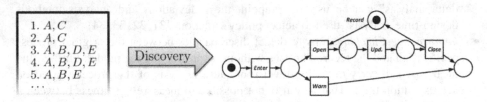

1. A, C
2. A, C
3. A, B, D, E
4. A, B, D, E
5. A, B, E
. . .

Fig. 4. Discovered process model

To analyze whether leaks can occur between parallel executions of the reconstructed process, the process is split into two "security levels": the high level, denoting the secret or confidential level and the low level denoting the public. The high level could be represented by process runs executed by physicians, whereas the low level could stand for process runs executed by nurses. The resultant model, depicted in Fig. 4, contains two process runs (for each level) and a shared resource (record).

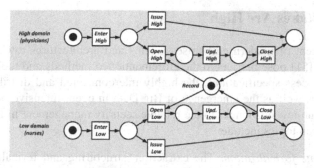

Fig. 5. Multi-level model

To analyze the model depicted in Fig. 4 for covert channels, properties must be defined to capture the interferences between the high and the low part. These properties are expressed in terms of patterns, which characterize places where information flows from high to low. Busi and Gorrieri [46] characterize the place-based non-interference, showing that the causal and the conflict places, if existing in a net, violate a non-interference property. Figure 5 shows these patterns. The rationale behind the causal place is that low can only fire if high triggers a prior activity. Hence, if low can fire his transition, then he/she knows that high has fired before. Similarly, in the conflict place both high and low compete for the control flow token. If low cannot fire, he/she can deduce that high as the token. Altogether, these patterns provably capture a very strict non-interference property, which in fact prohibit any flow of information between the security levels.

Fig. 6. Causal (left) and conflict (right) places *s*

Mechanisms, such as Seda and SWAT, exist to automatically detect these places. These tools reduce the verification problem to a reachability problem: is there a marking in which a causal and a conflict place contain a token? In the model in Fig. 4 the place "Record" stands for a conflict place, as well as for a causal place. Hence, information can leak across two instances of the process. Specifically, by observing the process execution, an attacker at the low level can profile the activity of a high subject (e.g. physician or senior physician), as well as deduce information about the patient's health state. Details about the analysis and tool support are given in [47, 48, 49, 50].

7   The Stakes Are High

The development of modern enterprise systems and architectures towards the BPM "in the large" [51] pose barriers to the aforementioned methods and tools. In particular flexible process specifications, the highly interconnected and distributed process architectures as well as the emergence of Big Data in event-intensive process execution bring about new challenges to the reliable security management in BPM. Current relevant research topics include:

- *Requirements formalization.* The extraction, structuring and formalization of requirements are essential for the analysis. Current works aim at collecting systematically libraries of patterns that capture requirements at a higher level of abstraction. Concrete requirements specifications can be then derived by instantiating these patters for a specific analysis context. While this works well for structural characteristics (control flow), capturing the time component and involved data objects is a present challenge.
- *Corrective enforcement.* Monitors, as introduced above, have a detective-preventive operating mode. Currently, approaches are being investigated to equip monitors with corrective capabilities to ensure that a process execution meets its non-functional goals. The so-called corrective enforcement [52] is challenging insofar as the modified execution must furthermore achieve the objectives planned for the process while being only aware of the current prefix, not the whole process shape as in the case of process rewriting.
- *Data-aware reconstruction.* Process Discovery approaches currently consider individual perspectives of the process (control-flow, data-flow and role model) in isolation [53]. For a security analysis, though, the interdependencies between these perspectives are particularly relevant. One pressing research topic focused on the simultaneous detection of control-flow and data-flows, so that more expressive approaches to the detection leaks can be built.
- *Simulation.* One challenge when it comes to designing mechanisms and testing process mining techniques is the presence of representative event log files which controllably contain violations of selected properties [54]. That is because, with such event logs, developers and scientists can assert the "kill-rate" of their techniques, i.e. evaluate whether the devised approaches can in fact, e.g., detect the violations of properties, reconstruct process outliers or correct process models.
- *Resilient BPM.* Resilience is defined as a composition of two quite contradictory properties of systems: on one hand robustness and, on the other, adaptability in case of disruptions and incidents. Current studies investigate the transfer and implementation of these properties to processes in the so-called "Process-Aware Information Systems", which includes AristaFlow, SAP R / 3 and IBM Websphere. In particular, given a process architecture, a present challenge is to circumscribe resilience properties and design continuous, non-intrusive monitoring techniques to measure the resilience state of a process architecture [55].

8   Summary

The systematic consideration of security in BPM is a young research field that is still in development. It counts on a very active, fast-growing research community. In recent years numerous conferences were held on the subject. Moreover, the topic has been receiving increasingly attention in established conferences, a development which can be substantiated by the increasing number of publications lying at the intersection of computer security, process management and formal reasoning.

Despite the relevance of the topic, the integration and transfer of the results achieved so far into the field of "Enterprise Security" turns out sobered. Business process management systems do not offer native support for security analysis, and if they do so, the types of analysis are rudimentary. The lack of guarantees poses a great hurdle for the widespread adoption of business process. Therefore, the trend to be observed in the next years will encompass the incremental integration of existing scientific approaches into enterprise systems' software. This is certainly not only the case at the level of business process specification, but also at certifying the secure composition of services and their realization on a particular type of infrastructure.

References

[1] Wolf, C., Harmon, P.: The state of business process management. BPTrends Report (2010), http://www.bptrends.com/

[2] Website,
http://finance.fortune.cnn.com/2011/09/27/
the-fine-line-between-bad-luck-and-rogue-trades/

[3] Epstein, J.: Security Lessons Learned from Société Générale. IEEE Security & Privacy 6(3), 80–82 (2008)

[4] Simmhan, Y., Barga, R.S.: Analysis of approaches for supporting the Open Provenance Model: A case study of the Trident workflow workbench. Future Generation Comp. Syst. 27(6), 790–796 (2011)

[5] Website, http://www.google.com/patents/US6009410

[6] Website, http://www.google.com/patents/WO2012166878A2?cl=en

[7] Saat, J., Franke, U., Lagerström, R., Ekstedt, M.: Enterprise Architecture Meta Models for IT/Business Alignment Situations. In: EDOC 2010, pp. 14–23. IEEE (2010)

[8] Sandhu, R.S., Samarati, P.: Authetication, Access Control, and Audit. ACM Comput. Surv. 28(1), 241–243 (1996)

[9] Sandhu, R.S., Park, J.: Usage Control: A Vision for Next Generation Access Control. In: Gorodetsky, V., Popyack, L.J., Skormin, V.A. (eds.) MMM-ACNS 2003. LNCS, vol. 2776, pp. 17–31. Springer, Heidelberg (2003)

[10] Accorsi, R., Lewis, L., Sato, Y.: Automated Certification for Compliant Cloud-based Business Processes. Business & Information Systems Engineering 3(3), 145–154 (2011)

[11] Ramezani, E., Fahland, D., van der Aalst, W.M.P.: Where Did I Misbehave? Diagnostic Information in Compliance Checking. In: Barros, A., Gal, A., Kindler, E. (eds.) BPM 2012. LNCS, vol. 7481, pp. 262–278. Springer, Heidelberg (2012)

[12] Brewer, D.F.C., Nash, M.J.: The Chinese Wall Security Policy. In: IEEE Symposium on Security and Privacy, pp. 206–214. IEEE (1989)

[13] Botha, R.A., Eloff, J.H.P.: Separation of duties for access control enforcement in workflow environments. IBM Systems Journal 40(3), 666–682 (2001)

[14] Accorsi, R., Wonnemann, C.: Strong non-leak guarantees for workflow models. In: ACM Symp. Applied Computing, pp. 308–314 (2011)

[15] Roscoe, A.W.: Intensional specifications of security protocols. In: Computer Security Foundations Workshop, pp. 28–38. IEEE (1996)

[16] Weske, M.: Business Process Management - Concepts, Languages, Architectures. Springer (2012)

[17] Basin, D., Burri, S., Karjoth, G.: Optimal workflow-aware authorizations. In: ACM Symp. Access Control Models and Technologies, pp. 93–102 (2012)

[18] Wang, Q., Li, N.: Satisfiability and Resiliency in Workflow Authorization Systems. ACM Trans. Inf. Syst. Secur. 13(4), 40 (2010)

[19] Lowis, L., Accorsi, R.: Vulnerability Analysis in SOA-Based Business Processes. IEEE T. Services Computing 4(3), 230–242 (2011)

[20] Lowis, L., Accorsi, R.: On a Classification Approach for SOA Vulnerabilities. In: IEEE Computer Software and Applications Conf., pp. 439–444 (2009)

[21] Lohmann, N., Verbeek, E., Dijkman, R.M.: Petri Net Transformations for Business Processes - A Survey. T. Petri Nets and Other Models of Concurrency 2, 46–63 (2009)

[22] Lehmann, A., Lohmann, N.: Modeling Wizard for Confidential Business Processes. In: La Rosa, M., Soffer, P. (eds.) BPM Workshops 2012. LNBIP, vol. 132, pp. 675–688. Springer, Heidelberg (2013)

[23] Accorsi, R., Wonnemann, C., Dochow, S.: SWAT: A Security Workflow Analysis Toolkit for Reliably Secure Process-aware Information Systems. In: Conference on Availability, Reliability and Security, pp. 692–697 (2011)

[24] Accorsi, R., Höhn, S.: Towards a Framework for Process Rewriting. In: IFIP Symposium on Data-Driven Process Discovery and Analysis (to appear, 2013)

[25] Fdhila, W., Rinderle-Ma, S., Reichert, M.: Change propagation in collaborative processes scenarios. In: CollaborateCom 2012, pp. 452–461. IEEE (2012)

[26] Accorsi, R., Sato, Y., Kai, S.: Compliance monitor for early warning risk determination. Wirtschaftsinformatik 50(5), 375–382 (2008)

[27] Ni, Q., Bertino, E., Lobo, J.: Risk-based access control systems built on fuzzy inferences. In: ACM ASIACCS, pp. 250–260. ACM (2010)

[28] Brucker, A.D., Petritsch, H.: Extending access control models with break-glass. In: ACM Symp. Access Control Models and Technologies, pp. 197–206. ACM (2009)

[29] Accorsi, R., Ullrich, M., Van der Aalst, W.M.P.: Process Mining. Informatik Spektrum 35(5), 354–359 (2012)

[30] Van der Aalst, W.M.P.: Process Mining - Discovery, Conformance and Enhancement of Business Processes. Springer (2011)

[31] Accorsi, R., Stocker, T., Müller, G.: On the exploitation of process mining for security audits: the process discovery case. In: ACM Symp. Applied Computing, pp. 1462–1468 (2013)

[32] Accorsi, R., Stocker, T.: Discovering Workflow Changes with Time-Based Trace Clustering. In: Aberer, K., Damiani, E., Dillon, T. (eds.) SIMPDA 2011. LNBIP, vol. 116, pp. 154–168. Springer, Heidelberg (2012)

[33] Accorsi, R., Wonnemann, C.: Auditing Workflow Executions against Dataflow Policies. In: Abramowicz, W., Tolksdorf, R. (eds.) BIS 2010. LNBIP, vol. 47, pp. 207–217. Springer, Heidelberg (2010)

[34] Accorsi, R., Wonnemann, C.: Detective Information Flow Analysis for Business Processes. In: Business Process and Services Computing, pp. 223–224. GI (2009)

[35] Accorsi, R., Stocker, T.: On the exploitation of process mining for security audits: the conformance checking case. In: ACM Symp. Applied Computing, pp. 1709–1716. ACM (2012)

[36] Accorsi, R.: Automated Privacy Audits to Complement the Notion of Control for Identity Management. In: Conference on Identity Management, pp. 39–48 (2007)

[37] Accorsi, R., Stocker, T.: Automated Privacy Audits Based on Pruning of Log Data. In: Enterprise Distributed Object Computing Conference, pp. 175–182 (2008)

[38] DoD, Trusted computer security evaluation criteria (1983), Website: http://csrc.nist.gov/publications/histroy/dod85.pdf

[39] ISO/IEC, ISO/IEC Information Security Management System 27001 (2005), Website: http://www.27000.org/iso-27001.htm

[40] Gallegos, F., Senft, S.: Information Technology Control and Audit. Auerbach Publications (2004)

[41] Ristenpart, T., Tromer, E., Shacham, H., Savage, S.: Hey, you, get off of my cloud: Exploring information leakage in third-party compute clouds. In: ACM Conference on Computer and Communications Security, pp. 199–212. ACM (2009)

[42] Pearce, M., Zeadally, S., Hunt, R.: Virtualization: Issues, security threats, and solutions. ACM Comput. Surv. 45(2), 17:1–17:39 (2013)

[43] Chen, S., Wang, R., Wang, X., Zhang, K.: Side-channel leaks in web applications: A reality today, a challenge tomorrow. In: IEEE Symposium on Security and Privacy, pp. 191–206. IEEE (2010)

[44] Subashini, S., Kavitha, V.: A survey on security issues in service delivery models of cloud computing. J. Network and Computer Applications 34(1), 1–11 (2011)

[45] Shabtai, A., Elovici, Y., Rokach, L.: A survey of data leakage detection and prevention solutions. Springer (2012)

[46] Busi, N., Gorrieri, R.: Structural non-interference in elementary and trace nets. Mathematical Structures in Computer Science 19(6), 1065–1090 (2009)

[47] Accorsi, R., Lehmann, A.: Automatic Information Flow Analysis of Business Process Models. In: Barros, A., Gal, A., Kindler, E. (eds.) BPM 2012. LNCS, vol. 7481, pp. 172–187. Springer, Heidelberg (2012)

[48] Accorsi, R., Wonnemann, C.: Forensic Leak Detection for Business Process Models. In: Peterson, G., Shenoi, S. (eds.) Advances in Digital Forensics VII. IFIP AICT, vol. 361, pp. 101–103. Springer, Heidelberg (2011)

[49] Accorsi, R., Wonnemann, C.: Static Information Flow Analysis of Workflow Models. ISSS/BPSC 2010: 194-205 (2010)

[50] Accorsi, R., Wonnemann, C.: InDico: Information Flow Analysis of Business Processes for Confidentiality Requirements. In: ERCIM Workshop on Security and Trust Management, pp. 194–209 (2010)

[51] Houy, C., Fettke, P., Loos, P., Van der Aalst, W.M.P., Krogstie, J.: Business Process Management in the Large. Business & Information Systems Engineering 3(6), 385–388 (2011)

[52] Khoury, R., Tawbi, N.: Corrective Enforcement: A New Paradigm of Security Policy Enforcement by Monitors. ACM Trans. Inf. Syst. Secur. 15(2), 10 (2012)

[53] Accorsi, R.: Business Process as a Service: Chances for Remote Auditing. In: IEEE International Computer Software and Applications Conference, pp. 398–403 (2011)

[54] Stocker, T., Accorsi, R.: Security-aware Synthesis of Process Event logs. In: Workshop on Enterprise Modelling and Information Systems Architectures (to appear, 2013)

[55] Koslowski, T.G., Zimmermann, C.: A Detective Approach to Process-centered Information Infrastructure Resilience. In: ERCIM Workshop on Security and Trust Management (to appear, 2013)

[56] Accorsi, R.: Sicherheit im Prozessmanagement. Zeitschrift für Datenrecht und Informationssicherheit (to appear)

Mental Models – General Introduction and Review of Their Application to Human-Centred Security

Melanie Volkamer[1] and Karen Renaud[2]

[1] TU Darmstadt Darmstadt / CASED
Hochschulstraße 10, 64289 Darmstadt, Germany
melanie.volkamer@cased.de
http://www.secuso.cased.de
[2] University of Glasgow
18 Lilybank Gardens, Glasgow, United Kingdom
karen.renaud@glasgow.ac.uk
http://www.gla.ac.uk/schools/computing/

Abstract. The human-centred security research area came into being about fifteen years ago, as more and more people started owning their own computers, and it became clear that there was a need for more focus on the non-specialist computer user. The primary attitude fifteen years ago, in terms of how these new users were concerned, was one of exasperation and paternalism. The term "stupid user" was often heard, often muttered *sotto voce* by an IT specialist dealing with the aftermath of a security incident. A great deal of research has been published in this area, and after pursuing some unfruitful avenues a number of eminent researchers have started to focus on the end-user's perceptions and understandings. This has come from a realisation that end users are not the opponents, but rather allies in the battle against those carrying out nefarious activities. The most promising research direction currently appears to be to focus on mental models, a concept borrowed from the respected and long-standing field of Psychology and, in particular, cognitive science. The hope is that if we understand the end-user and his/her comprehension of security better, we will be able to design security solutions and interactions more effectively. In this paper we review the research undertaken in this area so far, highlight the limitations thereof, and suggest directions for future research.

1   Introduction

Over the last few years a number of different security mechanisms have been developed in order to protect users from different kinds of attacks eg. the SSL/TLS protocol. Some of these mechanisms have been formally proven to be secure and evaluated based on international security standards such as the Common Criteria or ISO 27001. However, a number of user studies [82,95], as well as the prevalence of attacks [61] successfully targeting the human end-user, demonstrate that

M. Fischlin and S. Katzenbeisser (Eds.): Buchmann Festschrift, LNCS 8260, pp. 255–280, 2013.
© Springer-Verlag Berlin Heidelberg 2013

many of these security mechanisms falter and fail as soon as the user is involved in the process. One big problem is the large number of security warnings users are confronted with. Users are habituated into ignoring these since they do not understand and thus perceive any risk [94].

This can either be attributed to the 'stupidity' of the user (not understanding what is secure and what is not) or the 'obtuseness' of the developers (not designing systems properly and not giving due consideration to the non-security related nature of the end-user's primary goal or task). As a solution, one could try to eliminate the end-user from the security mechanism's operation altogether. While this might work in a few cases (eg. virus scanners and firewalls), there are many contexts in which this is not advisable, for many reasons, as discussed in the paper 'Security Automation Considered Harmful?' [40]. For instance, in some cases eliminating the user could restrict functionality to such an extent that users will reject the mechanism (e.g. preventing users from visiting https web servers which do not possess extended validation certificates). There are also applications where user input is mandated by law. For example, the user has to be able to verify the correct processing of their vote when voting electronically.

Instead of eliminating the user altogether, one could try to force users to behave securely by defining corresponding policies or (long) lists of security and privacy rules and guidelines and try to compel the user to comply. In some cases users are punished for non-compliance, or at least threatened with sanctions [53]. This does not really work very well, at least in the way policies are designed nowadays [88]. For instance, policies often forbid actions that human nature almost compels eg. password sharing between colleagues in order to perform the primary goal/task effectively.

Wash and Rader [106] argue against all these options. It is far better, argue Wash and Rader and other researchers [106,19,5,1], for developers to align necessary security-specific user interactions, educational endeavours and risk communication efforts with users' mental models and capabilities. Users' actions and decisions are directed by their mental models so it is crucial for designers and developers to know and understand *their* models when developing and designing security mechanisms. The design should be aligned with the end-users' mental models but not, as is nowadays often the case, purely with the developers' and designers' mental models and based on their assumptions about end-users' mental models and capabilities. Risk communication, particularly, can only be effective if it does not only depend on the nature of the risk but also on the alignment between the conceptual model embedded in the risk communication and the users' mental model(s) related to the context and reality of the risk. Risk communication is an important aspect of Human-Centered Security as it is implicit in each warning.

Mental models also influence trust and acceptance of technology: An incorrect mental model can make users mistrust insecure technologies [21]. This constitutes yet another reason for paying attention to end users' mental models.

In this paper we focus on aligning interactions and risk communication with the users' mental models. The first goal of this paper is to provide security researchers with an introduction to mental models and lessons learned from other disciplines in which they have been successfully applied for many years. We also address mechanisms by which mental models are modified (either implicitly or explicitly). This is important because mental models are not static but rather change over time and differ between users or groups. The second objective is to provide an overview of existing security-related research on mental models in human-centred security. The mental models identified in the literature are summarized to inform the development of future security mechanisms and, in particular, security-related user *interactions* or the improvement of existing interactions including *risk communication*. Finally, we identify some limitations in the research, and the findings, and speculate about how these can be mitigated in future human-centred security mental model research.

2   Mental Models – Introduction and Overview

The term 'Mental Model' was first used by Craik [28] in 1943 in his book titled "The nature of explanation". Craik explained that a mental model was *a physical working model which works in the same way as the process it parallels*" (p. 4.2). Yet others described the concept of mental model before that, even though they used different terminology. Johnson-Laird [58] wrote a history of mental models and points out that Ludwig Boltzmann, writing in 1890, spoke about "constructing an image of the external world that exists only internally". The literature on mental models uses a range of terms and nomenclature, which makes the field somewhat challenging to investigate, with researchers using the following range of terms to refer to internal constructs of the world: analogy [29], metaphors [69,56], perception [43,97], theme [51], theory [98], internal concept [25] and reasoning [67]. For the purposes of this discussion we will use the term "mental model".

In 1983, two books with the words "Mental Model" in the title were published [57,48]. Gentner and Stevens [48] argue that a study of mental models is beneficial since it helps us to understand human knowledge about the world. The first chapter in their book is written by Don Norman, who explains that mental models provide both a predictive and explanatory power for understanding our interaction with our environment, with other people and with technological artefacts. He also emphasises that the models are always evolving, are not necessarily accurate but that they do have to be functional.

Rouse and Morris [91] provide a definition of mental models that seems generic enough to encapsulate the meaning of all the different terminologies: "*mechanisms whereby humans generate descriptions of system purpose and form, explanations of system functioning and systems states, and predictions of future system states*" (p.49).

In the same paper, Rouse and Morris warn that any research into mental models, while it might deliver an understanding of the *what* of the mental model, cannot deliver an understanding of *how* the human uses it. For example, even if a person possesses a comprehensive mental model, and knows how to act in a particular situation, it is particularly difficult to control for the impact of emotions [100]. Emotions are an imponderable which can make competent people (with perfect mental models) perform poorly. Consequently, even having identified mental models does not necessarily mean that they can be used to predict how people will deploy them in a particular situation.

The subsequent discussion will provide further insights into the intricacies of mental models while focusing on those aspects that are important for security researchers and developers to take cognisance of. This includes considerations of how mental models can be measured, developed and fostered.

2.1   Important Properties and Aspects of Mental Models

In 1985, Rouse and Morris [91] suggest that the two important aspects are (1) how the model is manipulated (ranging from implicit to explicit) and (2) how much discretion the human has in developing the model (ranging from none to full). They also identify a number of issues which beset mental model research. These are phrased as pertinent questions below, and a review of related literature provides a brief discussion of each:

To What Extent Is It Possible to Capture the Details of Mental Models? Norman [80] warn that mental models are incomplete and unstable, and sometimes confused with each other. Also Jones *et al.* [59] say that mental models are "inconsistent representations" which "adapt to continually changing circumstances and to evolve over time". Given that this is true, how can one know that a study has delineated one particular mental model, in its entirety, without corruption from others? Researchers generally attempt to gain access to mental models by asking people to verbalise their understanding of a particular concept. Intuitively this seems a reasonable approach. Bostrom [15] *et al.* used interviews to determine whether the lay public have an understanding of the dangers of radon. They justify their use of interviews but acknowledge the potential for cognitive entrapment to occur. Unfortunately, Payne [83] found that when people were asked to verbalise their mental models of a particular construct, in their case bank machines, the *ad hoc* nature of the description was striking. Payne noted that people deployed mixed explanation types, where they mixed pre-existing conceptions with new insights and generally expressed something that was heterogeneous and it often seemed to evolve as they spoke. So eliciting verbalisations might well not be the best way of capturing details of mental models accurately. Also Richardson *et al.* [89] showed that attempts to elicit mental models ran the risk of distorting them.

Rowe and Cooke [92] carried out a comparison of four elicitation mechanisms: think-aloud, laddering[1], relatedness ratings and diagramming in order to determine the link between the quality of the resulting mental model representation

---

[1] The interviewer gives the participant a series of problem statements and asks them to identify four relevant aspects of the problem and then asks in-depth questions in order to explore linkages between answers the participant gives.

and the troubleshooting ability of the participant. They found think-aloud to be particularly weak in this respect. They argue that laddering and rating appear to measure different aspects of the mental model, seeming to confirm the argument made by Staggers and Norcio [99] about the multi-faceted and dynamic nature of mental models including hierarchies of models or related models that are subtypes or specialisms of others. Rowe and Cooke also reported that the diagram quality was predictive of troubleshooting performance, suggesting that a sound mental model leads to a high quality diagram. Langan-Fox *et al.* [73] review a number of different techniques for eliciting mental models, and offer a methodology for choosing the optimal elicitation mechanism, depending on the research problem being investigated as well as the practical and theoretical considerations of the study.

Important to know: Accessing a mental model is challenging because one runs the risk of interfering with what one is attempting to measure. Even so, it is important to elicit mental models and, in the process, try to alter them as little as possible. Drawing diagrams where applicable seems to be the most promising approach.

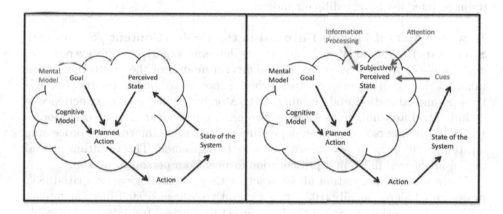

Fig. 1. Adapted from Richardson *et al.*'s [89] model

Do Cues Help to Draw out the Details of Mental Models? Since it is so challenging to elicit mental models it is worth considering the use of cues, both in elicitation and in assisting a user in correctly diagnosing a particular situation. Richardson *et al.* [89] apply a systems model approach[2] and claim the following primary mental model components: *intentions, perceptions, system structures* and *plans*. They see the mental model through the lens of a cybernetic loop, with the person interacting with the external system state and perceiving changes in order to close the feedback loop (Figure 1 - left). They then extend their model, as shown on the right of Figure 1, to include the presence of cues, as

---

[2] A conceptual model of a system which attempts to depict causatives and influences leading to outcomes.

mediating between the actual state of the system, and the perceived state of the system, showing the importance of cues that the human can interpret in order to understand the state of the system. They explain that cues are the signals that humans pay attention to, and argue that it is critical to understand how people use strategies and tactics are employed, and how such cues are interpreted. Dörner [35] explains that cues help to complete the feedback loop, thus helping people to build a more accurate mental model.

Even though cues are important it seems difficult to guarantee that cues will be efficacious. For example, visual cues are often missed, due to attentional limitations [96]. Even if someone notices a cue, sometimes they don't respond to it because they don't have the prior knowledge to understand it [87]. Even if they do have the necessary understanding, they may not react correctly due to emotional issues [39]. Finally, even if they notice the cue, know how to react, and their emotional state is such that they are able to react correctly, they may still misinterpret the cue if they do not trust its source [54].

Important to know: Cueing is simple and understandable in well-known contexts such as the theatre. When the cueing context is a human operating within a complex system, provision of a cue that will be noticed, interpreted and acted upon as intended is very difficult indeed.

How Are Mental Models Encoded in the Brain (Content /Structure/ Specificity)? Mental Models develop as humans experience life events [93], and are also impacted by the collective mental models of the group the person inhabits [59]. Such experiences can include exposure to media reports [70], culture [8] and education and training [104]. Moreover since human experience is as individual and unique as humans themselves, it is clear that mental models will differ from person to person depending on individual information processing strategies [90] even in people with similar backgrounds. The resulting mental model might well differ in representation format from person to person.

There is some speculation about whether they are analogical or spatial [83], or organized hierarchically [102] or as a cognitive collage [103]. Since the human brain can be considered one of the final, as yet uncharted, frontiers, it is possible that any attempt to pin down the precise mental model storage mechanism will be futile.

There is likely to be a difference in specificity too, especially when one compares novices and experts. Staggers and Norcio [99] provide some references to show that experts' mental models are more abstract and richer than those of novices [112]. They refer to the expert model as a *macro* model, that represents a higher level of functioning than novices' models. In addition to being more abstract, Thatcher and Greyling [102] also report that experts have more complete and detailed mental models. It is clear then that novices have an understanding of a particular problem situation, but do not possess sufficiently abstract mental models which enable them to extract core principles from such an understanding. This means they cannot necessarily apply their models to solve problems which are different in specifics but generically similar.

Important to know: Mental models are based on past general and educational experiences. Moreover, it seems that the process which makes someone an expert also results in an abstracting process, so that they develop generic mental models which can be applied to a wide variety of contexts.

How Should the Development of Correct Mental Models Be Encouraged?

If one wishes to impart a mental model one always has to contend with the person's prior knowledge and experience. In terms of topic-specific prior knowledge there are three possibilities:

The topic is completely new to the learner. How best ought we to develop mental models in the minds of learners? It is important to ascertain people's understandings and assumptions, and to ensure that one matches educational delivery to these [110]. If the new material is not 'pitched correctly the lecturer runs the risk of either boring or confounding the listener. If this happens the intended mental model will not develop.

The learner has a correct but limited mental model of the topic. One should try to elicit some sense of what the person already knows [26] and then try to provide links to this knowledge so as to ease extension of the pre-existing mental model [101].

The learner has an incorrect mental model of the topic. Chapman and Ferfolja [23] outline the consequences of poor learning, and consequent imperfect mental models. They term the outcome as disruptive. It is especially damaging if further learning needs to build on a mental model that is faulty since the misunderstanding then impacts the entire learning process. Bain [7] explain that we need to prove that mental models are incorrect nefore they will realise that they need to change the way they see something. Having understood this, people can be helped to unlearn concepts, but it seems that this needs to be facilitated explicitly: one cannot merely present people who hold an incorrect understanding with the correct information [22]. This is time-consuming and effortful and so this situation should be avoided if possible.

One of the most common examples of incorrect models is demonstrated by shared, incorrect yet commonly held *folk models* [30]. Some examples include beliefs about diabetes by Bangladeshi British (incorrect eg. believing living in Britain caused diabetes) [49], shared models of infection in an English suburban community (incorrect eg. feed a cold, starve a fever) [55], understanding about heat loss from houses (incorrect eg. closing off rooms preserves energy) [65]. The danger of incorrect folk models is that they resist scientific checks which would highlight their flaws and are too easily generalised to situations they were never meant to be applied to [31]. A primary example of the resistance of folk models to change is the MMR controversy in the UK, where many parents believed that the vaccine would lead to autism, despite scientific evidence to the contrary [18]. It is difficult to challenge folk models, since they are so resistant to change.

People's mental models are formed by life experiences, and informed by educational activities. As security researchers and practitioners, we need to understand how to design our educational endeavours so as to maximise the development of correct mental models. Morey and Frangioso[77] present six principles of effective learning: (1) acknowledging the person's existing mental models (2) fostering an understanding of the complexity of human-machine systems (3) challenge unthinking assumptions (4) listen to learners in order to understand where they are coming from (5) observe, assess, design and then implement (6) let learners teach others. These seem to deal with all of the possibilities mentioned above.

Important to know: In First Aid, the golden rule is "Do No Harm". In teaching, that should be our mantra too [17]. An effective teacher will first determine pre-existing assumptions, then challenge those that are faulty, then pitch the new material in order to match pre-existing knowledge, and finally facilitate peer to peer discussions to ensure that the new material is emphasised and remembered [23].

In Conclusion. Mental models are obviously complicated multi-faceted mental entities, they are dynamic, based on individual experiences, are not easily described, combine many different modalities, are different between different groups and in particular between lay persons and experts, and are very difficult for researchers to capture. Finally, any attempt to capture the details of mental models are likely to change such models, and researchers have to be sure that they do not affect mental models during the process of measuring them. Moreover, when we become aware of an incorrect mental model we need to confront the holder(s) with the inconsistencies and imperfections of this model, so that we can guide them towards a correct model.

2.2   Example Applications of Mental Models

Mental model research has been deployed successfully in a variety of contexts.

Application Areas. For example, Littman *et al.* [75] instructed programmers to develop two different kinds of mental models of software: the first was a systematic strategy and the second the as-needed strategy. Using the first, the programme attempts to trace the data flow all the way through the code where the latter focuses attention only on local code without consider antecedents. They found that the former strategy was by far the more effective in leading to better software maintenance. Pfeffer [84] explains how best to change the mental models of senior managers in organisations. Converse *et al.* [27] describe how best to foster shared mental models so that teams operate more effectively. Finally Nemire [79] relates how an incorrect mental model of how roller coasters operate led to a fatality, and argue for the need for specific educational efforts in order to address incorrect mental models.

Risk Communication vs. HCI. Fischoff *et al.* [44] write about the importance of aligning risk communication with people's perceptions, understandings and assumptions. They also explain that emotions and social processes play a role in how people make use of mental models in making risk decisions.

They acknowledge these confounding factors but argue for the need to get the cognitive aspects right, which *can* be controlled. An example where emotions and outrage overcame the best efforts of risk communicators is the MMR vaccine controversy [18].

In the human-computer interaction (HCI) field mental models have also been used to support design. Three examples serve to indicate the range of work in this area. Bates [11] designed an information searching interface that modeled a 'berry picking' metaphor which is better aligned to human information searching behaviour than traditional information retrieval interfaces. Khaslavsky [66] attempted to design an interface which was sensitive to cultural mental models but does not report on an evaluation of the interface. Finally, Donker *et al.* [34] determined that blind users had different mental models of an interface from sighted users, and designed an interface specifically for them.

3   Mental Models in Security Research

The only way to develop security mechanisms that effectively protect users against attackers is to align the design of security-related interactions, educational efforts and risk communication with users' mental models and capabilities. Users only protect themselves if their mental model includes some concept vulnerability to attacks. It might be necessary to attempt to adapt, extend or modify the target users' mental models in order maximise the possibility that users will act, and act correctly, to protect themselves from potential attacks.

The common knowledge about mental models presented in the previous section offers a suitable launching pad for a discussion of their use in security-related research. In particular, security researchers and developers of security mechanisms and security critical applications need to understand how end users mentally model their systems, and the security thereof (with or without taking additional security-related actions), and how they understand the effects of their actions on these systems.

As a consequence, security researchers have been conducting mental model research from different perspectives and using a variety of different methods to elicit their details. Many different methods have been applied, including: different types of interviews, user studies, and card sorting. Some researchers carry out the research in order to design security-related interactions more effectively, some to communicate risk, some to tailor educational efforts, and some to evaluate whether specific user interactions and interfaces lead to appropriate mental models. Such models are crucial in leading users to make informed security decisions.

Here we provide an overview of the most important research that has been conducted in the area of human-centred security, presenting findings and limitations. We also present and discuss research on mental model research which is intended to improve security interfaces, such as firewalls, anonymous credentials and Single Sign-On. We commence this discussion by presenting concrete security-related models identified in literature and continue by summarizing the general findings.

3.1   Concrete Mental Models of Threats and Security Mechanisms

Jean Camp [19] identified mental models of security and privacy based on a literature review, i.e. mental models that are currently indirectly or directly being used to communicate security and privacy issues. These are:

- *Physical Security Model* (e.g. because of the lock metaphors)
- *Medical Model* (e.g. because of the phrase 'infected by a virus')
- *Criminal Model* (e.g. being arrested if you hack into a system)
- *Warfare Model* (e.g. because of intrusion detection and firewall tools)
- *Market Model* (e.g. because of people losing money)

Camp discusses these in detail, outlining their positive and negative impacts on user behaviour, with respect to security decisions. She explains that each of the models only covers parts of the security and privacy problems and thus only helps to protect against a subset of possible attacks. For instance, the *Physical Security Model* helps to protect the computer hardware but does not preserve privacy in terms of how much information one should provide using different services. Camp *et al.* [76] validated these five models using card sorting and interviews while distinguishing between experts and lay-persons. They found out that many of the studied security risks were either assigned to the Physical Security or the Criminal Models. However, they also conclude, that none of the five mental models "fit the understanding of the impression of the related risk" [76]. The Physical Security Model was also indirectly supported by results of Furman *et al.* , [46] as they show that end-users have most trust in banks (due to their physical protection) and and shops that they know in the physical world.

   Rick Wash published a interview-based study [105] in which he indentifies folk models of security threats that are used by north American home computer users to decide what security software to use and which security advice to follow. He derived eight distinct folk models and explained how people used these to justify ignoring security advice. The mental model they ascribed to influenced whether or not they made backups, whether they installed anti-virus software and whether they were open to advice about their security-related actions or not. For example, some believed that one could only catch a virus if you visited a bad part of the Internet and thus did not think they were vulnerable. Some believed that they themselves were too insignificant to be worthy of a hacker's attentions, and thought they were therefore not at risk. Some also believed that hackers were only after large databases or companies and that they, as individuals, would not be attacked. Wash points out that efforts to explain how virus-protection software works to protect computers might encourage end-users to use it. Some advice was routinely ignored but not by all mental models. In summary, he identified the following models (four for risks and four for attacker types):

- *Virus – generally bad* but only high level understanding.
- *Virus causes mischief* i.e. annoying problems with computer/data.
- *Virus intentionally downloaded in buggy software* computer will misbehave in time: e.g. crash or does not boot any more.

- *Virus supports crime* e.g. by stealing personal/financial information.
- *Hackers perceived as geeks* who want to impress friends.
- *Hackers are criminals* who target big fish (rich and important people).
- *Hackers support crime*, looking for large databases of info.
- *Hackers are burglars* who steal personal/financial information.

Kauer et al. [62] replicated this study in Germany. They found, in total, eleven folk models, with the eight from Rick Wash being mentioned with small differences. In addition, the authors identified: *Viruses are Governmental software, Hackers are Governmental Officials* referring to the Bundestrojaner and Staatstrojaner (engl. Federal Trojan horse) as well as *Hackers are Stakeholders with individual and opportunistic purposes* such as Anonymous[3] and the Chaos Computer Club[4]. With respect to the first two extra models, all participants said that the government only acted if there were suspicion of some crime. With respect to the latter model, participants did not fear being targeted, since these stakeholders are perceived to be more interested in media-effective targets. The government was also mentioned throughout the interviews conducted by Furman et al. [46]. Considering the question about whom or what users need to protect themselves from, their participants also mentioned colleagues and scammers (in addition to hackers, bad guys, and criminals).

Dourish et al. [36] also probed mental models of security using interviews. They identified four classes of threats:

- *Hackers* cause mischief and harm; vandalism; the least commonly identified threat.
- *Stalkers* get information online but can also continue their activities offline.
- *Spammers* advertise by means of unsolicited messaging which is perceived as a type of denial-of-service attack.
- *Marketers* invade individual privacy by surreptitiously collecting information about activities, purchasing patterns, etc.

Weirich and Sasse [108] also conducted interviews to explore perceptions of who they thought tried to get into other people's accounts and whom they targeted. The identified models for the first question were:

- *Kids*: only want to prove that they can do it, but do not cause serious harm;
- *Vandals*: plain mad, who cause serious harm;
- *Criminals*: only attack online-banking;
- *Vengeful people*: disagree with individual or organization;
- *Others*: industrial spies, terrorists, and jokers.

Friedmann et al. [45] identified mental models related to secure connections. The authors conducted 72 interviews (including drawings) and, in particular, asked how the participants to decided whether a connection was secure or not. They also asked what a secure connection meant to them. They identified five

---

[3] http://du-bist-anonymous.de/
[4] https://www.ccc.de/en/

strategies with respect to the first question: (1) HTTPS; (2) lock/key icon; (3) point in transaction (main pages are usually not secured); (4) type of information requested, and (5) type of webpage. The last two are particularly interesting as this means that people assume that when sensible data is requested (like passwords or credit card information) the server ensures that the connection is secure. Similarly, they assume that organisations such as banks take due care. Regarding the second question (meaning of secure connection), the authors identified three mental models:

- *Transit*: protecting the confidentiality of information while it moves between entities.
- *Encryption*: specific mechanism for encoding/decoding messages.
- *Remote Side*: protecting data once it arrives at recipient. From the drawings, one can conclude that people with this mental model assume that the connection is either always secured, secured by default secured or unsecured since it there is no way for anyone to interfere with the transmitted message.

Benenson *et al.* [12] studied smartphone users with respect to privacy concerns and awareness. They identified two different mental models: the *Android* and the *iPhone* mental model. They confirmed that Android users did seem to be more privacy-concerned and -aware, as they significantly more often mentioned data privacy as an important factor for choosing an application. iPhone users thought that if applications needed the required access they would not necessarily ask. Moreover, iOS users were mostly unaware of application data usage. These results are explained by the different ways Apple and Google chose to inform users of applications' data usage. Furthermore, the researchers found that technical features were an important factor in informing smartphone choice. People are more likely to own an Android phone if they cared about technical features. People who were interested in technology were more likely to own an iPhone.

3.2   Mental Models to Improve Interactions in Concrete Applications (HCI)

HCI Sec researchers also studied mental models in the context of Interface design. We report here about five different types of applications.

Anonymous Credentials. There is a great deal of technical research on anonymous credential and privacy-friendly identity management. Wästlund *et al.* [107] evaluated three different interfaces and three different corresponding metaphors in a user study to explore the ideas behind anonymous credentials namely card-based, attribute-based and adapted card-based metaphors. The objective was to test whether the built mental models of anonymous credentials are correct for any of these approaches. Therefore, they asked the participants to explain which information flows exist and who could violate anonymity. While the adapted card-based technique performed best, further improvements are necessary.

Firewalls. Raja *et al.* [85] base their research on firewall warnings on the physical mental model described in [19] after having identified a number of misconceptions with personal firewalls in [86] . They use this model to visualize the functionality of a firewall. Their comparison with the Comodo personal firewall shows that the visualization of a fireproof wall separating parts of a building helps users to develop better mental models of firewalls. While most users preferred the warnings designed by Raja *et al.*, some participants in their study mentioned that they would take these new warnings less seriously than warnings from Comodo.

Web Single Sign-On (SSO). Sun *et al.* [52] investigate, using interviews and drawings, end-users' perceptions of web SSO technology for authentication and found many misconcepts causing mistrust in the whole approach and raising privacy and security concerns. They re-designed the interfaces of a SSO solution and show that many more participants would use and trust this "new" approach.

Password Manager. Chiasson *et al.* [24] studied two different password managers and found that most of the identified problems were caused by inaccurate or incomplete mental models of the software's operation. For instance, most users do not even have a high-level understanding of how password managers function. The authors recommend more visibility (vs. more transparency) to enhance the usability of password managers.

Secure E-Mail Products. Whitten and Tygar [109] conducted a cognitive walkthrough evaluation and a user study on PGP 5.0. From the cognitive walkthrough many problems were identified including those related to mental models and metaphors e.g. the lack of differentiation between private and public keys. The results from the user study showed that most users had difficulties with key management and with understanding the underlying PKI concept, leading to errors such as sending the secret unencrypted or encrypted with the wrong key.

3.3   General Findings

In this subsection, we provide the general findings on mental models in human-centered security independent from concrete mental models. We first report about a couple of general misconcepts and then about differences between different groups of users.

Problems with Diverse Antecedents. Computer users face different types of problems: common computer problems (e.g. buggy program or disk failure) and security / privacy[5] problems. Gross and Rosson [51] conducted interviews to better understand end-user security management based on the hypothesis that

---

[5] Note, although the focus of this subsection is on security some researchers have studied both aspects together and thus those results are included here if they are not only privacy specific.

people have difficulty distinguishing between these two kinds of problems. Their findings in [51] seemed to confirm this. However, in a later paper [50], Gross and Rosson showed, based on an online survey with 368 participants, that end-users did indeed make a distinction between these problems. Furthermore, they showed that end-users are more concerned with security and privacy problems than with general computer failure.

Focus on Confidentiality. Furman et al. [46] interviewed 40 people about their perceptions of security. They found that although the participants mentioned *confidentiality*, few mentioned the other two cornerstones of information security: *integrity* and *availability*. This was confirmed by other researchers. For example, in [45], the authors found out that people only considered confidentiality and encryption in their definitions of secure connections. When it comes to smartphone security, the situation is slightly different. When talking about these devices, users do actually mention availability [78] but more in terms of high mobility devices being more likely to be lost or stolen, thus referring to the availability of the device rather than the data on the device.

Others Being Responsible. Gross and Rosson [51] attempted to understand end-user security management in companies and organizations. Most of their participants attributed responsibility for security to the IT staff or to the organization or both but most emphatically not themselves. Furman *et al.* [46] report a similar result in the home environment: people assign the responsibility to third parties such as the government, software companies, credit card companies, banks, or IT professionals. Also, Dourish *et al.* [36] showed that end-users were in favour of delegating responsibility, namely to technology, knowledgable colleagues, family members, room mates, organizations and institutions such as banks. According to an AOL/NCSA study[6] from 2005, most people felt that the responsibility laid with the government and with big companies (while at at least 15% felt that they themselves were responsible). Hence, it is not surprising that many studies [16,46,36] confirmed that users are rather relaxed when it comes to online banking as they assume the bank will take care (including ignoring warnings for such pages [16]).

In the smartphone context, King [68] shows that end-users trust applications because they assume that an in-depth evaluation has been carried out by Apple and Google before these are provided in their stores. This was confirmed by Kelley *et al.* [64] for Android users. The same observation was made in [81] within the context of electronic voting: people assume that observers will make sure that votes are properly tallied.

An opposing finding is reported by Furnell *et al.* [47], who found that 90% of respondents (strongly) agreed with the phrase 'It is my responsibility to protect my computer from online attacks'. The difference might be caused by the question being concrete about one aspect while the other studies posed more broad-ranging open-ended questions. Moreover, it is possible that the direct question

---

[6] http://www.bc.edu/content/dam/files/
offices/help/pdf/safety_study_2005.pdf

asked by Furnell *et al.* might have been misunderstood by the participants in the sense that it is not clear to them what is implicitly required; maybe they thought it just meant that they should install a virus scanner. This mismatch is a more general finding and is addressed in the next paragraph.

Lack of Awareness/Misunderstandings/Misapplication. Wash and Rader [106] explain that users often believe they *were* doing what was necessary to protect their computers. They argue that users were motivated to take necessary actions but obviously only against those attacks or threats they were aware of. Gross and Rosson [51] also found that their participants believed themselves to be able to sufficiently protect themselves, e.g. they know about Phishing, and know that they should check for urgency-related terms and typos in emails. Unfortunately, their knowledge is often out of date. Similarly, they are aware that software and passwords should be regularly updated and changed. However, regularly, to them, means six monthly. The authors of [6,46,47] drew similar conclusions. For instance, most of the participants in [46] could identify the evaluated trust and security seals/icons but did not understand what they meant. Different researchers [64,41,74] report that smartphone users, in particular Android users, express doubts about the permissions applications might have, but explain that they do not understand the explanations about permissions provided at installation time.

Researchers could also show that there are situations in which the end-users know very well how to behave but still decide, for arbitrary reasons, not to behave securely: (1) When one considers smartphone authentication, the participants in [78] mentioned that they knew they ought to use an authentication mechanism but stated that the mechanism available on the phone was inappropriate. They felt it was overly stringent if they wanted to access innocuous information such as the weather forecast. Similar findings were reported by [68] with respect to general security and privacy concerns on smartphones. (2) In [108], the authors identified social aspects in organizational environments in the context of password security as one reason: not being a nerd but being a team player; and not being paranoid. Similarly [78] reported that people explained that they did not lock their smartphones because they did not want to type passwords in front of their friends. To them, this made it look as if they wanted to hide something. Some also felt that such an action would send a message that that they did not trust their friends and this might compromise the relationship.

Misconcept of Hackers' Targets. Weirich and Sasse's [108] participants mentioned the following targets: security-conscious organisations, high-profile organizations, people with important information, and anyone who had annoyed an attacker. Wash [105] confirmed their findings with his different folk models as well as others. This suggests that many people consider themselves too insignificant to be attacked.

Dourish *et al.* [37] found that many of the non-specialist participants they interviewed in their study reported feeling that their security efforts were futile. They reported that they were unable to protect themselves from the efforts of faceless and nameless attackers. Also participants in the interviews of [108] argued that hackers could always find a way in despite their efforts. Wash [105] reports that some of his participants felt powerless to protect themselves from the efforts of hackers and therefore did not take any action to protect their computers. Others believed that all they had to do was to make it harder for hackers to get into their accounts than into those of other home computer owners, what Wash refers to as the 'speed bump' theory. These findings were confirmed by [36]: people referred to the 'unknown other who will always be one step ahead'.

Differences between Novices and Experts. Several security researchers [76,4,47] confirmed the results reported by Staggers and Norcio [99] that experts' mental models are more abstract and richer than those of novices (see also Section 2.1). For instance, Asharpour et at. [4] showed that the mental model of security risks strongly correlate with their level of expertise in security. Liu et at. [4] even tried to measure the distance between the mental models of experts and lay-persons. Experts are defined in this paper as those who know all the technical definitions of the security-related words. In a later paper, the authors [4] changed the definition to "One who has at least five years expertise in security as a researcher, student or practitioner" (so called security specialist)[7]. The authors reported differences between the security specialists and 'the others'.

Bravo-Lillo *et al.* [16] report a mental model study on how people (experts and lay persons) decide whether to ignore or follow security warnings. Experts are defined by having completed at least one year's security or privacy course or at least a one year security or privacy project. They report very clear differences between novice and expert users and thereby confirmed previously mentioned research. For example, they found that experts actively looked for vulnerabilities and considered multiple factors when they encountered a potentially risky situation. Novices, on the other hand, performed fewer security checks. Novices can either relate a warning to viruses or they consider that there is actually no problem. Furthermore, novices tended to assess the safety of an action after they performed it whereas experts tended to be more cautious and assessed the safety of a task (considering recent actions, sensitivity of information and consequences) before they embarked on it. They also found that novices were more likely to trust in the ability of large corporations to protect them. Accordingly, they expected online banking to be secure because banks traditionally have good security. They confirmed the findings of [111,45,63] that novices made decisions based on the look and feel of a website. On the other hand, experts agreed that bank websites with warnings were usually not trustworthy. They also identified a misconcept with respect to file storage. Novices believed that storing a file was more dangerous than opening it They thought that opening the file provided them with a safe preview but saving the file on their desktop was akin having a time bomb on their computer: the file posed a danger due to its presence.

---

[7] The second definition is far more appropriate since it represents a higher level of understanding in terms of Bloom's [2] well-known taxonomy of the cognitive domain.

Bartsch and Volkamer [9] study in a qualitative card sorting study how lay and expert users assess risks connected to Web sites. Their results indicate the diversity of mental models, both between the two groups and between individuals, particularly related to their preferences (e.g. concerning privacy or financial consequences). Bartsch and Volkamer conclude that it is not enough to distinguish between experts and lay persons.

Cultural and Demographic Differences.   Most of the research on mental models has been conducted in the USA. Moreover, there are only a few studies considering cross-cultural issues in mental models of security and privacy. For instance, Diesner *et al.* [33] studied mental models of data privacy and security in India – a country without data protection laws at this point in time. They conducted interviews with 29 participants and used Network Text Analysis and map analysis techniques including Auto map [20,32] to evaluate the transcripts and NetDraw [14] to visualize the results. They found that personal information, identity, and knowledge were the central contents while security-related terms are not central in people's minds. Later, Kumaraguru *et al.* [72] conducted an exploratory study in India and in the USA to compare privacy perceptions and concerns in the two countries. They suggest that Indians and people from the US differ with respect to level of concern about privacy, and people from the USA are more privacy-aware when it comes to new technologies. Kauer *et al.* [62] replicated Wash's folk model study [105] in Germany. They identified some differences. In particular, they considered the government and the 'Bundestrojaner' as possible threats as well Anonymous and the Chaos Computer Club. These were not mentioned by participants in Wash's study.

There is very little research into demographic differences in the context of mental models. Dourish *et al.* [36] identified some differences between younger and older participants in terms of who might threaten them on the Internet: young people are more likely to identify big organizations as threats than older ones. Sheng *et al.* [95] studied the susceptibility of different demographic groups with respect to phishing. The results may lead to the conclusion that women are, in general, more susceptible than men and that people between the ages of 18 and 25 will be more susceptible than the older computer users.

Different Environments.   The authors of [51] concentrate on the organizational environment. They do not state this explicitly but it seems as if there is a mental model considering IT Staff as being responsible for security within organisations. Thus, it could be expected that people have different mental models about security at home and at work but this is not specifically addressed in this paper.

Muslukhov *et al.* [78], in common with other researchers, noticed differences between the mental models of *home computer* and *smartphone* security. People assume their smartphone to be a less secure device on which to store sensible data. They attribute this to the mobility of smartphones which made them more likely to be lost or stolen. In general, their main concerns are the smartphone being lost or stolen combined with losing data such as their address book or someone dialing expensive numbers on their account.

Trust and Adoption. Researchers who studied privacy critical applications have in common that they (indirectly) showed that misconception can lead to distrust and thus lead to end-users not adopting the proposed technology (e.g. in [107,52,60]). Thereby the confirmed the research from Castelfranchi and Falcone [21] who studied the relation between mental model in trust in general.

4   Conclusion

To conclude this overview paper, we present some pertinent limitations of mental model research and suggest directions for future research.

4.1   Limitations

This review of mental model research, with a particular focus on human-centred security, has revealed uncertainty about the following:

Validity of Findings: the findings of current mental model research may not be applicable in other contexts, because existing research has been carried out with limitations on:

- *Methodology:* Most of the studies on mental models are based on self reported data about security behaviour which might or is very likely to be inaccurate (participants want to be seen as more security and privacy aware and concious than they actually are) and depends on the context.
- *Heterogeneity:* Most of the studies have been carried out within one country within a limited area. We could not find any comparative study which contrasts or compares the findings from one country to that of others. Clearly the findings of such homogeneous studies need to be replicated in different contexts in order to confirm, validate, and if possible generalize their findings.
- *Time:* Much of the research seems rather dated in 2013. Mental models are extremely dynamic, and findings from some years ago are now of questionable validity.

Design Methodology: no attempt has been made to design interfaces to accommodate an understanding of mental models. This might be because these will probably differ slightly across user populations, and change during the lifetime of the system.

Measuring Mental Models: The question of how to identify individual mental models still remains an open question. A number of techniques have been deployed, as explained in Section 2, but none has so far been identified as being the best mechanism. Furthermore, it would be necessary to identify the mental models of a particular user before start using a security mechanism; and this should not take too much time. In addition, it is unclear how to handle the dynamic characteristics of mental models.

Identifying Experts: There is no clear way to distinguish novices from experts. It is important to be able to do this, since one might want to use different metaphors to train them [56] or provide them with different interfaces [3]. One cannot simply ask people if they are expert, since humans are notoriously bad at judging their own abilities [38,71]. One could ask people how long they have been using a particular technology, but people make use of their time differently, so time, being a weak indicator, is not a reliable predictor. The traditional way of assessing knowledge is to set a task to determine whether the person is able to complete it. That is probably an infeasible approach outside an educational setting, and, moreover might suffer from the same problems mentioned in the discussion related to accessing mental models in Section 2.

4.2   Future Work

To move forward we need to ascertain how to utilise this knowledge about the different aspects of mental models. A number of aspects are particularly prescient:

1. How do we *identify* which mental model(s) a particular user ascribes to in order to adopt risk communication, education and interfaces accordingly? Does it mean that we have to ask the user a set of questions before launching an application? Should the system learn about the mental model(s) by 'observing" the user's behaviour as he/she uses the system? How do we detect changes in the user's mental model(s)?
2. On the other hand, it might be possible to design *generic* risk communication techniques, educational efforts and interfaces. It could be that we ought to design in such a way that is independent of the different concrete mental model(s) users possess.
3. How do we *predict* user behaviour based on our understanding of their mental models? There is some research proposing different approaches, such as Blythe and Camp in [13] who model Wash's folk models [105] in software agents according to Gentner and Stevens [48] in a type similar to STRIPS [42] or in terms of causalities [10] or in terms of a graph [16]. Clearly more research is needed to consolidate the findings and generate guidelines for informing design.

4.3   In Closing

This paper has presented an overview of mental model research with particular application to human centred security. We do not claim this review to be exhaustive but it does give a flavour of the applicable research in the area and highlights areas that require more attention.

Acknowledgement. Karen Renaud carried out this research while on a visit to Darmstadt and funded by a KIVA grant (Kompetenzentwicklung durch interdisziplinäre Vernetzung). The authors would like to thank Steffen Bartsch for his valuable input.

References

1. Adams, A., Sasse, M.A.: Users are not the enemy. Communications of the ACM 42(12), 40–46 (1999)
2. Anderson, L., Krathwohl, D., Airasian, P., Cruikshank, K., Mayer, R., Pintrich, P., Raths, J., Wittrock, M.: A taxonomy for learning, teaching, and assessing. In: Anderson, L., Krathwohl, D. (eds.) A Revision of Bloom's Taxonomy of Educational Objectives, Complete Edition, pp. 212–218. Longman (2001)
3. Appelt, W., Hinrichs, E., Woetzel, G.: Effectiveness and efficiency: the need for tailorable user interfaces on the web. Computer Networks and ISDN Systems 30(1), 499–508 (1998)
4. Asgharpour, F., Liu, D., Camp, L.J.: Mental models of security risks. In: Dietrich, S., Dhamija, R. (eds.) FC 2007 and USEC 2007. LNCS, vol. 4886, pp. 367–377. Springer, Heidelberg (2007)
5. Asgharpour, F., Liu, D., Camp, L.J.: Mental models of computer security risks. In: WEIS: Workshop on the Economics of Information Security, Carnegie Mellon University, June 7-8 (2007)
6. Aytes, K., Connolly, T.: Computer security and risky computing practices: A rational choice perspective. Journal of Organizational and End User Computing (JOEUC) 16(3), 22–40 (2004)
7. Bain, K.: What the best college teachers do. Harvard University Press (2011)
8. Bang, M., Medin, D.L., Atran, S.: Cultural mosaics and mental models of nature. Proceedings of the National Academy of Sciences 104(35), 13868–13874 (2007)
9. Bartsch, S., Model, M.: Effectively communicate risks for diverse users: A mental-models approach for individualized security interventions. In: Informatik Jahrestagung (to appear)
10. Bartsch, S., Volkamer, M., Theuerling, H., Karayumak, F.: Contextualized web warnings, and how they cause distrust. In: 6th International Conference on Trust & Trustworthy Computing, London, United Kingdom, June 17-19, pp. 205–222 (2013)
11. Bates, M.J.: The design of browsing and berrypicking techniques for the online search interface. Online Information Review 13(5), 407–424 (1989)
12. Benenson, Z., Gassmann, F., Reinfelder, L.: Android and iOS users' differences concerning security and privacy. In: CHI 2013 Extended Abstracts on Human Factors in Computing Systems, CHI EA 2013, pp. 817–822. ACM, New York (2013)
13. Blythe, J., Camp, L.J.: Implementing mental models. In: IEEE Symposium on Security and Privacy Workshops, pp. 86–90. IEEE Computer Society (2012)
14. Borgatti, S.P., Everett, M.G., Freeman, L.C.: UCINET for Windows: Software for social network analysis. Analytic Technologies, Harvard (2002)
15. Bostrom, A., Fischhoff, B., Morgan, M.G.: Characterizing mental models of hazardous processes: A methodology and an application to radon. Journal of Social Issues 48(4), 85–100 (1992)
16. Bravo-Lillo, C., Cranor, L.F., Downs, J.S., Komanduri, S.: Bridging the gap in computer security warnings: A mental model approach. Security & Privacy 9(2), 18–26 (2011)
17. Buchmann, M.: Teaching knowledge: The lights that teachers live by. Oxford Review of Education 13(2), 151–164 (1987)

18. Burgess, D.C., Burgess, M.A., Leask, J.: The mmr vaccination and autism controversy in united kingdom 1998–2005: Inevitable community outrage or a failure of risk communication? Vaccine 24(18), 3921–3928 (2006)
19. Camp, L.J.: Mental models of privacy and security. IEEE Technology and Society Magazine 28(3), 37–46 (2006)
20. Carley, K., Palmquist, M.: Extracting, representing, and analyzing mental models. Social Forces 70(3), 601–636 (1992)
21. Castelfranchi, C., Falcone, R.: Trust is much more than subjective probability: Mental components and sources of trust. In: Proceedings of the 33rd Annual Hawaii International Conference on System Sciences 2000, p. 10. IEEE (2000)
22. Cegarra-Navarro, J.-G., Eldridge, S., Gamo Sánchez, A.L.: How an unlearning context can help managers overcome the negative effects of counter-knowledge. Journal of Management & Organization 18(2), 231–246 (2012)
23. Chapman, J.A., Ferfolja, T.: Fatal flaws: the acquisition of imperfect mental models and their use in hazardous situations. Journal of Intellectual Capital 2(4), 398–409 (2001)
24. Chiasson, S., van Oorschot, P.C., Biddle, R.: A usability study and critique of two password managers. In: Proceedings of the 15th Conference on USENIX Security Symposium, USENIX-SS 2006, vol. 15. USENIX Association, Berkeley (2006)
25. Clegg, S.R.: Ten propositions concerning security, terrorism and business. Global Business and Economics Review 10(2), 184–196 (2008)
26. Conrad, D.: Building knowledge through portfolio learning in prior learning assessment and recognition. Quarterly Review of Distance Education 9(2), 139–150 (2008)
27. Converse, S.A., Cannon-Bowers, J.A., Salas, E.: Team member shared mental models: A theory and some methodological issues. In: Proceedings of the Human Factors and Ergonomics Society Annual Meeting, vol. 35, pp. 1417–1421. SAGE Publications (1991)
28. Craik, K.J.W.: The nature of explanation. Cambridge University Press (1967)
29. Dagher, Z.R.: Review of studies on the effectiveness of instructional analogies in science education. Science Education 79(3), 295–312 (1995)
30. d'Andrade, R.G.: The development of cognitive anthropology. Cambridge University Press (1995)
31. Dekker, S., Hollnagel, E.: Human factors and folk models. Cognition, Technology & Work 6(2), 79–86 (2004)
32. Diesner, J., Carley, K.M.: Automap1.2 - extract, analyze, represent, and compare mental models from texts. Technical report, CMU (2004)
33. Diesner, J., Kumaraguru, P., Carley, K.M.: Mental models of data privacy and security extracted from interviews with Indians. In: 55th Annual Conference of the International Communication Association (ICA), New York, May 26-30 (2005)
34. Donker, H., Klante, P., Gorny, P.: The design of auditory user interfaces for blind users. In: Proceedings of the Second Nordic Conference on Human-Computer Interaction, pp. 149–156. ACM (2002)
35. Dörner, D.: On the difficulties people have in dealing with complexity. Simulation & Gaming 11(1), 87–106 (1980)

36. Dourish, P., Delgado De La Flor, J., Joseph, M.: Security as a practical problem: Some preliminary observations of everyday mental models. In: Proceedings of CHI 2003 Workshop on HCI and Security Systems, Fort Lauderdale, Florida, April 5-10 (2003)

37. Dourish, P., Grinter, R.E., De La Flor, J.D., Joseph, M.: Security in the wild: user strategies for managing security as an everyday, practical problem. Personal and Ubiquitous Computing 8(6), 391–401 (2004)

38. Dunning, D., Johnson, K., Ehrlinger, J., Kruger, J.: Why people fail to recognize their own incompetence. Current Directions in Psychological Science 12(3), 83–87 (2003)

39. Easterbrook, J.A.: The effect of emotion on cue utilization and the organization of behavior. Psychological Review 66(3), 183 (1959)

40. Edwards, W.K., Poole, E.S., Stoll, J.: Security automation considered harmful? In: Proceedings of the 2007 Workshop on New Security Paradigms, pp. 33–42. ACM (2008)

41. Felt, A.P., Ha, E., Egelman, S., Haney, A., Chin, E., Wagner, D.: Android permissions: user attention, comprehension, and behavior. In: Proceedings of the Eighth Symposium on Usable Privacy and Security, SOUPS 2012, pp. 3:1–3:14. ACM, New York (2012)

42. Fikes, R.E., Nilsson, N.J.: Strips: a new approach to the application of theorem proving to problem solving. In: Proceedings of the 2nd International Joint Conference on Artificial Intelligence, IJCAI 1971, pp. 608–620. Morgan Kaufmann Publishers Inc., San Francisco (1971)

43. Fischhoff, B.: Risk perception and communication unplugged: Twenty years of process1. Risk Analysis 15(2), 137–145 (1995)

44. Fischhoff, B., Bostrom, A., Quadrel, M.J.: Risk perception and communication. Annual Review of Public Health 14(1), 183–203 (1993)

45. Friedman, B., Hurley, D., Howe, D.C., Felten, E., Nissenbaum, H.: Users' conceptions of web security: A comparative study. In: CHI 2002 Extended Abstracts on Human factors in Computing Systems, pp. 746–747. ACM (2002)

46. Furman, S.M., Theofanos, M.F., Choong, Y.-Y., Stanton, B.: Basing cybersecurity training on user perceptions. IEEE Security & Privacy 10(2), 40–49 (2012)

47. Furnell, S., Bryant, P., Phippen, A.D.: Assessing the security perceptions of personal internet users. Computers & Security 26(5), 410–417 (2007)

48. Gentner, D., Stevens, A.L.: Mental models. Lawrence Erlbaum, Hillsdale (1983)

49. Greenhalgh, T., Helman, C., Chowdhury, A.M.: Health beliefs and folk models of diabetes in british bangladeshis: a qualitative study. BMJ: British Medical Journal 316(7136), 978 (1998)

50. Gross, J.B., Rosson, M.B.: End user concern about security and privacy threats. In: Cranor, L.F. (ed.) SOUPS. ACM International Conference Proceeding Series, vol. 229, pp. 167–168. ACM (2007)

51. Gross, J.B., Rosson, M.B.: Looking for trouble: understanding end-user security management. In: Proceedings of the 2007 Symposium on Computer Human interaction For the Management of information Technology, p. 10. ACM (2007)

52. Gupta, S., Bostrom, R.P.: Theoretical model for investigating the impact of knowledge portals on different levels of knowledge processing. International Journal of knowledge and Learning 1(4), 287–304 (2005)

53. Harris, M., Furnell, S.: Routes to security compliance: be good or be shamed? Computer Fraud & Security (12), 12–20 (2012)
54. Helm, R., Mark, A.: Implications from cue utilisation theory and signalling theory for firm reputation and the marketing of new products. International Journal of Product Development 4(3), 396–411 (2007)
55. Helman, C.G.: "feed a cold, starve a fever" folk models of infection in an english suburban community, and their relation to medical treatment. Culture, Medicine and Psychiatry 2(2), 107–137 (1978)
56. Hsu, Y.: The effects of metaphors on novice and expert learners performance and mental-model development. Interacting with Computers 18(4), 770–792 (2006)
57. Johnson-Laird, P.N.: Mental models: Towards a cognitive science of language, inference, and consciousness, vol. 6. Harvard University Press (1983)
58. Johnson-Laird, P.N.: Mental models and thought. In: Holyoak, K.J., Morrison, R.G. (eds.) The Cambridge Handbook of Thinking and Reasoning, pp. 185–208. Cambridge University Press (2005)
59. Jones, N.A., Ross, H., Lynam, T., Perez, P., Leitch, A.: Mental models: an interdisciplinary synthesis of theory and methods. Ecology and Society 16(1), 46 (2011)
60. Karayumak, F., Kauer, M., Olembo, M.M., Volk, T., Volkamer, M.: User study of the improved Helios voting system interface. In: 1st Workshop on Socio-Technical Aspects in Security and Trust (STAST), pp. 37–44. IEEE Digital Library (2011)
61. Kaspersky. The evolution of phishing attacks: 2011-2013 (2013), http://media.kaspersky.com/pdf/Kaspersky_Lab_KSN_ report_The_Evolution_of_Phishing_Attacks_2011-2013.pdf
62. Kauer, M., Günther, S., Storck, D., Volkamer, M.: A comparison of American and German folk models of home computer security. In: Marinos, L., Askoxylakis, I. (eds.) HAS 2013. LNCS, vol. 8030, pp. 100–109. Springer, Heidelberg (2013)
63. Kauer, M., Kiesel, F., Ueberschaer, F., Volkamer, M., Bruder, R.: The influence of trustworthiness of website layout on security perception of websites. In: Current Issues in IT Security 2012, May 7-11, vol. (18), pp. 215–220. Duncker & Humblot (2012); 5th MPICC Interdisciplinary Conference on Current Issues in IT Security, Freiburg i Breisgau, Germany
64. Kelley, P.G., Consolvo, S., Cranor, L.F., Jung, J., Sadeh, N., Wetherall, D.: A conundrum of permissions: installing applications on an android smartphone. In: Proceedings of the 16th International Conference on Financial Cryptography and Data Security, FC 2012, pp. 68–79. Springer, Heidelberg (2012)
65. Kempton, W.: Variation in folk models and consequent behavior. In: American Behavioral Scientist; American Behavioral Scientist (1987)
66. Khaslavsky, J.: Integrating culture into interface design. In: Cconference Summary on Human Factors in Computing Systems, CHI 1998, pp. 365–366. ACM, New York (1998)
67. Kindberg, T., Sellen, A., Geelhoed, E.: Security and trust in mobile interactions: A study of users perceptions and reasoning. In: Mynatt, E.D., Siio, I. (eds.) UbiComp 2004. LNCS, vol. 3205, pp. 196–213. Springer, Heidelberg (2004)
68. King, J.: How come I'm allowing strangers to go through my phone? - Smartphones and privacy expectations (2013), http://jenking.net/mobile/
69. Klimoski, R., Mohammed, S.: Team mental model: Construct or metaphor? Journal of Management 20(2), 403–437 (1994)
70. Kozma, R.B.: Will media influence learning? Reframing the debate. Educational Technology Research and Development 42(2), 7–19 (1994)

71. Kruger, J.: Lake wobegon be gone! the "below-average effect" and the egocentric nature of comparative ability judgments. Journal of Personality and Social Psychology 77(2), 221 (1999)
72. Kumaraguru, P., Cranor, L.F., Newton, E.: Privacy perceptions in India and the United States: An interview study. In: In The 33rd Research Conference on Communication, Information and Internet Policy (TPRC) (September 2005)
73. Langan-Fox, J., Code, S., Langfield-Smith, K.: Team mental models: Techniques, methods, and analytic approaches. Human Factors: The Journal of the Human Factors and Ergonomics Society 42(2), 242–271 (2000)
74. Lin, J., Amini, S., Hong, J.I., Sadeh, N., Lindqvist, J., Zhang, J.: Expectation and purpose: understanding users' mental models of mobile app privacy through crowdsourcing. In: Proceedings of the 2012 ACM Conference on Ubiquitous Computing, UbiComp 2012, pp. 501–510. ACM, New York (2012)
75. Littman, D.C., Pinto, J., Letovsky, S., Soloway, E.: Mental models and software maintenance. Journal of Systems and Software 7(4), 341–355 (1987)
76. Liu, D., Asgharpour, F., Camp, L.: Risk communication in security using mental models (2008), Usable Security Website:
http://usablesecurity.org/papers/liu.pdf
77. Morey, D., Frangioso, T.: Aligning an organization for learning-the six principles of effective learning. Journal of Knowledge Management 1(4), 308–314 (1997)
78. Muslukhov, I., Boshmaf, Y., Kuo, C., Lester, J., Beznosov, K.: Understanding users' requirements for data protection in smartphones. In: 2012 IEEE 28th International Conference on Data Engineering Workshops (ICDEW), pp. 228–235. IEEE (2012)
79. Nemire, K.: Case study: The wrong mental model can kill you. In: Proceedings of the Human Factors and Ergonomics Society Annual Meeting, vol. 51, pp. 554–558. Sage Publications (2007)
80. Norman, D.: Some observations on mental models. In: Gentner, D., Stevens, A. (eds.) Mental Models. Erlbaum, Hillsdale (1983)
81. Olembo, M.M., Bartsch, S., Volkamer, M.: Mental models of verifiability in voting. In: Heather, J., Schneider, S., Teague, V. (eds.) Vote-ID 2013. LNCS, vol. 7985, pp. 142–155. Springer, Heidelberg (2013)
82. Orgill, G.L., Romney, G.W., Bailey, M.G., Orgill, P.M.: The urgency for effective user privacy-education to counter social engineering attacks on secure computer systems. In: Proceedings of the 5th Conference on Information Technology Education, pp. 177–181. ACM (2004)
83. Payne, S.J.: A descriptive study of mental models. Behaviour & Information Technology 10(1), 3–21 (1991)
84. Pfeffer, J.: Changing mental models: HR's most important task. Human Resource Management 44(2), 123–128 (2005)
85. Raja, F., Hawkey, K., Hsu, S., Wang, K.-L., Beznosov, K.: Promoting a physical security mental model for personal firewall warnings. In: CHI 2011 Extended Abstracts on Human Factors in Computing Systems, CHI EA 2011, pp. 1585–1590. ACM, New York (2011)
86. Raja, F., Hawkey, K., Jaferian, P., Beznosov, K., Booth, K.S.: It's too complicated, so I turned it off! Expectations, perceptions, and misconceptions of personal firewalls. In: Proceedings of the 3rd ACM Workshop on Assurable and Usable Security Configuration, pp. 53–62. ACM (2010)
87. Rao, A.R., Monroe, K.B.: The moderating effect of prior knowledge on cue utilization in product evaluations. Journal of Consumer Research, 253–264 (1988)

88. Renaud, K.: Blaming noncompliance is too convenient: What really causes information breaches? IEEE Security & Privacy 10(3), 57–63 (2012)
89. Richardson, G.P., Andersen, D.F., Maxwell, T.A., Stewart, T.R.: Foundations of mental model research. In: Proceedings of the 1994 International System Dynamics Conference, pp. 181–192 (1994)
90. Robertson, I.T.: Human information-processing strategies and style. Behaviour & Information Technology 4(1), 19–29 (1985)
91. Rouse, W.B., Morris, N.M.: On looking into the black box: Prospects and limits in the search for mental models. Psychological Bulletin 100(3), 349 (1986)
92. Rowe, A.L., Cooke, N.J.: Measuring mental models: Choosing the right tools for the job. Human Resource Development Quarterly 6(3), 243–255 (1995)
93. Rumelhart, D.E., Norman, D.A.: Representation in memory. In: Cognitive Science Laboratory, Center for Human Information Processing, University of California, San Diego (1983)
94. Schechter, S.E., Dhamija, R., Ozment, A., Fischer, I.: The emperor's new security indicators. In: IEEE Symposium on Security and Privacy, SP 2007, pp. 51–65. IEEE (2007)
95. Sheng, S., Holbrook, M., Kumaraguru, P., Cranor, L.F., Downs, J.: Who falls for phish?: a demographic analysis of phishing susceptibility and effectiveness of interventions. In: Proceedings of the SIGCHI Conference on Human Factors in Computing Systems, CHI 2010, pp. 373–382. ACM, New York (2010)
96. Simons, D.J., Levin, D.T.: Change blindness. Trends in Cognitive Sciences 1(7), 261–267 (1997)
97. Slovic, P.: Perception of risk. Science 236(4799), 280–285 (1987)
98. Spears, J.L., Barki, H.: User participation in information systems security risk management. MIS Quarterly 34(3), 503–522 (2010)
99. Staggers, N., Norcio, A.F.: Mental models: concepts for human-computer interaction research. International Journal of Man-Machine Studies 38(4), 587–605 (1993)
100. Staw, B.M., Barsade, S.G.: Affect and managerial performance: A test of the sadder-but-wiser vs. happier-and-smarter hypotheses. Administrative Science Quarterly, 304–331 (1993)
101. Taber, K.S.: Mediating mental models of metals: Acknowledging the priority of the learner's prior learning. Science Education 87(5), 732–758 (2003)
102. Thatcher, A., Greyling, M.: Mental models of the internet. International Journal of Industrial Ergonomics 22(4), 299–305 (1998)
103. Tversky, B.: Cognitive maps, cognitive collages, and spatial mental models. In: Campari, I., Frank, A.U. (eds.) COSIT 1993. LNCS, vol. 716, pp. 14–24. Springer, Heidelberg (1993)
104. Vosniadou, S., Brewer, W.F.: Mental models of the earth: A study of conceptual change in childhood. Cognitive Psychology 24(4), 535–585 (1992)
105. Wash, R.: Folk models of home computer security. In: Proceedings of the Sixth Symposium on Usable Privacy and Security, p. 11. ACM (2010)
106. Wash, R., Rader, E.: Influencing mental models of security: a research agenda. In: Proceedings of the 2011 Workshop on New Security Paradigms Workshop, NSPW 2011, pp. 57–66. ACM, New York (2011)
107. Wästlund, E., Angulo, J., Fischer-Hübner, S.: Evoking comprehensive mental models of anonymous credentials. In: Camenisch, J., Kesdogan, D. (eds.) iNetSec 2011. LNCS, vol. 7039, pp. 1–14. Springer, Heidelberg (2012)

108. Weirich, D., Sasse, M.A.: Pretty good persuasion: a first step towards effective password security in the real world. In: Proceedings of the 2001 Workshop on New Security Paradigms, NSPW 2001, pp. 137–143. ACM, New York (2001)
109. Whitten, A., Tygar, J.: Why Johnny Can't Encrypt. In: Proceedings of the 8th USENIX Security Symposium, vol. 99, p. 1. McGraw-Hill (1999)
110. Willingham, D.T.: Why don't students like school: A cognitive scientist answers questions about how the mind works and what it means for the classroom. Wiley. com (2009)
111. Wu, M., Miller, R.C., Garfinkel, S.L.: Do security toolbars actually prevent phishing attacks? In: Proceedings of the SIGCHI Conference on Human Factors in Computing Systems, pp. 601–610. ACM (2006)
112. Ye, N., Salverndy, G.: Expert-novice knowledge of computer programming at different levels of abstraction. Ergonomics 39(3), 461–481 (1996)

Author Index